Flush

BRYN NELSON, PhD

Flush

The Remarkable Science of
an Unlikely Treasure

GRAND CENTRAL
PUBLISHING

NEW YORK BOSTON

Grand Central Publishing
Hachette Book Group
1290 Avenue of the Americas, New York, NY 10104
grandcentralpublishing.com
twitter.com/grandcentralpub

First edition: September 2022

Grand Central Publishing is a division of Hachette Book Group, Inc. The Grand Central Publishing name and logo is a trademark of Hachette Book Group, Inc.

The publisher is not responsible for websites (or their content) that are not owned by the publisher.

The Hachette Speakers Bureau provides a wide range of authors for speaking events. To find out more, go to www.hachettespeakersbureau.com or call (866) 376-6591.

Library of Congress Cataloging-in-Publication Data

Names: Nelson, Bryn, author.
Title: Flush : the remarkable science of an unlikely treasure / Bryn Nelson, PhD.
Description: First edition. | New York : Grand Central Publishing, 2022. | Includes bibliographical references and index.
Identifiers: LCCN 2022019468 | ISBN 9781538720028 (hardcover) | ISBN 9781538720035 (ebook)
Subjects: LCSH: Feces. | Feces—Social aspects.
Classification: LCC QP159 .N45 2022 | DDC 612.3/6—dc23/eng/20220512
LC record available at https://lccn.loc.gov/2022019468

ISBNs: 9781538720028 (hardcover), 9781538720035 (ebook)

Printed in the United States of America

LSC-C

Printing 1, 2022

For Mom and Dad with love—from your #2 son

CONTENTS

INTRODUCTION

I T TURNS OUT THAT SENDING YOUR CORPSE TO A COMPOSTING
facility may not be for everyone. After death, we still routinely bury
or burn our dearly departed. But from a biological standpoint, we're a
fairly complete plant food once we fully decompose into rich soil: cadav-
ers offer all of the minerals and nutrients generally deemed essential for
flora, except for nickel. As author Caitlin Doughty described in *From
Here to Eternity*, the "recomposition" movement—literally returning our
remains to the earth and embracing the freedom of a "body rendered
messy, chaotic, and wild"—is, well, gaining ground.

I vividly remember my visit to the Body Farm in Knoxville, Ten-
nessee, where people had donated their corpses to science so forensic
experts could learn *how* we decompose under a variety of natural and
sinister circumstances: in the ground, on the ground, in a car trunk, in
a trailer. I found the place utterly fascinating, and oddly touching. In
death, donors were helping scientists learn more about the inevitable
conclusion to life and helping forensic sleuths solve murders and bring
killers to justice. When I wrote about it, though, a photo editor was so
disgusted by the mere idea of the place that my pictures taken inside the
two-and-a-half-acre compound, sans any visible bodies, were rejected
as inappropriate.

Donating your newborn's umbilical cord may not be for everyone
either. After a birth, we still routinely dismiss the roughly two-foot-
long tether as medical waste and throw it in the trash. But from a biological standpoint, the cord is packed with stem cells and progenitor
cells that give rise to oxygen-carrying red blood cells, infection-fighting
white blood cells, and clot-forming platelets that can help treat or cure
more than seventy conditions ranging from leukemia to sickle cell disease. With more than 40,000 cord blood transplants performed around
the world, that movement is making headway as well. For another story,

I described how a leukemia patient was saved by a double cord-blood transplant. The procedure used umbilical cords from two anonymous babies (nicknamed Amelia and Olivia after their respective A and O blood types) to effectively reseed his bone marrow after his own had been obliterated by chemotherapy and radiation. Despite the lifesaving potential, many hospitals don't even give new parents the option of donating the cord after its removal.

When I began writing this book, a question kept popping into my head: *What has value?* Very few people would question the value of donating blood, an organ, sperm, or eggs. Saving a life or helping to bring a new one into the world is a celebrated act of altruism. But once we've branded something as useless or worthless—or let's face it, icky—it's often hard to see it in a new light. Which brings me to donating your poop. Sure, we're often eager to be rid of it as soon as possible. But from a biological standpoint, the normal by-product of digestion can be utterly transformative for both plant and human life. I'd like to think it's something we can all get behind. Before you make a face or go searching for the perfect gift bag, though, there's a good reason why we should care about this diamond in the rough that we work so hard to make from birth until death.

Poop is there at the beginning, usually as a greenish-black and tar-like but nearly odorless output that babies normally produce within a day or two of birth (though sometimes while still *in utero*). Meconium, its technical name, contains mucus, bile, water, shed intestinal cells, amniotic fluid, downy fetal hair, and other bits that the fetus swallowed during gestation. Its expulsion from the body christens the intestinal tract and clears the way for a newborn's digestive system to begin disposing of the leftovers from breast milk or baby formula. Poop is often there at the end as well, when the dregs of our lower digestive system sigh out of our bodies, the inner and outer anal sphincters relaxing one last time. Between those two seminal events, an adult at the midpoint of a fairly large distribution curve defecates eight or nine times every week. One compilation of studies put the weekly yield at about two pounds, or roughly the weight of a store-bought pineapple, though a newer global model more than doubled that estimate. Given the planet's current population of eight billion people, some back-of-the-envelope math suggests that our annual production equals, well, a crapload of pineapples.

As a former microbiologist maybe I'm biased, but it seems a bit strange that we consistently shun a natural substance that we make so much of (OK, maybe not John Waters—or Germans). As a homeowner, I was both mortified and amused when a neighbor compared the brown primer a crew had just sprayed onto our house to "an unmentionable bodily function." Shit brown. She had meant it as an insult, and yet her sense of propriety prevented her from saying the words. As a writer working on this book, I was disturbed to discover that the transcription service I used for some interviews had trained its artificial intelligence algorithm to censor objectionable slang like "shit" and "piss" and "butthole" in the recording. In an effort to be rid of them, we have even erased the words.

In *History of Shit*, French psychoanalyst Dominique Laporte's irreverent smackdown of the Western world's delusions of grandeur, he observes, "We dare not speak about shit. But, since the beginning of time, no other subject—not even sex—has caused us to speak so much." Consider for a minute how often parents (and pet owners) obsess over poop and what it might mean. We share vivid recollections of the colors, sounds, smells, or freakishly large volumes from a diaper-changing episode. We revel in the satisfied or even joyful exclamations of a properly pooping tot (not to mention the relieved parents). These moments are often remembered as celebrated steps in a child's development: The first poop! The first poop that isn't weirdly black or green! The first mostly solid poop! The first poop in a toilet!

Later in life, we often obsess over its *absence* and what it might mean, especially after an accident, illness, or surgery. Its reappearance can elicit joyful exclamations from the properly pooping patient and loved ones alike: the body's systems are recovering and life is getting back up to speed. My first normal bowel movement after a surgeon removed my gallbladder and a gallstone the size of a grape definitely sparked joy.

The loss of my gallbladder was a timely reminder that there's still much to learn about what body parts like the squishy pear-shaped organ actually do. Many doctors call it "expendable," by which they generally mean that you can live a normal life without it despite its role in concentrating, storing, and distributing fat-dissolving bile made by the liver. I reluctantly agreed to have mine removed after six years of dithering over sporadic episodes of punch-to-the-chest pain that initially felt like a heart attack. I wasn't particularly sad to see the organ whisked away in pieces like green wet wipes through a hole in my belly button. (OK, I didn't really see that happening, but it's a fun mental image along with the tan, multifaceted stone that had exceeded an inch in diameter by then and had to be popped out through the same portal and would have been a rather nice addition to the Ball jar of antique marbles that I keep on a bookcase in my office except that surgical centers are apparently unaccustomed to such requests for souvenirs.)

The surgeon, anesthesiologist, and nurses, unfortunately, all neglected to explain just how much the routine outpatient procedure and multiple accompanying drugs can wreak havoc on a person's indoor plumbing. (To be clear, there are many, many ways to muck up your intestinal tract.) In the post-surgery recovery room, a discharge nurse warned me against straining too hard while on the pot, lest the effort cause some serious collateral damage; in rare cases, severe constipation even kills. Death by bowel movement can happen through a defecation-associated pulmonary embolism when a blood clot in a deep leg or pelvic vein suddenly breaks free and blocks blood flow in an artery within the lungs. Alternatively, excessive strain can send a patient's blood pressure soaring. That pressure buildup can cause blood vessels in the brain or abdomen to bulge and rupture, leading to a life-threatening stroke, aneurysm, or heart attack. Elvis Presley, in fact, likely died on the toilet (poor Elvis). Due to a long-standing addiction to opiates, he had become severely and chronically constipated and developed a dramatically enlarged colon. The medical condition is called megacolon, and yes, it's pretty much what you'd imagine. One plausible hypothesis suggests that Elvis strained too hard and keeled over from a massive heart attack. So severe constipation may have dethroned the King.

In the end, my own Great Poop Vigil only lasted fifty-two hours after the surgery—far earlier than many other poor souls have reported. As a middle-aged man, I still felt a bit odd celebrating and telling my parents about my toilet triumph. But that first splashdown was a moment of pure relief that suggested things were on the mend. My surgery and recovery forced me to pay far more attention to how my body works and what goes in and comes out of it. The episode also brought home how much we still have to learn about the push and pull of the teeming masses that call the intestinal tract home. We're still accustomed to thinking that we live apart from the rest of nature, but an entire world in miniature is literally inside of us: we can't escape the fact that its fate is closely intertwined with ours. Understanding what it is and what it does may help us appreciate our own inner power and learn how to live in better sync with a remarkable ecosystem that has coevolved with us. By extension, that mental shift may help us understand how working with nature instead of trying to overwhelm or dominate it can help us avoid a lot of pain.

The truth is that we're out of step with the rest of the planet at an *"Oh, shit!"* moment in history. Climate scientists are telling us that we must act now to avert the worst outcomes of rising sea levels and air temperatures linked to our reliance on fossil fuels. A wholesale shift to other energy sources will require some creative reimagining of postindustrial society, of what is most important and meaningful in life. There's no single cure-all for what ails us, but one of the best places to start creating solutions is with our own humble number two.

If we understand that we are intrinsic members of the natural world, that we are literal conduits between the inner ecosystems within our guts and the larger ones all around us, it may help us embrace our essential roles as caretakers of both. Becoming better stewards may require new ways of thinking about conventional regulations and infrastructure that no longer fit the changing landscape. It also may require rethinking how we measure and talk about value and progress.

Multiple cultures have long told creation stories and folktales that emphasize the essential nature of regenerative cycles in which one creature's body or by-products give rise to others. One tale from the Chukchi people of Siberia and recorded in the *Chukchi Bible* by Yuri Rytkheu describes how the world was created from the stomach and bladder of a raven, the First Bird. "A raven, flying over an expanse. From time to time he slowed his flight and scattered his droppings. Wherever solid matter fell, a land mass appeared; wherever liquid fell became rivers and lakes, puddles and rivulets. Sometimes First Bird's excrements mingled together, and this created the tundra marshes. The hardest of the Raven's droppings served as the building blocks for scree slopes, mountains, and craggy cliffs."

These traditional narratives act as counterweights to our contemporary tales, which are more likely to feature polluted waterways, preventable diseases, and shit going where it shouldn't. In our words, in our images, in the very architecture of our spaces, "we're not taught to attach any meaning to what's thrown away," Shawn Shafner, a Washington, DC–based artist, educator, and activist, told me. To help break the taboo of what we've deemed unmentionable or insignificant, Shafner founded The POOP Project—short for People's Own Organic Power. Through art, theater, and humor (one of his songs includes the words, "I'm a pooper. Yes, I doo"), he disarms audiences into having more serious discussions about how to have healthier relationships with their bodies and the planet.

Despite our protestations that it's someone else's problem, we could all use a reality check. Towns and cities from Norway to Canada still pump their wastewater directly into the sea. New York City's famously dysfunctional combined sewer system regularly dumps raw sewage into canals and waterways after heavy rains. Cities in Florida, Texas, and other states have likewise been inundated with raw sewage after hurricanes. Even in my own city, Seattle, the enviable wastewater treatment infrastructure still occasionally fails, like when a January 2021 power outage at a pump station not far from my house dumped 2.2 million gallons of storm water and sewage—likely including some of my own—into Lake Washington.

We will always be vulnerable to natural disasters. But our inattention to the fate of a "useless" product is disproportionately harming the most vulnerable and increasing the risk to all. "Our shit is pointing to a place where we've rejected that regenerative cycle for a myth of waste, and we've rejected our own body's capacity to be a regenerative agent in that cycling," Shafner told me. Our very existence depends upon

Flush

reclaiming that responsibility, writes author Jenny Odell in *How to Do Nothing.* "Even if you cared only about human survival, you'd still have to acknowledge that this survival is beholden not to efficient exploitation but to the maintenance of a delicate web of relationships. Beyond the life of individual beings, there is the life of a place, and it depends on more than what we can see, more than just the charismatic animals or the iconic trees. While we may have fooled ourselves into thinking we can live cut off from that life, to do so is physically unsustainable, not to mention impoverished in still other ways."

Like it or not, we are linked to this web and bound to our own less-than-charismatic crap. It's true that our complicated antihero can raise a stink and reveal only glimpses of potential amid the peril, kernels of Jekyll amid the Hyde. But if we take more than a passing look, we spy a veritable cornucopia of possibilities for human protection, innovation, and transformation. We see lifesaving medicine and sustainable power. We see compost and fertilizer we can use to restore eroded, depleted, or otherwise degraded land. We see a time capsule of evidence for understanding past lives and murderous ends. We see ways of measuring human health from the cradle to the grave, early warnings of community outbreaks like COVID-19, and urgent indicators of environmental harm.

Clever prospectors have created extraordinary opportunities to plumb its depths for water, fuel, and minerals. It can grant mobility and independence. Poop is the ultimate multitasker and could even help sustain and protect astronauts while fueling their way to Mars. Human feces, in short, is *the shit*, and it's high time we break the taboo and talk about its many merits. Our vastly underrated output contains a crystal ball of narratives about the past and future, about our pitfalls and possibilities. What we see within it reflects our own prejudices and mistakes, as well as the power to learn and change our fate.

Let's revel for a moment in the great leveler, the democratic reminder that, as Japanese author Tarō Gomi famously illustrated, everyone poops. There is humor and delight and optimism in that childlike wonder, a welcome reminder of our link to whales and elephants and the rest of the animal kingdom. There are also far bigger stories to tell. Not only

7

warnings of damage to our bodies and the world around us, but also hopeful tales of our untapped potential to rebuild and restore. We've heard far too little about the intricate factories in our guts, the ingenious applications for their products, and the innovators who are connecting the dots. This book aims to change that: consider it your reintroduction to a universal if unloved companion and an invitation to discuss something that's both unmentionable and a common fixation.

Elevating the status of our number two may help us reconnect with the world. It may also create a new dilemma that will require us to reexamine our values and morals: How do we prevent it from becoming a commodity that perpetuates decades of forcible extraction from exploited communities and further concentrates the wealth in others? How do we ensure that the benefits are distributed responsibly and equitably and that the risks aren't disproportionately shouldered by those who are most vulnerable to the potential harm?

As one of our most underrated and complicated resources, it's clear that poop has given us plenty to talk about. So think of this deep dive into the world's most squandered and misplaced natural asset as a *cri de coeur* (or cri de colon?) for the vast, hidden potential in the "waste" and the common, obvious, and ugly things that we overlook every day. Beyond our umbilical cords after birth and our bodies after death, we have vast scientific and economic potential rumbling under our belly buttons throughout life, just waiting to toot out into the open air (sorry, gallbladder, not you). But the *why* and *how* of taking poop seriously are just as important to the future of our species as the *what* of its considerable contents.

Unlocking poop's enormous potential will require us to overcome our shame, disgust, and indifference, embracing our role as both the physical producers and the moral architects of a more just and habitable planet. More than that, to become the standard-bearers our feces deserve, we will need to change our collective Western mind-set about what has worth, what moves us forward, and what it means to live in balance. A world that values and elevates the importance of our everyday output is one that no longer prioritizes the new and shiny as default options to solve climate change and other daunting challenges. It's one that resists the siren song of disruptive, exploitative, and proprietary innovation and embraces a future of progress through imaginative retrofitting and reinvention. It is a world where we no longer require simple or pretty answers, but ones that offer more lasting solutions. Our poop is substance, not style. Form, not flash. But there is undeniable strength in its numbers, and the whole of our ample product is greater than the sum of its many parts.

First a splash. Then ripples. Then waves. This book is a series of interconnected stories highlighting the momentum that can grow from unexpected sources and help us transition to a more circular economy in which we discard nothing and abandon the fantasy that we exist outside of ancient cycles of life and death, growth and decay. It is a book about how to stop wasting and stop pretending that there's anything *left* to waste. To maximize poop's power, we'll need to come to better grips with an agent of change that can elicit strong reactions. It is nurturing and noxious, funny and temperamental, benevolent and explosive—and it's only beginning to reveal its many mysteries. As one researcher aptly told me, *We don't know shit.*

It's high time we did. Come, let's get more acquainted with our inner champion. Locked within us is a medicine cabinet, a mound of fuel briquettes, a bag of fertilizer, and a biogas pipeline. Because of us—and what comes out of us—a dying mother recovers. The lights come on. The crops grow. A bus accelerates. Sometimes hope arrives in surprising packages.

CHAPTER ONE

Matter

I MAGINE THAT YOU'RE A GIANT short-faced bear living in North America about 20,000 years ago. You would be massive, reaching roughly ten feet in height when standing on your hind legs—no small feat considering that you'd weigh more than a ton. Although paleontologists haven't fully reconstructed your regular diet, you would likely be an omnivore and devour leafy greens and the carcasses of large animals. And when you shit in the woods? Well, that's where things get interesting.

Giant short-faced bears, like the continent's more carnivorous American lions, dire wolves, and saber-toothed cats, were the apex predators in a complex food web during the Pleistocene Epoch. An exotic menagerie of lumbering plant-eaters provided a diverse source of prey. Western camels and large-headed llamas. Long-horned bison and stag-moose with their massive antlers. Harlan's ground sloths that rivaled the bears in size. Vulnerable young of the Columbian mammoths that towered above them all and swung their curving tusks across most of the continent.

These colossal beasts had diets to match, and their digestive systems transformed the masses of vegetation they consumed into vast piles of manure. When they were eaten themselves, their fats and proteins and other nutrients sustained the lions and wolves and cats. Giant short-faced bears likely sniffed out half-eaten carcasses and helped themselves to the rest, leaving their own scat littered across the landscape.

Plants and trees grew in the fertilized patches, new grazers appeared, and the cycle repeated.

Some 20,000 years later, the process looks rather different for the planet's second most prolific poopers (after cattle). Many of us send our own output hurtling through a miles-long odyssey from the toilet into a complicated sewer network that connects to a wastewater treatment plant. We screen, filter, and aerate the incoming sewage; digest it with microbes; treat it with chlorine or other disinfectants; and pump the effluent through more pipes until it discharges into a nearby lake or river or sea. We extract the solids and haul much of them off by trucks or trains to be burned in incinerators or buried in landfills.

Modern sanitation is a luxury for much of the world: with a simple flush, our poop disappears. But have you ever wondered what exactly is swirling down the drain? Unlike bears or whales or birds, we expend an inordinate amount of effort to sequester our by-products from the rest of the natural world. And in so doing, we're effectively wasting one of the planet's most versatile natural resources.

I know what you're thinking. *Seriously, poop?* Yes, the object of disgust, the butt of jokes, the rump of puns—and a dangerous substance to boot—is far more than meets the eye (or nose). To understand what we're missing out on, though, we need to know why we should care about it, how we make it, and what it contains. One particularly instructive example of why we should give a crap dates back to the heydays of the Pleistocene.

As rocks slowly weather and erode, they release phosphorus into soils, where plant roots can absorb it. Plants use the element—one of the fourteen soil-derived nutrients they require (fifteen if you include cobalt)—to produce and store energy from the sun, and to construct DNA, RNA, and cell membranes. Animals get phosphorus from eating plants and use it to store energy and to make DNA, RNA, membranes, teeth, bones, and shells.

Phosphorus, in other words, is essential for life. To increase its availability, we've learned how to mine it and add it to fertilizers. But phosphorus leaches from soils, washes into streams and rivers, and eventually sweeps into the ocean, where it sinks to the bottom and gradually accumulates in sediments. And that presents a big problem: we've already tapped most of the accessible deposits and don't have thousands of years to wait for geological uplifting from the ocean floor to expose more phosphorus-rich rocks. So how else can the element be redistributed to help replenish soils?

Chris Doughty is an earth system scientist who views our planet as one integrated system. In particular, he studies how large-scale ecological patterns such as nutrient cycles are influenced not only by wind, water, and plants, but also by animals. That means he spends a lot of time modeling and calculating how animals help complete the cycling of elements like phosphorus. After taking in nutrients over a lifetime, a bear or whale or elephant can return them to other living things when it dies and its body decomposes. But to Doughty's surprise, his research has suggested that a much more important contributor—by several orders of magnitude—is the periodic release of nutrient-rich bundles in the form of animal dung. That makes sense when you consider how long and far the producers can roam. For broad distribution by both aquatic and terrestrial fauna, Doughty determined, the bigger the better. "Big animals move more than small animals; they're key," he told me. That's because bigger animals are more likely to move into an area with limited nutrients.

Focusing on Pleistocene megafauna weighing at least ninety-seven pounds, Doughty and colleagues developed a model that suggested they were key players in a complex phosphorus transport chain that moved the element from the deep ocean back up to the vast interior of continents. Whales carried phosphorus from the depths and, upon surfacing to catch their breath, dispersed it in the shallows and across the surface through plumes of floating fecal slurry. Vast flocks of seabirds and schools of migratory fish such as salmon ferried the nutrient to the shore and up rivers and streams. And a succession of carnivores and herbivores completed the relay to forests and plains, mountains, and meadows.

The complex predator-prey interactions created what Doughty and other scientists call "landscapes of fear," where carnivores relentlessly stalking herbivores kept them both on the move. "That has a huge impact on where they poop and how the elements are incorporated into the ecosystem," he explained. Over time, the animals re-dispersed phosphorus fairly evenly across the landscape. Those regular deposits, in turn, left behind extensive trails of food for others. Poop, in other words, helped make the living world go 'round.

It still does. Filter-feeding whales in Antarctic waters can convert iron-rich krill into bright orange feces that fertilize the surface for iron-dependent phytoplankton, the microscopic algae that feed a vast array of sea creatures. On the sunbaked African savanna, elephants can disperse seeds up to forty miles from a parent plant and nearly double the amount of soil carbon through their dung, thereby enriching a common grass that feeds other herbivores like gazelles. In North America, research by Canadian ecologist Wes Olson suggests that microbes taken in by a bison's snuffling nose or mouth help break down the cellulose in grass, while each resulting pile of dung can support more than a hundred insect species. In *Beloved Beasts*, science journalist Michelle Nijhuis describes the profound impact of this "bison patty ecosystem" and "bison snot ecosystem" on prairies. When bison abounded, the clouds of insects in turn fed a community of birds and small mammals. "Without bison—without bison snot, bison crap, and everything in between—the prairie is a smaller and quieter place," she writes. It's no wonder that some researchers call these kinds of habitat-creating animals "ecosystem engineers."

But Doughty and his collaborators believe a massive die-off of land giants during the Late Pleistocene and early Holocene Epochs—peaking in what other researchers have called "a geologic instant" between 14,000 and 11,000 years ago—decimated the global recycling system. In North America, the collective loss is strikingly apparent at the La Brea Tar Pits in Los Angeles, where I marveled at the jumbles of fossils still being pulled from the bubbling asphalt and reassembled into a ghostly zoo of exquisitely preserved predators and prey. Researchers strongly suspect that human hunters, climate change, or maybe both contributed to these mass extinctions. For the survivors, more recent human roadblocks have sharply limited their ability to travel across ecosystems, whether due to freeways that carve up the habitat of panthers and bison or dams that impede the upstream migration of salmon. As a consequence, Doughty's group calculated that land mammals in Eurasia,

Australia, and the Americas have retained less than 5 percent of their former capacity to distribute nutrients. The nutrient-dispersing ability of whales and migratory fish has plunged as well. "Basically, animals used to be very key conduits of elements across landscapes, and right now they're not," Doughty said.

Humans and domestic animals are Earth's dominant megafauna now. In theory, we've taken over many of the ecological roles of the extinct or diminished giants: humans as the carnivores and our livestock as the herbivores. But instead of dispersers, we act as concentrators; animals that no longer live in landscapes of fear tend to poop in the same place. Consequently, the output piles up in some areas and drains from others. Or as Doughty observed, "The rich get richer and the poor get poorer." Danish businessman and philanthropist Djaffar Shalchi put it even more memorably: "Wealth is like manure: spread it, and it makes everything grow; pile it up, and it stinks."

Albeit on a smaller scale, we're not much help in redistributing phosphorus when we die either (it makes up about 1 percent of body mass). Our dead are most often cremated or embalmed and entombed in clusters of wooden or metal boxes. While the nutrients in our poop have mainly ended up in landfills or ocean sediments, those in our remains mainly feed cemetery microfauna or the garden flora that receive a commemorative sprinkling of ashes (the recomposition movement, though, is working to expand the list of beneficiaries).

In her book, *Braiding Sweetgrass*, Robin Wall Kimmerer writes about how *wiingaashk*, the sacred sweetgrass of the Anishinaabe Indigenous peoples of North America, can teach us about the necessity and beauty of a balance between taking and giving.

In the Western tradition there is a recognized hierarchy of beings, with, of course, the human being on top—the pinnacle of evolution, the darling of Creation—and the plants at the bottom. But in Native ways of knowing, human people are often referred to as "the younger brothers of Creation." We say that humans have the least experience with how to live and thus the most to learn—we must look to our teachers among the other species for guidance. Their wisdom is apparent in the way that they live. They teach us by example. They've been on the earth far longer than we have been, and have had time to figure things out.

In disrupting ancient cycles, we're unwittingly dumping vital nutrients where they're least useful. Whether in life or death, our equilibrium between consumption and production is seriously out of whack. And as the inexperienced arbiters of life in the Anthropocene Epoch, we can create imbalances that have a way of coming back to haunt us. Disorder in our inner ecosystem can harm our health through disease and antibiotic resistance. It shouldn't be a surprise that on a larger scale the same kind of asymmetries can threaten the health of the entire planet.

● ● ●

Phosphorus, if undeniably vital to Earth's well-being, is just one of many things that pass through us over a lifetime and retain their utility at the other end. For the giant short-faced bear, the incoming nutrients may have regularly taken the form of a rotting stag-moose or tender greens. I'm far more partial to a medium-rare burger on a sesame seed bun with cheddar cheese, tomato, avocado, and dill pickles from a local burger joint. It may not be the healthiest option, but a burger can provide a useful glimpse into how we as modern omnivores acquire and process a range of carbohydrates, proteins, fats, fibers, vitamins, and minerals from the plants and animals we consume. And just as extinct megafauna have helped us understand how we can disperse useful raw materials over vast distances, more contemporary species are helping us understand how we disassemble complex foods into building blocks that can nourish and harm us, alter the balance of our inner ecosystem, and reshape the flora and fauna all around us.

Digestion really begins when we start grinding up food by chewing it and softening it with enzymes in our saliva. Even here, at the front end of a tube that runs from the mouth to the anus, we still don't fully understand our inner workings. In 2020, stunned researchers in the Netherlands documented their discovery of a "previously overlooked" set of salivary glands set deep in the back of the throat behind the nose. Their report, in turn, set off a fierce debate about whether nineteenth-century anatomists may have actually discovered the glands and whether they really aid digestion or play a more obscure physiological role. This much we do know: from multiple locations, we can produce up to two wine bottles' worth of saliva every day. That spit helps us mash up each bite of, say, a sesame seed bun, into a manageable ball, or bolus. Swallowing that compact bolus, in turn, is one of the most complicated actions in the human body. Some experts suggest that about thirty muscle pairs and a half-dozen cranial nerves might be involved, while others say the true number of muscles may be closer to fifty.

Once the bolus moves from the throat into the esophagus, a top-to-bottom contraction of muscles acts like a conveyor belt to carry the mash through a sphincter into the churning vat of acids we know as the stomach. In 1824, a British physician and chemist named William Prout created a stir by isolating hydrochloric acid (also called muriatic acid) from the stomach of a rabbit—the first proof that the gastric juices described by researchers experimenting on animals as varied as kites and bullfrogs contained the potent acid. Prout wrote that he had found the acid "in no small quantity" in the stomach of a hare, horse, calf, and dog.

Nine years later, a US Army surgeon named William Beaumont confirmed Prout's findings and opened a new window onto the digestive process—literally—when he chanced upon French-Canadian fur trapper Alexis St. Martin. The young man had miraculously survived a grisly musket wound that left him with a hole in his left side that extended into his stomach. Beaumont nursed his patient back to health but then took full advantage of the opening and besieged St. Martin, who became both his live-in servant and his guinea pig, with hundreds of invasive experiments. In one, Beaumont tied multiple pieces of beef, pork, bread, and raw cabbage to a silk string and then coaxed them into the hole in St. Martin's stomach before fishing them out at regular intervals to time how long it took to digest each morsel. Beaumont's *Experiments and Observations on the Gastric Juice, and the Physiology of Digestion*, if a landmark in new gastrointestinal insights, was a low point in medical ethics.

From these and other observations, we know that salivary and pancreatic enzymes, not gastric ones, are responsible for breaking down a bun's high starch content into sugars like maltose and then glucose, delivering the first burst of energy from a hamburger. White bread lacks many of the nutrients and fiber of wheat's bran and germ layers, though, which is one reason why it's often considered a poster child of "empty calories" among dieticians.

Cheddar cheese and ground beef contain abundant calories, too, though almost all in the form of cow-derived proteins and fats. In the stomach, the hydrochloric acid–containing gastric juices begin to denature the proteins like the unfolding of an origami crane. The complicated three-dimensional shapes smooth into simpler forms that can be more easily torn apart. The stomach's chemicals, in essence, can partially "cook" beef; it's the same principle behind adding weaker citric acid to milk to form cheese or making ceviche by curing raw fish or shrimp in acidic lime juice.

Cooking our food ahead of time can ease digestion even more. Grilling or frying beef, for instance, can break down proteins like the collagen in connective tissue, making the meat more tender and easier to chew. To understand more about the general physiological process of digestion, a 2007 study used Burmese pythons as human stand-ins (or rather, slither-ins). In the wild, Burmese pythons are solidly in the megafauna category and famous for swallowing things like goats and pigs and even alligators. In a lab at the University of Alabama, sixteen youngsters dined instead on lean eye of round beef from South's Finest Meats in Tuscaloosa.

For the experiments, the researchers compared the more typical python meal of an adult rat to equivalently sized raw steak, microwaved steak, raw ground beef, and microwaved ground beef. The pythons needed the most energy to digest the raw steak—just as much as they needed to digest the rat. Grinding the meat decreased the digestive energy requirement by about the same amount as cooking it. And the pythons used the least amount of energy to digest beef that was both ground and cooked. We grind meat when we chew it. And for our human ancestors, learning how to outsource more of the digestive effort by cooking meat might have given them an evolutionary leg up on the competition by freeing up more energy for other activities.

Even so, breaking down the fats in beef and cheese requires the additional dissolving power of enzymes made in the pancreas and bile produced by the liver (the latter normally doled out by the gallbladder). Not far from where the stomach empties into the small intestine, the common bile duct and major pancreatic duct converge at a small sphincter that controls the delivery of what German physician and writer Giulia Enders has likened to detergent. "Laundry detergent is effective in removing stains because it 'digests out' any fatty, protein-rich, or sugary substances from your laundry, with a little help from the movement of the washing-machine drum, leaving these substances free to be rinsed down the drain with the dirty water," Enders writes. In the gut, the enzymatic action breaks fats into glycerol and fatty acid building blocks, carbohydrates into simple sugars, and proteins into amino acids to enable their mass absorption into the bloodstream.

Further details of our gastrointestinal tract have come from comparisons with a multitude of lab animals. Pigs, in particular, share similar organ structures and arrangements within the gastrointestinal system, and have become a favored model for intestinal injuries and diseases. In both pigs and humans, the lining of the small intestine is studded with a vast fractal network of villi and microvilli projections that resemble the threads of a shag carpet and collectively act like a huge sponge. Intricate capillaries connected to the villi soak up amino acids, sugars, glycerol, smaller fatty acids, and water-soluble vitamins and minerals, while a mirror web of lymphatic vessels collects larger fatty acids and fat-soluble vitamins. This is how we pluck the nutrients, energy, and building materials we need to fashion our own fats, proteins, and other molecules from liquified food as it flows through the tract.

The regular appearance of undigested food in a stool may suggest that the normally ultra-efficient small intestine is struggling to properly absorb these nutrients. Patients with short bowel syndrome, for example, can lose much of the energy in food through malabsorption, while a foul-smelling and greasy or oily stool may suggest poor fat absorption due to a deficiency in bile or pancreatic enzyme production. Bile can also provide color indicators: after doing its duty to help digest fats, bile's pigments slowly degrade into a chemical called stercobilin that turns the greenish-yellow intestinal remnants brown on their way toward the exit doors. A rushed departure, though, can prevent the bile from fully breaking down, leaving the output more yellow or green.

In people with lactose intolerance, the small intestine can't make enough lactase, the enzyme that helps it digest the main sugar in dairy products. Like roughly two-thirds of adults around the world, I've become partially lactose intolerant and can get abdominal cramps, gas, and diarrhea after drinking too much milk. People with celiac disease instead mount an abnormal immune response against the gluten protein in wheat, like in that hamburger bun, which can damage the lining of the small intestine and cause either constipation or diarrhea, among other symptoms.

Within the broad range of a typical transit time, about 10 percent of a burger might pass through my stomach within an hour of eating it, while emptying half of it into my small intestine might take upward of two to three hours, on average, and somewhat longer for women. The arrivals to the small intestine might spend another six hours or so wending their way through up to sixteen feet of twists and turns until they reach the large intestine, or colon. And then, through the final five feet of the digestive pathway (shaped like a corkscrew in pigs but a question mark in humans), the pace drops down to a leisurely crawl. That's a good thing, because the colon has plenty of work to do, such as absorbing some of the remaining minerals like calcium and zinc from the cheddar cheese, breaking down some last bits of food, and regulating the gut's balance between electrolytes and water.

The colon's workload, in turn, is made far easier by a dense population of tiny helpers. Based on recent estimates, the trillions of bacterial colonists inhabiting our gastrointestinal tract, mostly in the colon, roughly equal the sum of our own cells. In the human gut, a thriving ecosystem that can support up to several hundred bacterial species—rivaling the complexity of a rain forest—has coevolved with us. This microscopic jungle constantly shifts and adapts in response to what we eat, where we live, whether we're sick or have taken antibiotics, and other environmental influences. All told, research suggests that several thousand bacterial species have colonized the guts of people around the world.

Scientists have compared the entire community of microbial inhabitants—our inner microbiome—to a "hidden metabolic organ." So far, they have found that this "organ" aids in tasks like digesting food through breaking down plant fibers and other carbohydrates (like in that avocado and tomato), synthesizing nutrients like vitamin K and all eight B vitamins, and balancing the immune system to recognize genuine external threats without overzealously attacking our own cells. Researchers have linked an out-of-balance microbiome, known as dysbiosis, to everything from inflammatory bowel disease and high blood pressure to diabetes and obesity.

We're only beginning to understand what the bonanza of bacteria and their products can do, but a few microbial groups have become particularly well-known. The most common genus in both the human gut and human feces in many parts of the world, *Bacteroides*, feeds its microbial neighbors by breaking down complex carbohydrates and protects us against pathogens; studies have suggested that it's a major regulator of the human immune system. In the relatively rare instance that it gains access to a vulnerable spot, though, *Bacteroides* can become an opportunistic pathogen and invade our cells. *Escherichia coli* is another opportunist. In one form, it's a relatively benign gut resident that can aid digestion and vitamin production and is a favorite experimental organism of labs like the one where I received my doctoral degree; in more toxic versions, or strains, it can be a deadly assailant that invades through contaminated food or water.

Another widely known genus, *Bifidobacterium*, includes dozens of species that specialize in fermenting plant fibers and carbohydrates in the gut while those in the genus *Lactobacillus* release lactic acid as a product of fermenting carbohydrates in foods such as breast milk. By releasing the acid as well as antibacterial peptides and hydrogen peroxide, lactic acid bacteria can aggressively protect their home turf in the gut (and the vagina, where they dominate) by making the surrounding environment inhospitable for pathogenic microbes.

Bifidobacterium and *Lactobacillus* abound in the infant gut, where they play a critical role in development and infection control. Thousands of years ago, our ancestors learned how to exploit the fermentation strategy used by these bacteria and by yeast cells to convert goat, sheep, camel, cow, horse, and buffalo milk into early versions of kefir and yogurt. Lowering the milk's pH through fermentation gave the foods a sour but not unpleasant taste while preserving them from spoilage by other microbes. We've since branched out to ferment thousands of foods and beverages like kimchi, kombucha, miso, sauerkraut, sourdough, salami, beer, wine, cheese, and some pickles (pickling in acidic brine is a separate process). More natural preservatives are added when the yeast or bacteria produce ethanol and antimicrobial proteins. Many of the fermented foods we love today, in other words, are made with the descendants or variants of microbial specialists that originally thrived in our ancestors' guts and were expelled in their poop.

When eaten regularly, fermentation expert Robert Hutkins said, the live microbes in fermented yogurts (the focus of most research) can outcompete gut pathogens and shift the balance toward a more favorable intestinal mix by unleashing products that kill off other microbes. They can digest complex fibers and alleviate the gas and bloating from lactose intolerance. Yogurt, Hutkins said, still contains plenty of lactose; the fermenting microbes only consume a fraction of the sugar within each cup. So how does it aid lactose intolerance in those of us who have it? When we eat yogurt, he said, the accompanying microbes effectively supply our small intestine with the missing lactase enzyme and help break the more complex sugar into the more readily absorbed simple sugars glucose and galactose. Little or no intact lactose remains to cause trouble in the large intestine by overfeeding other gas-producing bacteria. Yogurt-derived bacteria contained within a capsule can achieve the same thing if swallowed with a glass of milk.

Perhaps even more impressively, the microbes may help pacify the immune system by continually training it. The immune system regularly conducts friend-or-foe inspections to distinguish between safe and unsafe substances. Incoming fermenters normally pass the test, but by triggering the immune system's screening process, Hutkins said, they keep it from looking for trouble on its own and inadvertently attacking things it shouldn't—like the gut.

One of the many other reasons to keep our gut microbes happy is that they help intestinal cells produce a remarkable 95 percent of the body's serotonin molecules. These neurochemicals can influence both our mood and the muscle contractions needed to move undigested left-overs toward the rectum. These regular peristaltic movements can be stimulated by more moisture, more bulk, or both, and tend to be a bit faster in men than in women. That difference, in turn, may help explain why men seem to have faster gut transit times than women.

If you're curious about the timing in your own gut, the sesame seeds on a bun can provide a rough marker. That's because certain seeds pass through the human digestive tract more or less intact. Plants like rasp-berries, blackberries, tomatoes, and peppers make particularly good use of this adaptation: they can disperse their next of kin directly within the rich growing medium left behind by bears, birds, and other animals. Drinking a tablespoon of sesame seeds mixed in a glass of water and noting when they begin reappearing in a stool can provide a rough estimate of the normal journey time. Incidentally, I can report that my own sesame seed vanguard reappeared after twenty-five hours, though laggards kept arriving over the next thirty-five hours. Sweet corn also works since the indigestible cellulose coat around each kernel often remains intact. Because I'm partial to corn, I tried that timing method as well, and found that the kernels' journey averaged a bit less than sixteen hours over four trials.

No conversation about poop and regularity would be complete without fiber, and it's true that the fiber in foods like avocados and tomatoes helps to keep things moving. The small intestine doesn't digest dietary fibers well, making them available to help sweep out the large intestine, and it's here in the depths of our guts that roughage really shines. Brooklyn-based registered dietician Maya Feller told me that in addition to speeding the transit time, insoluble fiber can slow sugar absorption and improve blood sugar levels. It can help remove cholesterol and other lipids from the bloodstream—as well as bind up carcinogens and toxins formed during digestion—and flush them out with the rest of our waste.

That means poop can team up with fiber to literally clear away potentially harmful by-products.

By feeding fermenters like *Bacteroides*, soluble fiber can help diversify the gut microbiome. Thriving colonies of fiber eaters may at least partially offset the potential for excess fatty acids from foods like hamburger meat to decrease the gut's diversity and abundance of beneficial microbes. Feller also noted fiber's role in helping to tamp down inflammation in the gut, the body's biggest immune mediator, meaning that it may reduce inflammation throughout the body. Like a reverse Las Vegas, what happens in the gut doesn't stay in the gut. "So if we see people are having GI distress, or they're constipated, or they have diarrhea on the other side, or inflammation in the gut, then we're thinking, 'Okay, well, what's happening in the rest of the system?'" Feller said.

At the end of our winding intestinal corridor, the rectal exit is actually a series of two portals, the inner and outer anal sphincters. Our unconscious nervous system relaxes and contracts the inner portal in response to inputs like rising pressure and volume, while we have more direct control over the outer one, which fortunately cuts down on accidents. When the rising volume and pressure from the remains of a burger send an unmistakable message that it's time to open the double doors, some researchers say the act of letting go can be literally pleasurable.

Gastroenterologist Anish Sheth, coauthor of *What's Your Poo Telling You?* refers to this sensation as "poo-phoria." Pooping, especially the evacuation of a sizable stool, can clearly relieve the buildup in pressure. But it also may distend the rectum enough to fire up the vagus nerve (which extends from the brain to the colon) and release a flood of endorphins, the same chemicals released during an orgasm. The sweet release can drop the heart rate and blood pressure, triggering a feeling of happy light-headedness. The momentary pooper's high, Sheth suggests, could even become addictive. Other doctors have suggested that the pleasurable sensation could involve a nerve in the anus and anal canal called the pudendal nerve. And in men, some experts have speculated that the movement may massage the prostate gland, a well-known G spot more associated with anal sex.

Flush

• • •

Whew! So after getting a little high on the pot, what do we really deliver? Here again, we can look to other species for guidance. Instead of megafauna or lab animals, many of the most intriguing answers have come from the tiny microbes we shed in our feces and from the by-products of their own metabolism and decay. Their world in miniature is likewise one of predators and prey, competition and complexity, and varied niches that shift with the changing conditions in each one of us.

Human poop is slightly acidic, and in healthy adults, about three-fourths of it is water. To determine the constituents of the remaining fourth, researchers around the world have documented volunteers' intake or put them on diets and closely inspected their output. One 1980 experiment in the United Kingdom, for instance, examined the poop of nine men in their twenties and thirties who ate a "standard British-type diet" for three weeks (at the time, that meant Weetabix whole grain cereal, milk, sugar, orange juice, biscuits and jam, meat, vegetables, fruit, white bread with butter, and tea and coffee). The complicated process of sorting through their feces started with freeze-drying to remove the water, crushing the solids with a rolling pin, mixing them with detergents in a plastic pouch called a stomacher, and then filtering what remained.

One discovery from this experiment was just how widely the fecal mass and transit time can vary from person to person despite an identical diet. One volunteer passed his food in about two days, on average, while another took nearly five days to complete the gastrointestinal circuit. A 1996 *Gut* study by Mayo Clinic researchers confirmed not only highly variable gut transit times between individuals, but also significantly faster journeys in men than in women, especially through the colon. The observation fits with studies suggesting that women are more prone to constipation. It's also a good reminder of why studies should include diverse populations: unrepresentative participants such as similarly aged men from the same country eating the same things can easily skew what counts as "normal."

The same is true for socioeconomic diversity. A 2015 review of studies from more than two dozen countries calculated that the daily median stool weight of people from low-income countries—who ate more plant-based, high-fiber diets rich in foods like yucca, lentils, and black beans—was double the weight of poop from their counterparts in high-income countries. The minimum and maximum stool weights also varied far more in richer countries, perhaps due to the wider variety of diets available to their residents.

Surprisingly, the 1980 British study has remained one of the few to sketch out the main constituents of human fecal solids. For the nine men, at least, the report found that insoluble plant fiber accounted for about 17 percent of the solids, by weight. Soluble plant fiber and other material such as undigested proteins, fats, and carbohydrates contributed another 24 percent. And the big winner? Bacteria, making up more than half. Other studies have put the bacterial fraction somewhat lower, though diet can play a significant role.

As with fiber, people who eat more resistant starches such as the less digestible carbohydrates in beans, brown rice, green bananas, and lentils tend to have bulkier stools with more microbes. That's due to the frenzy of gut bacteria fermenting the plant material that remains in the colon after our own cells have taken a crack at it. Think of these bacteria like herbivorous long-horned bison grazing a Pleistocene grassland. One study suggested that nearly 50 percent of the bacterial cells in human feces are still viable if they find a suitable home beyond the gut. Some are aerobes, like us, and need oxygen to grow. Others are anaerobes, and don't require it or even die in its presence. Until a few decades ago, though, we knew little about how to keep most of our gut bacteria alive in a laboratory version of their native habitat, leading to vast undercounts before DNA sequencing began revealing their identities.

Some members of a vastly different domain of gut microbes, the archaea, also dine on carbohydrates but are more akin to stag-moose or reindeer that can withstand extreme conditions. Some archaea produce methane gas through the breakdown of carbon-containing molecules in the absence of oxygen. Anaerobic bacteria first ferment carbohydrates into their building blocks, and then the methanogens take over, using substrates like hydrogen plus carbon dioxide to create methane. This multi-step process, anaerobic fermentation, can be quite useful in producing human-derived biogas. Not everyone is a methane producer but a 1972 study in *The New England Journal of Medicine*, titled "Floating Stools—Flatus versus Fat," helpfully suggested that one way to tell is whether your stools float instead of instantly sinking to the bottom of the toilet bowl. Mine usually float, I can confidently state.

Like a biomedical version of "You sunk my battleship!" the researchers even degassed volunteers' previously floating stools and watched them plunge to the watery depths. Of course, this simple demonstration has a larger point: excess fat in the stool can be an indicator that it's not being properly absorbed in the gut, which is a clinical feature of several intestinal conditions. But the study's results suggested that gas and methane in particular—not fat—is behind the buoyancy and that doctors shouldn't rely on floating stools as a sign of trouble.

For as many bacteria and archaea that our bodies host, we likely harbor just as many viruses, based on recent estimates, meaning that the combined population of gut microbes probably outnumbers our own cells by a factor of at least two to one. Viruses like Ebola, polio, influenza, and SARS-CoV-2, cause of the devastating COVID-19 pandemic, are unquestionably dangerous. Fortunately, the vast majority of tiny viral particles that normally make up our inner collection—the human virome—are harmless, at least to us. Like dire wolves ganging up on bison, many viruses known as bacteriophages instead stalk and kill gut microbes. Others, like the pepper mild mottle virus, infect plants and are mostly incidental reflections of our diet. The pepper mild mottle virus, in fact, is one of the most abundant viruses in human poop and a useful marker of where we've been in the environment.

Just as archaeologists can examine the remains of a fossilized landscape to infer the overall health of a prehistoric ecosystem, physicians inspecting our poop can surmise plenty about our own internal environment. Some of these clues can come from the gut cells that break free from the intestinal lining—a regular exfoliation that provides the source material for noninvasive biopsies. Pathologists can examine the shed cells or extract their genetic information to seek out abnormalities associated with colon cancer and other diseases. Additional DNA, RNA, or whole cells from the legions of expelled bacteria, archaea, viruses, and fungi such as yeast can provide indicators of disease or a census of the gut inhabitants. So can the cells or eggs of parasites like tapeworms, *Giardia*, and *Cryptosporidium.*

After we've gone (to the bathroom or to the great beyond), the bouquet of odors in our poop can tell other stories about us. Blame the organic compounds released as we and our microbial inhabitants digest our food. Many of these compounds are quite useful: a chemical class called the polyamines, which includes cadaverine, putrescene, spermidine, and spermine, for instance, is believed to aid biological processes such as helping our cells grow, mature, and proliferate.

It's just that many of these compounds are, well, stinky. Gut microbes use the amino acid arginine to make putrescine, the smell of rotting flesh. A further breakdown process can transform putrescine into spermidine, which is—you guessed it!—also quite plentiful in sperm. Spermidine, in turn, can become spermine. The latter two molecules are behind the distinctive fishy odor that might make you do a double-take during a springtime walk. As science writer Kiki Sanford helpfully notes, some trees also give off the "spunky odour" (which makes sense if you think of pollen as the equivalent of sperm). And cadaverine's odor is, well, pretty much what you'd expect from its name.

Another malodorous constituent, skatole, was named for scat despite some researchers' insistence that the pure compound smells more like mothballs. The organic molecule forms when bacteria break down the amino acid tryptophan. It's naturally present in beets, one of my least favorite foods, and can spoil pork products through an unfortunately named process known as "boar taint." In small amounts, though, it can smell flowery-sweet and is partially responsible for the lovely smell of the jasmine flowers in our yard. A synthetic version is even used in ice cream and perfumes, which makes me wonder if *eau de toilette* is a little too on the mark. Skatole's close relative, indole, is an odor twin that is likewise described as pleasantly flowery in low doses and unpleasantly mothball-ish and musty or putrid in higher doses.

But perhaps beauty is in the nose of the beholder. A group of

Norwegian researchers argued in their 2015 paper, titled "Indole—the scent of a healthy 'inner soil,'" that the compound's odor is an underappreciated sign that everything is working as it should. "Indole is an example of a microbe-generated signal substance that has positive effects on its host as well as the microbiome, and normal-smelling faeces may be an underestimated health indicator," they write.

It might surprise you to learn that the true contributors to poop's natural tang have been hotly debated. In a 1987 study, a trio of researchers from Utah claimed that they had isolated and characterized the chemicals responsible for specific scents that might otherwise "be dominated by the offensive, foul odor of feces and remain undetected by human olfaction." The scientists, pooh-poohing the reputed contributions of skatole and indole, assembled an "odor panel" of six women and four men to help sniff out the essence of three other chemicals they had isolated. The results led them to conclude that members of the methyl sulfide chemical family are probably the main culprits behind the "foul and disagreeable odor of feces."

Whether our own output's true nose is more redolent of rotten cabbage or fetid death, it seems clear that poop, like coffee, is brimming with organic chemicals that may add their own little flair to the aroma. Other researchers at the UK's University of Bristol identified nearly 300 separate odor-producing compounds from the feces of healthy volunteers and a comparison group of patients with the inflammatory bowel disease ulcerative colitis, or with *Campylobacter jejuni* or *Clostridioides difficile* infections.

Since microbes are behind the production of many of these chemicals, an imbalance or collapse or boom of certain populations might alter the resulting odor in telltale ways. One study found that children who can't absorb nutrients well, likely leading to more fermentation of their partially digested food by gut microbes, tend to produce more putrescine and cadaverine in their feces. A separate study suggested that adult patients with irritable bowel syndrome marked by diarrhea likewise had higher signals from both compounds. Identifying these chemical signatures, whether by laboratory equipment or disease-sniffing dogs, could point out the trouble and prove lifesaving.

As a species, and even as individuals, we can leave behind distinctive calling cards based upon a host of chemical and genetic markers released through metabolism and digestion, to the increasing interest of archaeologists and forensic scientists. If that seems surprising, consider that the presence and description of animal droppings can already tell us plenty about how things are going in the natural world. Florida panther scat in Everglades National Park, identifiable as soft black cords of meat waste mixed with hair and bones, helped pinpoint white-tailed deer as the endangered cats' favored prey, supplemented by marsh rabbits and raccoons. In Arizona's Kartchner Caverns, I learned that guano from common cave bats is sometimes called liquid "sunshine" due to its critical role in sustaining a unique ecosystem that includes a tiny fungus-feeding mite, other predatory mites, fly larvae, crickets, and spiders. A grizzly bear shits in Alaska's woods, and the large brown clump of slightly fruity scat packed with seeds elicits a thrill of recognition in wildlife biologists and citizen scientists tracking the animal's territory and population density.

And of course, we're learning how important our output can be as the input for other forms of life. In Amsterdam's Micropia museum, an interactive ode to microbes called "Tour de Poep" includes a small demonstration of how piles of dung from the Asian elephants at the adjacent ARTIS Amsterdam Royal Zoo are being composted through a process inspired by ancient farming methods in Korea and popularized as *bokashi* in Japan. The elephants no longer travel huge distances, distributing nutrients as they go. So the zookeepers have lent a hand by fermenting the dung with a select mix of eighty bacterial and fungal species under anaerobic conditions—not unlike the process used to make kimchi—to create a rich "supercompost" for growing herbs, vegetables, and other edible plants. The harvested plants then become food for the elephants and other zoo animals, thus completing the cycle.

You may not find your own poop quite as interesting as a bear's or elephant's, but it's no less useful. And as Earth's dominant animals, we still have the power to learn from other species and live up to our potential by rejoining natural cycles and recycling limited nutrients and resources. Our collective pump won't ever run dry as long as we stick around. So why are we continuing to plug it up and waste our waste, instead of reusing it for the common good?

The answer may have a lot to do with our inability to get past long-held misconceptions and ingrained disgust that have conspired against the know-how and technology we've already developed for principled innovation. Getting to the bottom of the deep aversion that has driven a wedge between us and our natural output, many scientists believe, may be critical to moving past the mental block, striking a better balance with the worlds inside and around us, and curtailing even more collateral damage to both.

CHAPTER TWO

Horror

I T WAS INFAMOUS, THIS PAINT-
ing: deemed so vulgar and blas-
phemous and offensive by some
that a protestor smeared white paint
on it and then–New York City mayor
Rudolph W. Giuliani singled it out as
"sick" when denouncing the provocative
1999 "Sensation" art show it was included in at the
Brooklyn Museum. In his fury, Giuliani froze city funds for the venerable institution and threatened to boot it from its longtime home beside
the borough's Prospect Park. The mayor lost an ensuing federal law-
suit filed by the museum, and British artist Chris Ofili's *The Holy Virgin
Mary* became a cause célèbre of artistic freedom. When I first saw it, I
expected to feel at least a twinge of disgust.

Ofili, who is of Nigerian descent, depicted a Black, hip-hop version
of the virgin adorned with gold glitter, a collage of female buttocks and
vaginas from porn magazines, and elephant dung—one piece forms the
virgin's right breast while two lumps support the base of the large can-
vas at either end. Tiny yellow pins embedded in the lumps spell out her
name. *Virgin. Mary.*

Although I was living in New York at the time, I didn't see *The Holy
Virgin Mary* until years later. When I viewed it from a distance for the
first time, Ofili's Black Madonna seemed to undergo a curious transformation. Rather than recognizable and scandalous body parts, the
porn cutouts appeared more like indistinct fleshy baubles in a dazzling
yellow-orange sky that sparkled from the swirls of paint and glitter and
the rays that seemed to emanate from a sun behind the virgin's head.
She looked directly outward at me, serene and confident in a flowing
blue dress that ended in large jewel-toned leaves. Her pose mimicked

those of more traditional religious icons like the ones Ofili saw in London's National Gallery, in which white, bare-breasted Madonnas were likewise suffused with sexual energy.

Ofili was inspired to incorporate elephant dung into his art after winning a scholarship to travel to Zimbabwe. In an early interview for an exhibition at the Walker Art Center in Minneapolis, he called his medium a "crass way of bringing the landscape into the painting," and a nod to the found objects that populated modernist art. A more abstract piece incorporating that medium atop golden rectangles and swirls, *Painting with Shit on It*, is just as unsubtle and just as compelling. In a later interview with the *New York Times*, Ofili said he also used the dung, a cultural symbol of regeneration, as a reference to his African heritage. "There's something incredibly simple but incredibly basic about it," he said at the time. "It attracts a multitude of meanings and interpretations." Sacred and profane. Earthy and ethereal. Disgusting and inspiring.

At its most basic, poop tends to top the lists of things that repel us. "Probably of all of the elicitors of disgust that we've collected from studies around the world, that's the most universal," Val Curtis, a self-described disgustologist and the late director of the Environmental Health Group at the London School of Hygiene and Tropical Medicine, told me in an interview before her death in 2020.

There's a good reason for that "yuck factor," she said. Researchers have identified abundant disease-causing microorganisms in our poop, from parasitic worms and protozoa to fungi, bacteria, and viruses. For our ancestors, Curtis hypothesized, revulsion to that potent reservoir of disease may have been a matter of self-preservation. Those who didn't stay away would have fallen ill more often, decreasing their chances of reproducing. "We've evolved a healthy sense of fear that keeps us away from big animals that want to eat us from the outside," she said, "but we've also evolved a healthy sense of disgust which keeps us away from tiny little animals that want to eat us from the inside."

As a focal point of an ancient mechanism dubbed a behavioral immune system that may have evolved to protect us from harm, excrement has helped us understand how the perception of vulnerability to death and disease can shape our aversions. As a window into our modern world, it has also illuminated how the mechanism can be amped up and exploited to sell us unnecessary solutions to be rid of "icky" things or turn us against outsiders and our own best interests by labeling them distasteful as well.

If poop is an imperfect hero, then disgust is its foil: enough revulsion at the right time can avert harm, while too much at the wrong time can backfire. In the pivotal years to come, commonsense hygiene will remain vital for public health, as we've learned with COVID-19. But so, too, will open-minded curiosity and practicality in the face of our own mess, our own bodies, and yes, the mess of others' bodies. Understanding why we're so repulsed by poop may help us reframe our output as a normal and useful product of nature. We may be able to get past the yuck factor when it does more harm than good and inoculate ourselves against bad-faith politicians, pseudoscientists, and other influencers who prey upon our emotions in order to deceive us.

It's not just poop. Disgust can target anything that comes out of another human body—blood, sweat, vomit, urine, semen, saliva. The common fear, Curtis said, is that someone else's secretions and excretions might end up in your own body. "Because other people are the prime sources of the diseases that might make you sick, we really don't want to have other people's stuff inside us." The same is true of our own secretions once they've been expelled. "Try spitting in a clean glass, then drinking it!" she said.

In her book, *Don't Look, Don't Touch, Don't Eat: The Science Behind Revulsion*, she argued that our inner defense mechanism likely extends to other signposts of disease, like the sweaty, unkempt appearance of someone who is unwell; the smell of rotten food; and the sight of rats, flies, and parasitic worms. In all, Curtis identified seven categories of things that humans find icky. All of them, she argued, can be linked to a defense mechanism that protected our ancestors from disease. Some psychologists have collapsed the categories into three main groupings: pathogen disgust, sexual disgust, and moral disgust. The latter two groups, though likewise centered on the idea of avoiding harm, tend to vary more in their strength and specificity. And a taboo in one culture, like premarital sex or blowing your nose in public or eating frog legs, isn't necessarily verboten in another. Overall, though, the disgust emotion is among the most powerful in our armamentarium.

• • •

What disgusts you? This question has become a bit of an icebreaker for me, and a way to marvel at my own oddly specific aversions and those of my friends. When I asked a bunch of them on Facebook, they listed some fairly common triggers: slugs, drool, snot, rotting fish, vomit, and skunk spray. But also vinegar, unvarnished wood utensils, cold tomato juice, fascism, frogs, fresh papaya, insincere compliments, and canned spinach.

Multiple research groups have put their own spin on disgust sensitivity tests and attached them to scientific studies or released them as printable surveys. The ones I've taken—which generally ask participants to choose how gross they find a list of items or behaviors such as chocolate shaped like poop, a bowel movement left in a public toilet, and anal sex—suggest that I have slightly less disgust sensitivity than the average man in the United States. I've spent a considerable amount of time in labs and hospitals, so I'm less disgusted by blood or death. And I routinely clean the bathroom of our backyard rental cottage after guests have departed, so I'm fairly inured to other bodily fluids.

But I'm absolutely revulsed by lutefisk, a dish in which dried cod has been soaked in a lye solution until it reaches the consistency of fish Jell-O—and which was a horrifyingly common entrée at Christmas dinners when I lived in Minnesota. I'm also disgusted by cockroaches, which seemed to seek me out by land or by air when I lived in New York; and beets, which to me carry an overwhelming stench of iron and rot and dirt and despair. Disgust, of course, evokes an *emotional* response. But why?

Psychologist Steven Taylor admitted that when he first had to change his newborn son's diaper, he gagged and almost threw up. "It was *so* gross," he said, laughing at how revolted he felt. As part of the admittedly imperfect behavioral immune system, his knee-jerk disgust may be tied to the body's reckoning of its own defensive shortcomings. "It's based on the idea that our biological immune system is not sufficient for us to avoid pathogens because we can't see them: viruses and bacteria are obviously too small," Taylor told me. Instead, the behavioral immune system elicits a disgust response against the things we can see or smell or taste or touch or even hear that might be sources of pathogens.

Flush

To help test whether disgust is indeed an evolved defense mechanism, biological anthropologist Tara Cepon-Robins and colleagues tested the emotion's costs and benefits in communities that face a high burden of disease. For their study, they worked with three Indigenous Shuar communities in southeastern Ecuador. From the regional market center of Sucúa, one rural community in the Upano River Valley is about an hour away by bus and has quickly transitioned to a more market-integrated economy. Many residents sell agricultural produce, work as laborers, and live in wood- or cement-floor homes with tin roofs. The other two communities, which lie east of the Cordillera de Cutucú mountains, are roughly seven to twelve hours away by bus and motorized canoe on tributaries of the Amazon River. The villagers there rely more on hunting, fishing, foraging, and horticulture and live in thatched-roof, dirt-floor homes.

Cepon-Robins and her colleagues asked a representative sample of seventy-five villagers from the three communities a list of nineteen questions adapted to their local environment. *How disgusted would you be by: Stepping in feces with your bare feet? Finding a cockroach in your food? Picking up a dead animal with your hands? Someone vomiting on your shoes? Drinking chicha (a fermented drink made by chewing the pulp of a manioc root) made by someone who has no teeth?*

To examine whether higher disgust sensitivity protected them, the researchers also measured blood-based markers of inflammation caused by bacterial or viral infections, or by helminths—parasitic worms that can be transmitted through feces-contaminated soil. Among villagers with higher disgust sensitivity, Cepon-Robins and her colleagues found less inflammation linked to bacterial or viral infections. Their findings, in fact, suggested that the more easily disgusted individuals were more likely to protect their entire households. "So if the people in your house had high disgust and were avoiding everything around them, then they were less likely to get infected and to pass that on to you as well," she said.

Shuar villagers living in the communities closer to the market town had a higher disgust sensitivity than their more distant counterparts. That makes sense, Cepon-Robins said, because villagers with greater access to piped-in water, cement floors, and separate kitchen spaces can more easily avoid pathogens, and shunning gross things could be useful.

People living in dirt-floor homes with no running water would have a harder time avoiding pathogens, so a higher level of disgust would be of little use to them.

Cepon-Robins was initially surprised that higher disgust didn't likewise protect villagers from the parasitic worms, which can often be seen with the naked eye. But after being deposited in the soil through feces, the worms' much smaller eggs take about three weeks to develop into embryos that can then live for months and re-infect humans. By that time, the contaminated site is often indistinguishable from the surrounding soil. Disgust can present trade-offs and in this case, it didn't seem to offer a clear advantage: the Shuar villagers still rely on growing food, and to be effective, their disgust of worms would need to encompass soil too, putting them at odds with their reliance on it for horticulture.

If evolution has given us a general capacity for revulsion, then, different cultural cues and even facial expressions can help fill in the blanks for what we find revolting. In China, a mother may signal a greater degree of disgust with her eyes to police a wayward child. In Europe, a mother may use more of her whole face in a wrinkled-nose grimace. I remember the fluorescent green Mr. Yuk stickers that made the same grimace when I was a young boy in the 1970s: my mom and dad used them to mark the cleaning supplies under the kitchen and bathroom sinks of our house in Ohio as icky and best avoided.

Without that guidance, young children may have a rather short list of things they find icky. Psychologist Paul Rozin (widely known as Dr. Disgust), led a particularly vivid exploration of the indiscriminate tastes of toddlers in a 1986 experiment. He and his colleagues found that more than half of their study subjects under the age of two-and-a-half readily ate a dish introduced as "doggie doo"—in reality peanut butter with Limburger and blue cheese. Older kids, however, were less likely to gobble it up, suggesting they had since learned what not to eat. My own family has delighted in the story of a cousin happily playing on his parents' farm in northwestern Minnesota as a toddler. The ingredients for his "mud pies" were in fact cow pies, to everyone's disgust but his own.

Veterinarian and epidemiologist David Waltner-Toews, author of *The*

Flush

Origin of Feces, told me that our specific response to poop may reflect a complicated and contradictory cultural history based more on geography. Whereas feces were traditionally associated with fertilizer in rural agricultural areas, he said, they took on a more sinister role in urban centers as officials emphasized the very real danger of diarrheal diseases. Call it the cow manure versus cholera dichotomy. "The further we get from the farm, the more we see only the threats and less the opportunities," he said. Some urban centers in Asia and South America established strong links with nearby farms, but for others, excrement became a problem rather than a solution.

The pendulum between problem and solution, revulsion and acceptance has likewise swung over time. In ancient Rome, respectable waste collectors called *stercorarii* collected human excrement and animal dung and sold it as fertilizer (cleaning the city's famed Cloaca Maxima sewer, however, was apparently considered more menial labor better suited for enslaved people and prisoners). In the first century CE, the Roman emperor Vespasian even imposed a tax on urine, *vectigal urinae*, that was paid by early launderers called fullers who bought the contents of public urinals (basically jars placed in front of stores where shoppers could relieve themselves). The fullers then used the urine as an important source of ammonia for washing clothes (tanners also used it for tanning hides). From the revenues that tax accrued, Vespasian is widely credited with coining the phrase, *pecunia non olet*. Money does not stink.

In the sixteenth century, Paris did, and successive edicts aimed at eliminating the common practice of dumping the contents of chamber pots from windows and doorways created the mirage (if not the *odeur*) of a city undefiled by filth. In *History of Shit*, Dominique Laporte writes that the directives essentially privatized waste management by moving it from public streets and alleyways into household privies and cesspools. The relocation failed to resolve the city's underlying sanitation issues, but it filled the coffers of the state through fines and taxes and birthed a capitalistic ecosystem of perfumers to mask the stench and professional haulers to whisk it away, whether to a river or a field.

Through agricultural alchemy, Laporte writes, the fruits grown with some of that urban waste reappeared in city markets. The "recollection of ancient customs buried under centuries of oblivion" may have sparked a revival of the concept that waste has value, particularly as a fertilizer. But it also prompted a curious distancing from contemporary customs:

> The investment of waste—in particular, human waste—with value is consistently marked by a feigned oblivion of recent practices. It is offered as a discovery, or better yet a *re*discovery, of ancient models. When the discourse of triumphant hygiene introduced the idea of profitable waste in the nineteenth century, not a single enthusiast argued for its agricultural benefits by pointing to the fresh example of its current use in the French countryside.
>
> Rather, they found justification for the *nec plus ultra* of agricultural technology in the diaries of travelers who had journeyed to China. This pattern of repetition and revival helps us better understand the oscillation of civilization's anal imaginary: that which occupies the site of disgust at one moment in history is not necessarily disgusting at the preceding moment or the subsequent one. There are even instances of microvariations, whereby the attitude toward waste reverses, reinstituting previous practices within the space of a few short years.

As with our own bodily fluids, our emotional response to death has likewise fluctuated over time. Before the rise of the modern funeral home industry in the mid-nineteenth century, author Caitlin Doughty contends, we were far more inured to the sight of death and less frightened of the potential "danger" of dead bodies, which, with the exception of a few communicable diseases, are not all that hazardous.

Context, of course, matters a lot. Seeing death or blood or poop in a hospital or on a nature documentary may be less unsettling than seeing it on a street corner. Gross-out scenes and disgusting people behaving badly have captivated television and social media audiences for years: revulsion from a safe distance can be entertaining. From an evolutionary perspective, though, we may be better off with an over-sensitive alarm than an under-sensitive one, Taylor said. Being easily disgusted may help someone avert more things that are truly dangerous (and truly avoidable), though it can trigger plenty of false alarms. A pandemic or other perceived threat, in turn, can turn up the disgust sensitivity. "We're more likely to be squeamish. We're more likely to avoid possible sources of contagion. We're more likely wary around other people," he said.

The intense emotion may have even played a role in successive waves of toilet paper panic-buying during the COVID-19 pandemic, or the #ToiletPaperApocalypse, as amused and horrified Australians called it on Twitter in March 2020. True, stores may have been initially wiped clean due in part to major shifts in where people began spending their weekdays, meaning that utilitarian single-ply rolls in shuttered offices and factories were sitting idle while socially isolating residents suddenly realized they needed more of the plush stuff at home. People were soon sharing hot tips on where to score some Charmin. A Mexican restaurant in California's Orange County began throwing in a free roll for food orders of $20 or more.

The excessive hoarding and occasional fisticuffs over a four-pack, though, prompted some deeper reflections on why we had become so emotionally attached to something with so little purpose in an emergency other than to keep our butts clean. Taylor, also the author of *The Psychology of Pandemics*, told me he viewed the panic buying as a way for people who were already highly anxious about the virus to avoid additional disgust. The emotion had been amped up by the threat of being infected, he suggested, so toilet paper became a way to tamp it down. Amid official recommendations to gather up two weeks' worth of necessities during lockdowns, a ready supply of bathroom tissue ensured that no poop would be left behind to gross us out in our time of need, while the thought of running out and the distress it would cause created an incentive to over-buy.

A separate personality trait called intolerance of uncertainty also likely played a starring role in the absurdity. No one likes uncertainty, Taylor said, but some people have a tougher time dealing with it. "And pandemics, by definition, are associated with a host of uncertainties," he pointed out. Viral videos of the ensuing chaos in store aisles only added to the frenzy and the general fear of missing out.

Disgust sensitivity, Cepon-Robins said, seems to calibrate to our environment and to what we can control. When we can control more of our surrounding environment—obsessively scrubbing our hands, hoarding stacks of toilet paper, or avoiding contact with other people, say—a mismatch between high disgust and low threat may become pathological. Disgust can delay when people go to the bathroom, particularly during work or travel and sometimes to their own detriment. It has been implicated in obsessive-compulsive disorder and a range of phobias like arachnophobia and an irrational fear of blood or injections. And multiple studies have suggested that people with a high disgust sensitivity are more prone to xenophobia, racial prejudice, and bias against people with even superficial signs of poor health.

Taylor explained that disgust sensitivity toward foreigners is tied to the perception of disease vulnerability: that avoiding contact with them may help people avoid new pathogens to which they have no previous immunity. When former US president Donald Trump disparaged Haiti, El Salvador, and African nations as "shithole" countries in 2018 and repeatedly referred to SARS-CoV-2 as the "China virus," in turn, he may have acted as a dis-inhibitory influence on others with racist or xenophobic tendencies. Disgust, in other words, can be transmissible and weaponized.

● ● ●

Psychological warfare may have factored into at least some fecal weapons throughout the ages, as historian and folklorist Adrienne Mayor recounts in *Greek Fire, Poison Arrows, and Scorpion Bombs: Biological and Chemical Warfare in the Ancient World*. The Scythians, a tribe of nomadic warriors who hailed from the steppes of Eastern Europe and Central Asia, were legendary for their deadly archers and gruesomely barbed arrows. They were also one of the first known cultures to use poop as a weapon. In the fourth century BCE, Scythians dipped their arrowheads in a concoction called *scythicon* that included a putrefied mix of human blood serum and animal feces, viper venom, and rotting viper carcasses. These Scythians, they didn't mess around. Nonfatal puncture wounds still provided fertile ground for gangrene and tetanus infections that subsequently killed or incapacitated their victims. As Greek historian and geographer Strabo observed, "Even people who are not wounded by the poison projectiles suffer from their terrible odor."

In twelfth-century CE China, a tricked-out catapult launched what historian Stephen Turnbull has dubbed an "excrement trebuchet bomb" of gunpowder, dried human feces, and poison in a ceramic container. The contraption, believed to have released toxic smoke upon impact, was literally a dirty bomb. And in the Middle Ages, European invaders initiated biological attacks by catapulting the corpses of bubonic plague victims, their feces, or both over the castle walls of their adversaries. The contaminated corpses were successful in spreading the Black Death in some cases, but medieval medical theory supposed that the potency in causing disease came from the foul stench of rotting organic matter rather than the bodies themselves.

The miasma theory that "bad air" caused disease through *miasmata* (a poisonous vapor released by decaying or infected matter), persisted for centuries. It no doubt contributed to the Parisian edicts of the sixteenth century and was still a popular rationale for cholera through the mid-nineteenth century. England's sanitary movement in the 1830s and 1840s, in fact, gained strength from *miasmists* who fervently believed that diseases could spread through "infectious mists or noxious vapors emanating from filth in the towns," according to one account by public health researchers. Preventing disease thus required new sanitary measures "to clean the streets of garbage, sewage, animal carcasses, and wastes that were features of urban living."

One of the great ironies of the nineteenth century was how a profound revulsion of urban smells—the smell of human excrement in particular—bolstered the miasma theory and actively contributed to multiple cholera epidemics wherein leading miasmists mistook the smells that warned of danger for the danger itself. In *The Ghost Map*, author Steven Johnson's account of the ferocious 1854 cholera outbreak that swept through London's Soho neighborhood, he describes how "the perseverance of miasma theory into the nineteenth century was as much a matter of instinct as it was intellectual tradition. Again and again in the literature of miasma, the argument is inextricably linked to the author's visceral disgust at the smells of the city."

What people commonly believed caused disease, however wrong, strongly influenced what disgusted them. That's important, because it suggests that new information that leads to a better recognition of true threats (water contaminated by *Vibrio cholerae* bacteria, in this case) can help shift the focus of disgust or dissipate its force against innocent victims. In densely populated Victorian-era London and New York, for example, some miasmists ascribed a constitutional failing or moral deficiency to the urban poor that made them more susceptible to poisoned air. Physician John Snow helped pinpoint a contaminated water pump as the outbreak's source, and French chemist Louis Pasteur's germ theory of disease, developed less than a decade later, bolstered the growing case against bad bacteria instead of "bad air." But the enduring grip of miasma theory also suggests how influencers can exploit public fear and confusion to redirect revulsion toward scapegoats or opponents.

In 2017, Michael Richardson, a media and cultural studies researcher in Australia, detailed how Trump, a widely acknowledged germophobe, exceled in using his and others' squeamishness to his own advantage. Richardson's article, "The Disgust of Donald Trump," describes how the visceral and often dramatically conveyed revulsion toward political targets became a shared and mutually reinforcing aversion that bound Trump and his supporters together. "I know where she went, it's disgusting, I don't want to talk about it. No, it's too disgusting," he said at a rally after presidential candidate Hillary Clinton took a bathroom break during a 2015 debate of Democratic candidates. Richardson argued that in doing so, Trump capitalized on an observation backed by several studies that people with conservative ideologies tend to be more swayed by disgust.

Revulsion as political theater, you might say. Trump's audiences may not have come into physical contact with a disgusting object, but Richardson said Trump was able to attach a shape and name to their anxiety or uncertainty or fear or rage and evoke the visceral sense that they were being contaminated. Women talking about their bodies. Immigrants walking over the border. Minorities threatening a way of life. Even with the collective recoil, Trump positioned himself as the one person who could ritually wash away the disgust that still clung to his supporters: his revulsion became theirs, just as his need to eradicate the offending object became theirs. The chanted soundbites like *Lock her up* and *Build that wall* were ready-made for the stenography of cable news. Even at a distance, social media became a kind of affect-amplifying machine, delivering short bursts of outrage or revulsion or fear in viral memes and slogans boosted by algorithms that prioritized rapid engagement over thoughtful reflection.

Multiple minority groups have been the targets of similar disgust-driven campaigns. After the SARS-CoV-2 virus spread beyond China's borders in January 2020, science journalist Jane C. Hu noted in *Slate* that multiple articles emphasized the "unusual" or "weird" foods sold in the country's markets. The wording insinuated that revolting food choices were to blame for the virus and reinforced stereotypes of Chinese people as carriers of disease; Asians across the world soon found themselves targeted with what Hu called "casual acts of racism." In 2020, while reported hate crimes dropped overall, a survey of sixteen major US cities found that crimes targeting Asians soared by 150 percent, an upsurge that continued into 2021.

For years, political campaigns against gay men have instead emphasized anal sex, feces, and a mythology branding them as pedophiles—major triggers in the categories of sexual, pathogen, and moral disgust. The overarching goal, argues one study, "is to trigger revulsion—to brand a group of humans as nothing more than 'disgusting' vectors of moral transgression and physical contamination." Recent political campaigns targeting LGBTQ+ people, for example, have used words like "grooming" to pair perceptions of immoral and deviant behavior with insinuations that they could harm innocent children. The emergence of HIV and AIDS compounded the effect, with escalating attacks on gay men as vectors of disease. Shame, often described as disgust turned inward, has further reinforced and amplified some of those aversions in the form of internalized homophobia.

In a 2014 paper, "Of Filthy Pigs and Subhuman Mongrels: Dehumanization, Disgust, and Intergroup Prejudice," psychologist Gordon

Hodson and colleagues argue that dehumanizing others by comparing them to animals, bodily functions and all, may make us more apt to view them with disgust. The strategy has been honed throughout history. European colonists and then the US government routinely referred to Indigenous people as "savages." Nazi propagandists dehumanized Jewish people as cockroaches and rats carrying disease. Racists have likened Black people to apes. Trump called some Latino immigrants "animals" while attacking the establishment of sanctuary cities.

But Hodson said strategies that shift public attitudes toward animals—emphasizing our similarities with them instead of our superiority over them, for example—can help re-humanize racial, ethnic, and other groups and reduce prejudice directed at them. Once animals are no longer seen as inferior, Hodson said, the social gain or value in dehumanizing other people seems to dissipate as well. In that light, Tarō Gomi's *Everyone Poops* becomes a reaffirmation of our common link to pigs, dogs, and elephants and a subtle countermeasure against dehumanization, disgust, and prejudice.

●　●　●

For some of his public talks, J. Glenn Morris Jr., an expert on emerging pathogens, uses two lasting images to illustrate the devastating impact of a well-established disease. The first is the opening stanza of Rudyard Kipling's 1896 poem, "Cholera Camp," which describes an adversary that easily overpowered the British infantry in India.

> We've got the cholerer in camp—it's worse than forty fights;
> We're dyin' in the wilderness the same as Isrulites;
> It's before us, an' be'ind us, an' we cannot get away,
> An' the doctor's just reported we've ten more to-day!

The second is a photo taken a century later at the cholera ward in the International Centre for Diarrhoeal Disease Research, Bangladesh, in Dhaka. Just as powerfully, it depicts how the disease might be outlasted: Women lie on their backs on cholera cots, each of which is covered by a rubber sheet and outfitted with a strategically placed hole. A plastic bucket beneath each hole catches the watery diarrhea, nicknamed "rice-water stool," that the patients effortlessly expel. At its peak, the bacterial killer can drain a quart of fluid from its victims every hour.

Nurses use rulers to periodically measure each bucket's contents, Morris told the audience at a talk I attended. For every quart lost, according to the rule of thumb, one and a half must be replaced through an oral rehydration solution or an intravenous catheter. The disease can kill quickly, if no one intervenes: a fully grown adult can die in as little as ten to twelve hours from circulatory collapse if not adequately rehydrated. But if started in time, the relatively simple rehydration therapy can keep a patient alive until the infection burns itself out.

For centuries, cholera has been endemic in Bengal, the densely populated region of South Asia that includes Bangladesh and India's state of West Bengal. The dreaded disease is so closely tied to Bengali culture that villagers make offerings to Oladevi, the goddess of cholera, to appease her and spare themselves from her wrath. Aided by wars, expanding trade routes, and poor sanitation, though, cholera epidemics struck London, Paris, New York, and other cities around the world in the nineteenth century.

What could limit the spread of such a catastrophic illness? Well-aimed revulsion, perhaps. In her book, Curtis writes, "Disgust is a voice in our heads, the voice of our ancestors telling us to stay away from what might be bad for us." A culturally reinforced aversion to poop or contaminated food might help ward off the dire consequences of a gastrointestinal disease readily dispersed by *Vibrio cholerae*–tainted diarrhea. But what happens when the person expelling so much of it is a loved one— your child, parent, or partner? I asked Morris. It's an intriguing question, he said. Disease transmission can sweep through a house when a caregiver tends to a family member suffering from diarrhea full of the pathogenic bacteria. "They're passing huge numbers of microorganisms that are hyper-infectious, and then it becomes fairly easy to get enough on your hands, your fingers, in the food, in the water sources, then you then proceed to infect everybody else in the household."

Other members of the community may steer clear, but close relatives are less likely to abandon family members in need. "I think you've probably got conflicting drives here," Morris said. "It is such a devastating disease that people recognize somebody's dying, and they want to be there." In this case, he said, love may indeed trump disgust.

A small but fascinating 2006 study led by psychologist Trevor Case, titled, "My Baby Doesn't Smell as Bad as Yours. The Plasticity of Disgust," backs the idea that close relatives may clear the revulsion hurdle more easily, especially mothers caring for their own infants. In a series of experiments, thirteen mothers compared the stink of their baby's dirty diaper to another baby's. They were consistently more likely to rank their own offspring's as less disgusting, even when the researchers deliberately mislabeled the diapers or didn't label them at all.

In a separate survey of forty-two moms within the same study, the majority said their reactions to their baby's dirty diapers had changed over time—specifically, the poop seemed less smelly and disgusting. In other words, they seemed to habituate. That's probably a good thing because, as Case and his colleagues write, "a mother's disgust at her baby's feces has the potential to obstruct her ability to care for her baby and may even affect the strength of the bond she has with her baby." Case, who prepared all of the diapers for the experiments, didn't seem to similarly acclimate to the smells. As the paper reported, he "found them similarly intense and overpoweringly unpleasant and depended solely on rigorous labeling procedures to prevent any confusion of the materials."

Taylor, fortunately, habituated to his own son's dirty diapers, just as other parents have acclimated over the millennia. From her own research, Cepon-Robins found that just as some forms of disgust may be more easily acquired than others, we can adapt to shifting realities as well, like a mother lacking soap and water to wash her hands or a hunter needing to touch dead animals. "If you can't avoid something, then you don't get as disgusted by it," she said.

Recall that studies have suggested people with high disgust sensitivity overall are more likely to be politically conservative; they may harbor biases that make them more inclined to avoid people in other social groups, historically seen as sources of contamination. Women often score higher on disgust scales than men, but they've become *less* conservative than men in several Western democracies, particularly in the US and UK. Some of the apparent paradox may be due to the type of disgust being measured or the willingness of participants to answer honestly, Cepon-Robins said. Because women bear children, there's also more at stake if they get sick, suggesting that they should be more sensitive to pathogen disgust when they're pregnant or most likely to conceive. That explanation could help explain strong aversions to certain foods, particularly during the first trimester of pregnancy. But Cepon-Robins said the hypothesis hasn't been well-tested beyond college-age students and women in wealthier countries.

Disgust also can be "suspended," Hodson said, particularly through repeated exposure and more intimate bonding. Gay men struggled to talk openly about poop and semen and the taboo of anal sex until the horrors of HIV and AIDS forced them to have fraught and frank conversations about protecting themselves and their partners. If disgust keeps us safe, perhaps love and resiliency keep us human. Reaching those who are terrified of being infected by others, who believe that they're especially vulnerable, and who already have strong biases may be considerably harder. Once we code others with symbolic significance, recalibrating our emotions toward them can require a significant cognitive effort. But Hodson's work suggests that interventions based on compassion, empathy, and trust-building can chip away at prejudicial attitudes as well.

In 2013, his lab developed a more specific measure called "intergroup disgust." The scale measures someone's revulsion toward a social outgroup: a separate group with which that person doesn't identify due to differences in race, ethnicity, religion, sexual orientation, or other characteristics. Intergroup disgust is most relevant to prejudice, Hodson said, and women and men score the same on it. A separate analysis of fifty years' worth of psychological studies found that men were actually slightly more likely to display prejudice than women. The analysis also found "no instance where women demonstrated more prejudice than men."

For the swayable majority, Taylor said, awareness and public education can help point out the links between disgust and discrimination, whether in creating a new self-awareness of unconscious bias or in exposing the tactics wielded by others to shape public perception. Amid the confluence of health, social, and political upheaval in 2020, the Black Lives Matter movement drew new attention to the long-term harms of systemic racism and discrimination, while the COVID-19 pandemic laid bare the disproportionate impact of health crises and insufficient resources on communities of color.

As for politics driven by fear and revulsion, the ever-escalating rhetoric can collapse under its own weight if given a good push, Australian researcher Michael Richardson told me. Scales that measure our moral disgust of behaviors like lying, cheating, and stealing have found little difference between women and men, though an upsurge in the emotion among US women in particular may have helped sway a few recent political elections.

Trump had presented himself as the relief valve for the revulsion he shared with his supporters, even as he cranked up the dial ahead of the 2018 midterms by accusing his opponents of supporting a migrant caravan contaminating the southern US border. "The paradox is that you are promising to rid the disgust, but your power resides in people continuing to feel the disgust or the shame or the fear or the mix of all of those things," Richardson said. It's a difficult balance to sustain, it's exhausting, and it often fails to materially benefit the people kept in a perpetual state of repulsion, he said. Several political analysts suggested that Democrats won back the House of Representatives in the elections (though they lost ground in the Senate) in part by refocusing some of the public fear toward a potential loss of affordable health care and by redirecting some of the disgust back at Trump.

Immigration policies decried as cruel and dehumanizing and the administration's disastrous response to the COVID-19 pandemic may have added to the backlash, and political reporters pointed to the disgust of suburban women as a key contributor to Trump's reelection loss in 2020. The pitfalls of his aggressive and polarizing strategy were perhaps best distilled by the widespread horror and condemnation of the events of January 6, 2021. After fiery speeches by Trump and supporters like Giuliani (then his personal lawyer, who called for "a trial by combat" to decide the already settled election), a violent mob stormed the Capitol. In the aftermath of the failed insurrection, multiple news accounts reported that the Capitol's cleanup effort required washing away human feces that had been smeared through the halls of Congress.

And so a former mayor disgusted by elephant feces befouling a sacred icon became an object of disgust himself for helping to incite a mob that used human feces to befoul another sacred icon. One form of revulsion overtook another. "Bipartisan Disgust Could Save the Republic," one headline hopefully proclaimed. That remains to be seen. But understanding how an evolutionary adaptation can be exploited may at least make us less susceptible to manipulation by those who wield it as a weapon. Poop, in other words, may be a useful teacher.

And what of the painting that once scandalized a city? *The Holy Virgin Mary* traveled to Tasmania and back before winning a permanent home at New York's Museum of Modern Art in 2018, to the rapturous delight of the museum's curators. Ofili, for his part, didn't forget about his other benefactors. For a decade, three female Asian elephants at the London Zoo—Mya, Layang-Layang, and Dilberta—had supplied him with dung for his artwork, including *The Holy Virgin Mary.* To thank them, he donated the US$105,000 auction price of another painting to help pay for an outdoor play area at the elephants' new home in what is now the ZSL Whipsnade Zoo north of London. Mya, now in her early forties, has since moved to Italy. Layang-Layang and Dilberta have passed on. But before her death, Layang-Layang bore four sons, two of which have survived and live in other European zoos.

The benefactors became the beneficiaries and Ofili's act of altruism helped to further the cycle of regeneration—of transformation—he so memorably symbolized in a once-reviled painting of a holy virgin. Profane becomes sacred. Earthy becomes ethereal. Disgust becomes inspiration. And the lowliest matter becomes an unlikely source of life.

CHAPTER THREE

Savior

M ARION'S DAUGHTER HAD been sick for more than four years with a severe autoimmune disease that had left her colon raw with bloody ulcers, and after multiple doctors and drugs failed, Marion was frantic. Then she heard about another option she had never considered, one that sent her daughter's disease into remission, virtually overnight, with a single act of love. "Who wouldn't do that for their daughter?" Marion asked me. It's like a miracle, she said. "An overnight magic wand."

Marion agreed to do it again—twice—for strangers. She'd seen firsthand how effective it could be and she felt so badly for the patients and their families. Had she donated blood or plasma, no one would blink. But this? She couldn't tell anyone else about *this* because of how they might react. She even implored me not to use her real name, and so we settled on Marion as a pseudonym.

There are more like her, anonymous donors who have given others a remarkable reprieve from two chronic and increasingly common conditions that can inflame and damage the gastrointestinal tract. Roughly three million people in the United States have some form of inflammatory bowel disease. Marion's daughter had been plagued by a type known as ulcerative colitis, which attacks the colon and rectum. The second, Crohn's disease, prefers the small intestine but can strike anywhere from mouth to anus.

There are many more donors who have cured patients of a bacterial scourge known as *Clostridioides difficile*. The hardy microbe can persist in a cocoon-like spore for years, impervious to nearly everything except bleach. It is also fast becoming resistant to multiple antibiotics.

Rough estimates suggest that 20 to 35 percent of patients now fail their first antibiotic treatment. Of those, 40 to 60 percent will have a second recurrence. C. diff infections, as they're commonly known, sicken about 460,000 people in the US and kill an estimated 15,000 to 30,000 every year. Although the burden has dipped over the past decade, infections acquired beyond clinics or hospitals now make up half or more of all cases.

Marion lived in the Tampa Bay area of Florida when I spoke with her. "I don't talk with anybody about it," she said of her daughter's therapy. "I've told people that we replaced her…" and she paused, "unhealthy bacteria with healthy bacteria. I didn't go into specifics." Here are the specifics: Marion was the donor in a fecal microbiota transplant. She gave her daughter her poop. And in doing so, she may have saved her life.

Human poop is a decidedly imperfect delivery vehicle for a medical therapy. It's messy. It stinks. It's antithetical to Western medicine's traditional approach of identifying and eradicating specific threats, it's hard to separate into precise doses, and it's a regulatory nightmare. But it can be incredibly potent. Because bacterial cells make up roughly half of its mass and may represent hundreds of distinct species, each fecal deposit contains a dizzying array of microbial proteins, carbohydrates, fats, DNA, RNA, and other cellular constituents. Archaea, viruses, and fungi can further add to the mix.

Every bowel movement releases a bit of this enormously complex gut ecosystem, acting as a biopsy to signal the health of the metabolic organ and a starter culture to partially replicate the teeming intestinal jungle under the right conditions. Viewed in this ecological light, a fecal microbiota transplant, or FMT, is nothing more than an attempt to reseed an intestinal tract with an approximation of its normal inhabitants, often after antibiotics have killed off the native flora that might have kept invasive species at bay. Think of it like densely planting a plot of land so that weeds have little room to grow.

The replacement flora can make their way to the right spot from either end. For years, desperate patients resorted to DIY fecal enemas using donor poop from friends or relatives. As the therapy has evolved and matured, the crude enemas have been joined by more effective (and less messy) doctor-administered deliveries via a sigmoidoscope to the lower colon or a colonoscope to the upper colon. From the other end, gastroenterologists can introduce donor poop directly to a patient's stomach via a nasogastric tube or to the small intestine via a nasoduodenal tube, while many doctors have moved toward safer and more palatable triple-coated pills.

Despite its relatively low-tech nature, no other medical therapy can claim such a high cure rate for a recurrent C. diff infection. One nurse likened the recoveries of desperately ill patients to what she and her medical colleagues saw with anti-HIV protease inhibitors in the mid-1990s. After the Mayo Clinic in Phoenix, Arizona, performed its first FMT in 2011, a patient who had been bedridden for weeks left the hospital twenty-four hours later. And in 2013, researchers in the Netherlands reported halting a landmark C. diff clinical trial early because the overall cure rate of 94 percent with donor feces was far outpacing the 31 percent cured with vancomycin, once considered "the antibiotic of last resort." When they ended the randomized portion of the trial, the researchers gave fecal transplants to eighteen patients who had relapsed after their vancomycin therapy and cured fifteen of them.

Yet few other medical interventions have elicited such revulsion, ridicule, and disgust. Chronicling a potential advance by a team of Canadian scientists, one newspaper account still warned readers to "Hold your nose and don't spit out your coffee." In a 2012 commentary, fecal transplant provider Lawrence Brandt called the "yuck" factor—"which seems more common in physicians than in patients"—a major impediment to the therapy's wider acceptance. Many of his patients' former doctors, he wrote, had been unmoved by any of the positive published data and had labeled FMT "quackery," "a joke," or "snake oil."

As our antibiotic defenses begin to crumble, the search for viable alternatives has run up against more than simple disgust. Fecal transplants have had to win over a medical establishment that has long struggled to accommodate solutions emphasizing balance over elimination. They have faced off against an inflexible bureaucracy that has long failed to accommodate biological imprecision—or to prioritize the public good over commercial profit. And they have had to break our habit of looking past ordinary but useful tools in favor of flashy solutions. But we've had glimpses of the potential for centuries.

The first known description of a fecal transplant dates back to fourth-century CE China, when a doctor named Ge Hong included several mentions in his ambitious collection of therapeutic formulas, *Zhou Hou Bei Ji Fang* (in English, translated as *Handy Therapy for Emergencies*). Ge dutifully described how to treat patients with food poisoning or severe diarrhea by feeding them a soup-like fecal suspension.

Chinese gastroenterologist Faming Zhang and colleagues have researched the method's history and write that the treatment was deemed a "medical miracle" that brought patients back from the brink of death. Zhang told me that the donor was generally a child and that the therapy was sometimes known by its true name, "fermented solution of stool," and sometimes euphemistically called "yellow soup" or "golden juice." Hong also reportedly referred to it as "Huang-Long" decoction, or Yellow Dragon syrup. In the sixteenth century, Chinese doctor and herbalist Li Shizhen recorded a variety of names for the practice in his influential compilation of herbal remedies, *Ben Cao Gang Mu* (or *Compendium of Materia Medica*). The list of applications had continued to expand during the Ming dynasty, as Zhang and colleagues discovered: "Li Shizhen described a series of prescriptions using fermented fecal solution, fresh fecal suspension, dry feces, or infant feces for effective treatment of abdominal diseases with severe diarrhea, fever, pain, vomiting, and constipation."

Similar methods began to catch on with European veterinarians and physicians. In the seventeenth century, Italian anatomist Girolamo Fabrizi d'Aquapendente (sometimes called Hieronymus Fabricius ab Aquapendente) described one such technique in ruminants like cows and sheep. The simple process, which he termed "transfaunation," required only the transfer of chewed food from a healthy animal to a sick one to treat gastrointestinal disorders. The therapy, of course, also transfers bacteria, protozoa, and fungi. Veterinarians now use an inserted or implanted tube and siphon to transplant the contents of a donor animal's rumen, its first stomach compartment, to a recipient, much like a motorist might siphon gas into an empty fuel tank.

Flush

In 1696, German physician Christian Franz Paullini penned his notorious and immensely popular *Heilsame Dreck-Apotheke* (or *Salutary Filth-Pharmacy*). In the initial book and multiple revised editions, Paullini compiled hundreds of eye-opening prescriptions from the medical literature and his own practice, all devoted to the curative powers of feces, urine, and other bodily secretions ranging from menstrual blood to earwax. A 1958 catalogue of German baroque literature described the work as "one of the filthiest books in world literature."

Paullini evidently had an eye for the exotic and based his remedies in part on excrement from an entire zoo's worth of animals summarized by US Army captain John Gregory Bourke in his 1891 tome, *Scatalogic Rites of All Nations*. Among them: camels, crocodiles, elephants, falcons, foxes, geese, owls, peacocks, squirrels, storks, wild hogs, wolves, a lioness, a black dog, and a red cow. Got vertigo, gout, or lovesickness? Paullini was on it. Many of the directives, such as using horse dung to soothe toothaches, hawk feces to remedy sterility, or the urine of young boys mixed with honey to aid earaches, have perhaps rightfully faded into obscurity. Among his many other recipes, though, the good doctor included one that used human poop to treat dysentery (known then as bloody flux) and another that addressed kidney diseases or stones with "the scrapings of chamber-pots taken in brandy."

One of the first documented indications that microbial therapy might help rebalance the human gut and resolve a bacterial infection arrived in 1910, when the *Medical Record* reported on a new technique for treating "chronic intestinal putrefactions." Gastroenterologist Anthony Bassler described how he had treated multiple patients for the intestinal disorder by injecting a mix of human gut-derived bacteria or pure cultures of *Bacillus coli communis*—now *E. coli*—into the rectum of each patient every four days. The injections led to a noticeable improvement in the patients' health and a shift in their resident bacterial populations.

Nearly fifty years later, a doctor named Ben Eiseman, then chief of surgery at the Denver Veterans Administration Hospital, used fecal enemas to cure three men and one woman of a life-threatening inflammatory condition called Pseudomembranous enterocolitis. Although Eiseman linked the intestinal disease to a well-known pathogen called

Staphylococcus aureus, scientists now suspect the true culprit was C. diff. Amid the dry clinical language of his case reports, Eiseman recorded remarkably similar outcomes for each in his 1958 write-up. "There was an immediate and dramatic response by this critically ill patient to a fecal retention enema," he noted for the first. Nevertheless, he concluded: "enteric-coated capsules might be both more aesthetic and more effective."

An ad hoc experiment involving capsules, in fact, was already underway on the East Coast at the direction of a surgeon who worried that the heavy use of preoperative antibiotics was disrupting the normal intestinal flora of his patients. In 1957, the surgeon instructed a young bacteriologist named Stanley Falkow to begin collecting stool samples from surgical patients admitted to an unnamed hospital. Falkow, a pioneer in the field of bacterial pathogenicity, or the study of how bacteria cause disease, never named the surgeon either but recalled in a 2013 blog post how he dutifully divided each patient's poop into twelve large gelatin capsules before stashing the collection in a refrigerated ice cream carton. The surgeon and a second doctor who agreed to try the same strategy sent each of their patients home with a prescription of two capsules per day in an attempt to reestablish their presurgical intestinal microbes.

Although anecdotal reports at the time suggested that enrollees in the uncontrolled clinical trial were faring better than other postsurgical patients, Falkow wrote that they likely never knew what they were ingesting. The experiment ended abruptly when the chief hospital administrator found out and accused Falkow of feeding patients their own shit. A more formal follow-up to Eiseman's suggestion would have to wait another half-century.

● ● ●

Even as FMTs began to seep into mainstream medicine, few doctors or patients knew about their potential to cure an infection that can turn deadly within days. On a Thursday morning in April 2010, Peggy Lillis was a healthy fifty-six-year-old kindergarten teacher studying for a master's degree in education. She had a bad back from decades working as a waitress, sure, and she smoked when she was raising two boys on her own in Brooklyn before quitting in her thirties. The chronic inflammation in her right shoulder occasionally flared and she had put on some pounds over the years, but she had normal blood pressure and was healthy overall. She was as tough as nails, her sons said.

Peggy didn't feel well after school, though, so she went to bed early. Then the diarrhea began, at 4:00 a.m., with a sudden urgency. She took her first sick day in memory. Maybe it was just a stomach virus. Her sons brought her Gatorade, and she was still herself on the outside, despite being weak and tired and pale. Only something horrible was happening on the inside, made worse by the strong antidiarrheal drugs that her doctor prescribed. Her intestines were steeping in a toxic stew released by millions of proliferating bacteria.

By Tuesday evening, she was in intensive care in a Brooklyn hospital. She was craving a Diet Pepsi, but her infection had progressed so far that she had developed septic shock and toxic megacolon: her large intestine was dying. An unapologetically frank surgeon removed it the following morning in a last-ditch attempt to save her life. Dozens of friends and members of her large Irish Catholic family crowded into the waiting room in a stunned vigil. It wasn't enough. By 7:20 p.m., just over six days since her first symptoms, Peggy was gone, leaving her reeling sons to wonder how she could have died so quickly from something they had never heard of.

Perhaps Peggy Lillis contracted the deadly infection in a nearby nursing home, where she periodically visited her godmother. Perhaps, as her autopsy report suggested, it began in a dentist's office, where she underwent a root canal and was given clindamycin, a broad-spectrum antibiotic that can profoundly alter the gut microbiome and clear the way for C. diff infections. The infection progressed so quickly that an FMT—even if her sons had heard of it—may not have helped. "But I think the key thing here," her oldest son, Christian, told me, "is that the disease itself and our ability to raise awareness of it and talk about it, identify it, prevent it, treat it, is all complicated by the fact that we don't want to talk about shit."

Later that year, Christian and his brother, Liam, cofounded the Peggy

Lillis Memorial Foundation to focus more attention on an infection that has long outgrown its reputation as a nuisance in hospitals and nursing homes. Christian subsequently ruffled feathers on an online C. diff support group by suggesting that descriptions of symptoms shouldn't be relegated to a folder labeled "TMI" and pushed so far down the message board that he likened it to the sub-basement. "If you were hemorrhaging from the eyes or if you had a weird rash, you'd be like, 'Here's a picture,'" he said. "Does the breast cancer support group say, 'We don't want you to discuss what your tits look like'?" Why should poop be any different, he said, especially when its description could provide crucial information about someone's condition?

Catherine Duff (now Williams), a C. diff patient who defied the odds, had to force herself to talk about it when it really mattered. Between 2005 and 2012, she had the infection eight times. The first six times, it responded to antibiotics. Then it didn't. "My colorectal surgeon gave me the choice of having my colon removed or dying," said Duff, then fifty-six. She had already lost a third of her colon and wasn't about to give up any more.

With every bout of diarrhea and vomiting, though, the mother of three felt the life draining out of her. "I had basically resigned myself to the fact that I was going to die." Then one of her three daughters came across the research of an Australian gastroenterologist named Thomas Borody. His group was recording astonishing results in C. diff patients with a therapy called a fecal microbiota transplant. Duff read everything she could and brought the printouts to her doctors. By then, she was seeing eight of them. Only two, an infectious disease expert and a gastroenterologist, had ever heard of the technique. Neither was willing to try it.

And so in the spring of 2012, two years after Peggy's death, Catherine became part of a burgeoning DIY movement born out of necessity. Her husband at the time, John, a retired submarine commander who frequently spent months at a time submerged with 180 men, readily agreed to be her fecal donor. "Nothing grosses him out," she said with a laugh. They convinced her gastroenterologist to at least have John's poop screened for pathogens, and then found a recipe and protocol on the Internet. According to some of the more detailed how-to videos on

YouTube, the typical shopping list for a DIY transplant could be easily procured at a neighborhood drugstore:

A plastic container for collection
Latex gloves
A dedicated smoothie blender
A metal sieve
A disposable enema bottle
0.9% saline solution
A large measuring cup
A plastic spoon

Duff recalled taking the enema at 4:00 p.m. "By seven o'clock, I felt good," she said. "It was almost miraculous how quickly I felt better." The next morning, for the first time in months, she took a shower and got dressed, put on makeup, and went downstairs for breakfast.

A few months later, Duff had to undergo emergency surgery to correct a spinal cord compression caused by an old horseback riding accident. She came down with her eighth case of C. diff before she ever left the hospital. But this time was different. This time, her colorectal surgeon agreed to try an FMT using a colonoscopy—delivering the poop through a long, flexible tube inserted the length of her colon—and she became the first patient in the state of Indiana to undergo the procedure at a medical facility. "By the time I woke up from the sedation, I felt fine," she said.

Officials at the US Food and Drug Administration were increasingly feeling otherwise after struggling over how to regulate the procedure. In an April 2013 letter, the director of the agency's Center for Biologics Evaluation and Research wrote that an FMT would be labeled as a biological drug instead of a transplant, a move that dismayed patient advocates and threatened to sharply curtail the fledgling field. Providers wanting to continue treating patients would have to file an investigational new drug application, an arduous and time-intensive requirement that most physicians were ill-equipped to meet.

A week later, the FDA hosted a public two-day workshop on the therapy in Bethesda, Maryland. Duff, who had just launched The Fecal Transplant Foundation to connect patients with providers and encourage more doctors to offer the treatment, was one of the meeting's 150 participants. When she scanned the list of attendees, she saw that she was the only patient. Midway through the meeting, Duff realized there would be no discussion about the impact from a patient's perspective, and knew she had to do something. She wrote an impromptu speech on her iPad over lunch and tearfully begged the moderator to let her talk during the Friday afternoon session.

Duff told me she was deathly afraid of only two things: public speaking and crickets. The mere thought of talking to the entire group made her hyperventilate. But the moderator signaled the audiovisual booth and then pointed to her, as he had agreed to do, and her microphone lit up. She began to tell her story, haltingly. She didn't finish, but by then it didn't matter. She received a standing ovation from the doctors in the room. Afterward, they lined up to introduce themselves and thank her. Within a few weeks, she had assembled the majority of her foundation's board of directors and board of advisors.

Despite the ubiquity of home remedies, including Duff's first DIY fecal transplant, she stressed that a medically supervised procedure was far preferable due to the potential for disaster from an improperly screened donor. A transplant that transmitted a bacterial or viral infection to a recipient could prove lethal. University of Minnesota gastroenterologist and immunologist Alexander Khoruts, who joined her board of advisors, said he worried about the potential long-term impacts as well, particularly in light of evidence linking the microbiome to obesity, diabetes, and allergies. "At this point, I think science tells us to be very cautious with this material," he said.

Perhaps Duff's testimony helped sway the FDA. Perhaps it was the deluge of bad press that caught the agency off guard as it struggled to balance the growing evidence of benefits with a highly variable and virtually unregulated practice that had flown beneath the radar. Either way, the FDA bowed to pressure and partially reversed course, agreeing to exercise "enforcement discretion" for FMTs used to treat C. diff infections that weren't responding to standard therapy. They wouldn't be FDA-approved, but neither would they be prohibited. For all other indications, such as ulcerative colitis, the agency pulled the plug unless the transplants were part of clinical research that received the agency's approval under an investigational new drug application.

After the small victory, Duff and her fledgling foundation continued to push for more FMT clinical trials aimed at other digestive disorders, more parity in research funding, and more public awareness and education. She found herself smiling more. Laughing even. Her disgust of poop had long since dissipated, and when dealing with endless diarrhea and a complete loss of modesty, she told me, it helps to have a sense of humor. Duff and a few like-minded board members began crafting slogans to adorn T-shirts or sweatshirts. Her favorites? "Poop is the Sh*t!" and "Give a sh*t. Donate to the Fecal Transplant Foundation." The site also featured an FMT awareness ribbon. Brown. "It is what it is," she said with a laugh. "There's no way around what we're talking about or what we're dealing with."

• • •

In the meantime, more clinics began discreetly offering FMTs and curing people around the world. On the periphery of the Wellswood neighborhood in Tampa, Florida, I visited gastroenterologist R. David Shepard, whose practice was one of the first in the Southeastern US to offer the therapy. Shepard began each multiday transplant procedure at the Tampa Bay Endoscopy Center, where he would insert poop from a screened donor into the sedated patient. Up one end through a colonoscope. Down the other through an endoscope that extended past the throat and stomach to the jejunum—the midsection of the small intestine.

The next day, Shepard would continue the procedure at RDS Infusions, his office in an unassuming one-story building with privacy windows made from glass blocks. Patients would lie on a bed that sloped downward toward the head, allowing gravity to help them retain a follow-up fecal enema. A day later, they'd get another enema. In one of the procedure rooms, a small, laminated poster hinted at the high anxiety of patients hoping for a breakthrough. A rainbow arced through a cloudbank above a single word repeated eight times in progressively bigger letters: "BREATHE."

Among the first sixty or so C. diff patients that had followed his directives, Shepard told me he hadn't had a single failure. One patient, the fifty-five-year-old daughter of a prominent local businessman, had racked up more than $150,000 in medical bills during a bout with a recurrent infection. At her wit's end, she came to Shepard and was cured after a single treatment. Until the FDA forced him to put his ulcerative colitis program on hold, he had achieved a success rate of about 70 percent for that condition, he said. Marion's daughter had been the first.

Shepard struck me as cautious but unfailingly polite, with a subtle Southern accent and a steady gaze. At a table in a small kitchen behind the practice's warren of rooms and hallways, he described his foray into a therapeutic field that he had initially dismissed. After talking to a half-dozen doctors, in fact, I began to hear some familiar refrains. Nearly all had brushed off the idea at first. "The thought of it was basically one of disgust and 'Oh, I'll never do that. You've got to be kidding,'" Shepard said.

Elaine Petrof, an infectious disease specialist in Ontario, Canada, told me that doctors in her specialty were often conditioned to associate infections with germs that must be eliminated. "Just conceptually speaking, pouring sewage into people doesn't seem like a good idea, right?" she said. "I confess that I fell into that category until I saw what this can actually do for people's lives."

The technique lingered on the margins of accepted medical practice for years because there simply wasn't a great need for it. That changed when more virulent strains emerged, including one linked to a particularly nasty 2002 outbreak in Canada's Quebec province, and C. diff became an epidemic. Doctors began to routinely encounter patients whose infections had stopped responding to all antibiotics, and some started giving FMTs a second look.

Up until then, a doctor refusing to offer the procedure might have had the last word. But the Internet changed everything in the aughts; patients began to search until they found somebody who would do it. They could be persuasive. For Petrof, the turning point came in 2009:

after antibiotics failed to resolve a woman's C. diff infection, she began bouncing in and out of the intensive care unit. Every day, the patient's relatives asked Petrof to consider a fecal transplant. "I thought, *this is crazy*," she recalled. Then they brought her a bucket of poop. She relented. "What completely floored me was the fact that within less than seventy-two hours, this patient who had been having over a dozen bowel movements a day, basically completely turned around and at the end of the week walked out of the hospital," Petrof said.

Even so, doctors like Khoruts marvel at the effectiveness of such a "medieval" intervention: "You plop a turd in the blender and draw it up in a syringe. Voilà! There's your transplant," he said. Many of the initial success rates hovered between 85 and 95 percent—consistent with published reports since then. "In medicine, it's pretty startling to have therapy that's that effective for the most refractory patients with that condition," Khoruts said.

Amid the astonishing success, the field faced another major challenge: finding enough qualified donors. When Khoruts screened a medical student who had responded to a flyer, she told him her medical school classmates had laughed at her interest. Nobody would giggle at a blood donor, he said in exasperation. That donation merits buttons and stickers and a collective sense of pride. But this? Even a medical student couldn't tell her classmates without feeling embarrassed, even though she might save more lives in the end.

OpenBiome, a nonprofit cofounded by two Massachusetts Institute of Technology students, set out to meet the growing demand by recruiting top-notch donors and providing low-cost, prescreened, filtered, and frozen poop: $250 a pop, plus shipping costs. When I visited OpenBiome in the fall of 2015, it had recruited twenty-seven donors. Volunteer coordinator Kelly Ling said the nonprofit had just passed a major milestone by sending out 7,000 treatment units in all (by the end of the year, its deliveries would reach more than 500 hospitals in four countries). Although tucked away in a suburban business park at the time, it took advantage of being near Tufts University and next door to a gym called Work Out World, which, as it turned out, was a perfect place to find new recruits.

Could you be a poop donor? Based on the most stringent selection criteria, you'd have to be screened for everything that would prevent you from donating blood, like HIV and hepatitis. You couldn't be a sexually active gay man or older than sixty-five. You couldn't have had a tattoo within the past six months. Or have recently traveled to a long list of countries. You couldn't have taken any antibiotics for the past three months. You couldn't have a history of autoimmune, neurologic, or gastrointestinal diseases. Or of metabolic syndrome, a cluster of conditions like hypertension and high blood sugar that raise the risk of stroke, heart disease, and diabetes. You shouldn't be overweight, ideally, and definitely not obese. Or squeamish.

In 2017, Canadian researchers reported spending more than US$15,000 to screen forty-six potential donors for a trial assessing the effectiveness of FMTs on diseases associated with metabolic syndrome. After excluding half based on their medical history or physical exam, doctors tested the remaining candidates for biochemical markers of health and thirty-one viral, bacterial, fungal, and protozoan pathogens. Ultimately, they found only a single eligible donor who met all of their criteria. Harvard and Stanford have better acceptance rates.

For the few donors who passed OpenBiome's similarly stringent screening process, staff established a routine to make deliveries as seamless as possible. After ringing a doorbell outside the first-floor lab, the donors had to pass a visual inspection to ensure that they looked well, and then do their business on-site in a blue lidded bowl or bring in a fresh sample from home. A new delivery appeared in a bright blue bag during my tour, and a lab technician weighed it to make sure it exceeded the minimum (roughly the weight of a tennis ball). It passed easily and joined what Ling called the "poop queue." For every accepted fecal sample, the donor received $40.

After waiting its turn, each sample received a squirt of buffer containing saline to maintain its pH and glycerol to protect its microbial inhabitants during their eventual storage at minus 112 degrees Fahrenheit. A "poop smasher" homogenized the buffered mix, which the lab tech then poured into a clear plastic bag divided lengthwise by a fine mesh filter. Fiber (a good sign) would remain on one side, while brown liquid would drain to the other. The solution could then be divided into separate delivery units, bar coded, and frozen for up to two years.

In a second-floor meeting room, OpenBiome research director Mark Smith told me how he had been inspired to cofound the nonprofit by a motivated friend with C. diff who had failed seven rounds of vancomycin. His friend's determined search had netted only a single doctor in the New York metropolitan region who was offering FMTs as a passion project. The next available appointment was in six months. So Smith's friend, who had been sick for eighteen months, instead cured himself with his roommate's poop, a margarita blender, and an at-home enema kit.

After mixing, freezing, and shipping its own more refined concoctions, Smith said the nonprofit stool bank had achieved an overall success rate of about 86 percent for curing C. diff. During my visit, I met one of the newer donors: a friendly, soft-spoken twenty-six-year-old named Joe who had been making regular deliveries for a little more than two months. He had heard about the nonprofit from his brother, applied, and passed the rigorous screening process. He initially thought he would just be part of a research study and earn some quick cash and was pleasantly surprised to discover that it was something much bigger.

It helped that Joe was fit and had a fairly healthy lifestyle with a diet that was high in fiber. He felt thankful to have a thriving microbiome and had developed what he called a "strange sense of pride" whenever he produced a morning stool that he knew would make the donor bank workers happy. "So I think, *wow, this is a great one!*" he said. Depending on his work and running schedule, he might deliver four of them every week.

Gastroenterologists readily understood that the remarkable effectiveness of FMTs from donors like Joe against one bacterial infection wouldn't necessarily translate to other conditions. With recurrent C. diff, Petrof explained, repeated antibiotic regimens essentially "torched the forest," killing off much of the bacterial diversity and opening up spaces for the C. diff microbe to take root and grow. Adding back almost any other flora—the equivalent of planting seedlings in the dirt—could help the ecosystem keep the pathogen at bay. For more complicated autoimmune conditions like ulcerative colitis, a basic FMT might not always be enough, at least not with typical donors from the Western world.

Scientists had also raised the idea that a rise in allergies and auto-immunity in industrialized nations may derive in part from reduced gut microbial diversity due to low-fiber diets and high antibiotic use. Similar to the high bar set by OpenBiome and other banks, Khoruts had dismissed about 95 percent of fecal donor applicants at the University of Minnesota. "It turns out that healthy people are rare," he told me. Healthy people who have never taken antibiotics are even rarer; he had yet to find a single one. Perhaps no Western donor could provide the microbes needed to fully reseed the gut. What then? Researchers were finding considerably higher gut microbiome diversity—and fewer allergies and autoimmune conditions—in rural African and Amazonian populations with non-Western diets and minimal exposure to antibiotics. Khoruts said it might be necessary to seek out our ancestral microbes—the ones we harbored before the advent of the antibiotic era—within these communities. "It's just a disappearing resource," he said.

Australia's Borody, a pioneer in the fecal transplant field, said he agreed with the principle but cautioned that any donor screening process would also have to account for endemic parasites and pathogens. Researchers, he added, still knew very little about the particulars of a complex and variable organ that might derive its power not only from diverse bacteria but also from fungi and viruses such as microbe-infecting bacteriophages. "The shortest way of saying it is, 'We don't know shit, man,'" he said.

● ● ●

Getting more acquainted with our inner inhabitants may require some better strategies for overcoming the all-too-familiar baggage of disgust. Scientists like Khoruts have urged a deemphasis on the poop part, preferring more technically accurate terms like "intestinal microbiota transplant." For donors, another way past the "yuck" factor has been to emphasize their curative powers. Borody, who cited Ben Eiseman's once obscure 1958 study as one of his early inspirations, said he and his staff

regularly notified donors whenever the center used their poop to successfully treat a patient. The news, he said, often brought them to tears.
The rest of us may need stronger tactics. Give it a better name. Dye it
blue. Disguise the smell with lavender or citrus or pine. Take it out of
its normal context and surround it with stainless steel and clean glass
surfaces. To minimize those ancestral voices of disgust, poop must be
made more palatable, advised Val Curtis, the disgustologist we met in
the previous chapter.

Curtis recalled working with clinicians who once told patients to
collect their fecal samples in old butter tins (akin to using Tupperware).
No wonder people wouldn't do it, she said. The same off-putting associations could hamper the delivery process, Curtis asserted, one reason
why methods that discreetly encapsulate the poop into pills may prove
more acceptable. "Our disgust system evolved for us to detect threats
outside our bodies that want to get inside our bodies," she said. That
system may spring into action if a parasite crawls across your arm.
But it won't necessarily stop you from swallowing a triple-coated pill
that releases its contents only when it has reached your intestinal tract.
"You're fooling the voices of your ancestors if you do put it into a capsule," she said.

Or a "crapsule," according to Borody's daughter. His own poop-in-
a-pill technique, she joked, could be delivered in a "shitment." It was the
therapy of the future, he and other researchers concluded, in large part
because it could replace a far more invasive colonoscopy. If only the shit
part could be a bit more refined. Just weeks after a Canadian research
team announced its own take on the feces-filled capsule method,
dubbed "poop pills" by the press, Shepard brought the new technique
to RDS Infusions in Tampa. He filled thirty-five triple-coated capsules
with donor poop and used them to successfully treat a nursing home
patient on her fourth bout of C. diff. Even then, he and others predicted,
the crude method would likely be fundamentally different in a few years.

Colleen Kelly, a gastroenterologist in Rhode Island, similarly predicted a rapid shift that hasn't fully materialized nearly a decade later.
Nevertheless, the field had begun moving away from whole-stool transplants toward other alternatives. "And it's not just because it's not

aesthetically pleasing, it's more just because it's really tough to try to find these donors," Kelly told me. Even if you've identified the perfect donor, how often can you really expect them to deliver? "We don't have farms with cows lined up in stalls," she said. Petrof was likewise conflicted about the "poop milk shakes" she was putting in her own patients. There had to be a better way to do it. Her angst yielded an infusion from a more defined subset of ingredients—essentially a mix of bacterial cultures. She called it a synthetic stool. Or "rePOOPulating" the gut. It smelled, well, different, she said. At least that's what the endoscopy nurses told her. Unpleasant, but not as bad as the real thing. The cloudy white liquid in a syringe just didn't seem as gross.

But the synthetic stool infusions faced their own obstacles. For one thing, the bacterial mix required a finicky anaerobic growing environment dubbed a "robogut": something akin to an artificial colon. Then there was the question of whether Petrof's team had stripped down its mixture of thirty-three bacterial strains a bit too much. Health Canada's regulatory officials put her therapy on hold pending the outcome of more tests and asked whether it was possible to simplify her slurry even more. The request seemed to crystallize the tension between the regulatory desire for simplicity and the scientific need for complexity. The principles of microbial ecology, Petrof said, pointed to the need for a robust bacterial community. And that likely meant more, not less, diversity. "We're trying to maneuver our way through this rather muddy water," she said.

She had plenty of company. The initial therapeutic failure of a widely touted mix of about fifty spore-forming bacterial species isolated from stool donors, known as SER-109, underscored the difficulty of replicating the gut microbiome's natural defense system against C. diff infections. The blend, by Seres Therapeutics, later rebounded with more promising results in a subsequent Phase III trial when researchers increased the dose. By early 2021, Petrof and colleagues found early success with their own encapsulated mix of forty species called MET-2 that they had isolated from the stool of a healthy donor and learned how to grow in the lab. In a small open-label trial, they reported resolving recurrent C. diff in fifteen of nineteen patients after a multi-day course of pills, and in all but one of the remaining patients after a higher-dose follow-up.

Clinical trials were continuing to expand the experimental repertoire of FMT therapies. They took aim at other stubborn bacterial infections and gastrointestinal complications. At autoimmune and inflammatory conditions. At metabolic, neurological, and psychological disorders. Amid all of the excitement, though, the field was rocked by another major setback, one that researchers had feared for years. In June 2019, the FDA issued an urgent safety alert after two patients in experimental trials at Massachusetts General Hospital in Boston contracted a multi-drug-resistant *E. coli* infection from the same fecal donor whose sample had been frozen for months. One patient with cirrhosis of the liver became seriously ill. The other, a blood cancer patient who had been immune suppressed in advance of a bone marrow transplant, died from a bloodstream infection.

The FDA had struggled for years over whether to regulate FMTs like tissue or blood donations, as advocates wanted, or as an investigational new drug, which would favor drug companies. In response to the infections, the agency temporarily halted multiple clinical trials until the sponsors could demonstrate adherence to beefed-up testing requirements. In March 2020, another FDA safety alert warned of a second round of *E. coli* infections in six C diff. patients. This time, investigators linked the infections to contaminated donor stools from OpenBiome. Stools distributed by the nonprofit were initially suspected in the deaths of two other chronically ill patients also being treated for C. diff. Follow-up tests, though, suggested that at least one of those deaths was unrelated to the transplant, while a doctor treating the other patient attributed that death to an underlying cardiac condition.

Christian Lillis told me that the FDA still could have put FMTs into their own, complex category, like it had for blood and blood products. But the *E. coli* cases escalated the fight that had pitted doctors and advocates pushing for that regulatory classification against pharmaceutical companies hoping to create alternative therapeutics under the more tightly controlled drug classification. The companies won the tug-of-war when the FDA retained its status quo designation of the microbiota transplant as an investigational new drug under an "enforcement discretion" policy for recurrent C. diff, opening the door to new proprietary blends. Poop was still free, Lillis said, but distinct formulations of its active constituents could become lucrative intellectual property.

In October 2020, a decade after their mother died, Christian and his brother, Liam, cohosted a virtual edition of the Peggy Lillis Foundation's annual gala: "C. diff Is a Drag." The theme could have summed up the entire year, except that it was hard not to smile when Cacophony Daniels, the gala's drag queen emcee who informed her Zoom viewers that she had been performing from her basement for the last seven months, belted out Cher's "Strong Enough" while bedecked in a blond beehive wig and glittery pink dress against a background of blue-silver streamers. The mood was corny, sweet, and defiant. Christian, who changed midway through the gala from a blue button-up and black blazer into a green T-shirt that read, "I give a" followed by the poop emoji, implored viewers with C. diff infections to come out and be visible. "We have a big taboo in the US about talking about poop," he said, repeating one of his common refrains.

They introduced a series of energetic survivors who had become committed advocates as well as honorees, including Teresa Zurberg, a canine scent-detection specialist at Canada's Vancouver Coastal Health, and her English springer spaniel, Angus. Zurberg, an expert in drug and explosives scent work, nearly died from C. diff in 2013. Then she trained Angus to sniff it out, helping to point out hotspots in Vancouver General Hospital as part of a team effort that cut the infection rate by nearly half within two years. Both of them appeared on Zoom, though Angus seemed more interested in a pink unicorn toy. Daniels neatly summed up the evening, "Doctors, dogs, and drag queens, oh my!" and then capped it off by singing, "Somewhere over the Rainbow" as family photos of Peggy Lillis flashed on the screen to her left.

Some hopeful signs have appeared in the decade since Peggy's death. In the US, hospitals began reporting C. diff infections to the CDC's National Healthcare Safety Network in 2013 and the overall number of cases dipped between 2011 and 2017. The estimated burden of community-associated cases remained unchanged, but health care–associated cases dropped by more than one-third. As COVID-19 raced across the planet, though, FMTs all but ground to a halt. The FDA required additional testing for the SARS-CoV-2 virus before it would authorize stool banks to continue sending out fecal samples. Khoruts's program at the University of Minnesota satisfied the requirement by developing an extensive screening protocol for donors over the summer. But OpenBiome, the nation's largest provider, instead sought out a test for the poop itself, only to see its application languish as the same FDA division charged with reviewing it waded through a flood of COVID-19 vaccine candidates.

The nonprofit finally won approval to resume regular shipments in May 2021. By then, it was too late. Tenuous finances forced it to begin phasing out its extensive stool banking program, sell off equipment and other assets, and focus on supplying hospitals in its core network. A subsequent collaboration with Khoruts's program in Minnesota offered a bit of a reprieve by helping the nonprofit meet the growing demand for C. diff FMTs until the arrival of the first FDA-approved alternative. Eventually, though, the slack would have to be picked up by start-ups such as Finch Therapeutics, whose chief executive officer, Smith, had been OpenBiome's research director. Seres and other companies with names like Ferring and Vedanta were likewise rising to the fore with promising but still unproven transplant alternatives based on their own blends of purified bacterial strains.

• • •

When Joe Timm and I talked again in 2021, the former OpenBiome donor was thirty-two and living in Boulder, Colorado. This time, he told me his full name and proudly recalled that he had donated at the stool bank for more than two years, one of the longer tenures. At one point, the nonprofit was also using his stool for a clinical trial on ulcerative colitis. Timm had recruited his roommate, who donated for nearly as long; both of them were broke and training for the Boston Marathon at the time. Timm had been excited to be on the forefront of medicine, though the $40 per donation certainly helped. He remembered feeling stressed about getting his deliveries in on time so that he wouldn't waste anything. Otherwise, it felt like flushing money down the toilet.

One day, Timm hoped to tell his grandkids about his early role in the medical procedure. His side gig had even become an icebreaker with some new acquaintances. "One of my fun facts is I used to sell my poop for cash," he said. "That begs the question, 'What? Tell me more!'" He realized that what he possessed had a power that wasn't fully understood but was nothing less than miraculous for someone struggling with repeated infections. It had value. The fact that he was paid for it, Timm said, had another curious benefit: it largely overcame others' initial disgust. "People are like, 'No way, you got to sell your poop?'" he said. "If I said I was just donating my poop for a good cause, not getting paid for it, people might think that's weird." But attaching a value to it shifted perceptions. "Because that's a considerable amount of money for going to the bathroom," he said.

The future of FMTs is far from settled: How do you regulate a complex gut ecosystem as an investigational new drug? If they prove their mettle, pills based on proprietary blends of gut microbes grown in the lab may help resolve some safety and donor availability concerns. Lillis and other patient advocates, though, worry that the push toward more palatable and marketable formulations could come at the expense of affordable and accessible treatments. This much is clear: the transition from throwaway by-product to precious commodity has been nothing short of remarkable. Within a few decades, FMTs have evolved from scorned folk remedy to desperate DIY treatment to highly effective and accepted medical therapy.

Despite all the doubt and ridicule, poop has affirmed its value and launched a search for gut microbes that might become medical marvels. Doctors are revisiting strategies based more on restoring community balance than on removing individual threats. Proud donors like Timm are no longer anonymous. Patients are breaking their silence and describing how a moment of grace can cause a lifetime of relief. We may not know *shit*, but at least more people are giving one.

CHAPTER FOUR

Memory

O N A SUNNY OCTOBER AFTER-
noon in 1977, a German shepherd named Crow waited with his
handler at one end of a football field
in Albany, New York. A camera operator had set up near midfield in the multi-
purpose Bleecker Stadium, built in 1934 as
part of the Depression-era Works Progress Admin-
istration, to record the proceedings. At the other end of the field, police
had erected five tall plywood barriers, each braced from the front and
labeled 1 through 5 in large black numerals. A man hid behind each one.

Kickoff. Crow and New York State trooper John Curry began walking
down one side of the field toward the hidden men as part of an unusual
experiment. The state's police force had acquired three German shepherds from the US Army in 1975 to launch its own K-9 unit, and initially
trained them to detect explosives. According to an account by anthropologist Denis Foley, Curry had spent several months training his dog
to accurately identify individuals by the scent of their feces. The trooper
called Crow the best tracker he had ever worked with. The next few
minutes, though, would determine whether Crow's remarkable sense of
smell could help sniff out a serial killer who had left behind a critical
clue after a sadistic double murder.

In Germany, a different mystery surrounded the partially mummified body of a young girl in an airless apartment. The malnourished toddler had short, dark hair, weighed just shy of fifteen pounds, and would
have stood two-feet-seven-inches tall when she was found near her bed
on July 10, 2000, still dressed in a light-colored tank top and badly soiled
diapers. She had lived with her twenty-year-old mother in central Germany until her mother stopped paying rent and left to go stay with an

uncle. "When asked by the police, the mother did not remember when she had seen the child for the last time," according to one account. The woman had been living with her uncle for about two weeks, and reportedly told him that the toddler was staying with a grandmother. "She also asked him how long a human could survive without food."

German police had two main questions about the child's tragic death: When did she die and how long had she been neglected? To solve the mystery, Rüdiger Lessig at the Institute for Legal Medicine at Leipzig University collected maggots from several parts of the girl's body: from her genitals and anus, and from her face. He killed the fly larvae in hot water and preserved them in a solution of 70 percent ethanol and then shipped them to a forensic entomologist sometimes known as Maggot Man.

● ● ●

Like other living things, we signal our presence through what we excrete, and observant scientists are piecing together remarkable stories from evidence that was long dismissed or ignored. Poop is not just a guardian that might save us from disease, but also a witness that can recall our human experience. There is vast memory in the DNA and odors and microbes and insects that populate what we leave behind. The scientific detectives who decipher the clues and fill in the blanks can help re-create a moment in time and solve cases to bring a measure of justice to the victims.

Extend it further back in time, and preserved poop can act more like a buried time capsule to tell us how people lived and migrated, where their ancient settlements rose and fell or persevered, and how they dealt with death, disease, and the world around them. We have long sought the story of our past in the written fragments and other artifacts of our ancestors: Dead Sea Scrolls, Rosetta Stone, volcano-buried cities, bits of long-lost epic poems. What we have not realized until the past sixty years or so is that civilizations often leave behind libraries far larger than Alexandria's, largely unburnt if a bit less grand. It turns out that the histories tucked away in those unassuming libraries can prevent the erasure of a person or a neighborhood or an entire metropolis.

In Albany, New York, Fifty Columbia Street is now a blank space in the city: a small parking lot sandwiched between a law firm and a deli. But

Flush

on the day before Thanksgiving in 1976, the John F. Hedderman Church Goods store that stood on the site was the scene of a shocking double murder. Robert Hedderman, the store's owner, was found that afternoon in a pool of blood in a small bathroom at the far back of the store. He was still wearing his overcoat, and his wallet held $110 in cash. He had been bound, and his throat slashed so deeply that he had been nearly decapitated. Margaret Byron, Hedderman's clerk, was found separately in a rear storeroom. She had been bound as well, perhaps with cord from a religious vestment. She had been strangled and stabbed repeatedly in her left breast. Her purse remained, but her watch was missing. Hedderman's seventy-eight-year-old father, who had gone to another room near the back of the store to nap, had unbelievably slept through it all and didn't recognize his own son when he awoke and called a police dispatcher to report finding a dead man.

Later that night, two detectives followed a trail of blood from the religious store to a trash can on Broadway near Columbia, in which they discovered a bizarre clue: a soiled priest's alb. It had been smeared with blood and feces, and the trail of blood continued north on Broadway, in the general direction of a furniture upholstery store called the House of Montague.

The investigation soon zeroed in on Lemuel Smith, a thirty-five-year-old janitor at the upholstery business. On the afternoon of the murders, Smith had been free on parole just shy of seven weeks, after spending eighteen of his last twenty years in prison. In 1969, he had assaulted and attempted to rape a woman in Schenectady, New York, and then assaulted, abducted, and raped a second woman on the same day. A decade earlier in 1958, he had assaulted a woman in Baltimore by beating her with a fifteen-inch iron pipe. Six months before that, he had been the main suspect in the murder of a friend's mother in Amsterdam, New York, when he was a sixteen-year-old basketball star, but the police there had been unable to gather enough evidence to indict him.

The Albany investigators used lie detector tests on witnesses and early suspects, but Smith refused to take one and his seventeen-year-old girlfriend claimed she had spent the whole day with him, giving him an alibi. Several witnesses saw someone matching Smith's description leaving the store around the time of the murders—and one saw the man throw something that looked like robes into the trash can on Broadway—but detectives doubted the strength of their testimony. More evidence emerged: a strand of hair on Smith's navy-blue wool sweater matched the brown-dyed gray hair of Byron, and black hairs on her body and in the bathroom near Hedderman's body were consistent with Smith's hair. Bloodstains on the alb matched Smith's type-O blood. At one point, out of desperation, the investigators brought in a psychic to help shed light on the case. And then an Albany detective remembered Trooper Curry and his talented detection dog. He reasoned that a scent lineup matching Smith to the feces and bloodstained alb might apply enough psychological pressure to extract a confession.

In Europe, scent lineups have figured in criminal investigations since the start of the twentieth century. Detection dogs first sniff an item collected from a crime scene, or an "evidence" scent, much like a bloodhound smells the clothing of a missing person. But instead of tracking an odor plume or footsteps to find the person, dogs in a scent lineup pass by a row of five to seven scents, including the "target" scent collected from the suspect and "comparison" or foil scents collected from other people. The dogs then choose which one, if any, matches the original evidence scent.

In 1903, in Braunschweig, Germany, a policeman named Bussenius and his detection dog made the first-known identification of a murder suspect using the tactic. A girl had been murdered on a farm and the leading suspect, a farmhand named Duwe, was included in the lineup. Accounts differ on what happened next: in one retelling, Bussenius asked Duwe and five others in a lineup to hold a pebble in their hand and then put it on the ground. The police inspector's trained dog, a German shepherd named Harras von der Polizei, first smelled a knife found at the crime scene and then matched it to the pebble that had been held by Duwe, who subsequently confessed to the murder. A different version of the story suggests that the dog first sniffed around the crime scene and then repeatedly lunged at the guilty farmhand in a lineup of all twelve farm employees.

Scent lineups didn't begin appearing in the United States until the 1970s and are still a controversial forensic technique sometimes dismissed by critics as "junk science," in part due to the method's perceived unreliability and potential for cross-contamination, bias, and abuse. Whether such evidence is admissible in court varies widely by jurisdiction. Given the training, resources, and procedures required, some researchers have suggested that the FBI may be one of the few agencies capable of producing admissible scent lineup evidence in a US court of law. But proponents have sought to refine and standardize the methodology to demonstrate its validity and have argued that rigorous detection dog training can yield highly accurate and reliable forensic evidence or help secure other useful information. "Interestingly, confronting the suspect with a positive identification often leads to confession," a French research team wrote in one study. From 2003 to 2014, the authors noted, positive identifications by dogs from what is now called the National Forensic Science Service helped to solve 120 out of 435 criminal cases that used the detection tool.

If the Albany lineup had been conducted today, Lemuel Smith and four or more men matching his age and ethnicity might have been asked to handle a sterile cotton square for ten minutes to transfer their scent onto it. Each square would have been placed in a separate jar or tin out of sight of both Curry and Crow, and three trials would have determined whether the dog could consistently match the odor from the blood-and-feces-stained alb to the cotton ball handled by Smith. The actual lineup used four men as the comparisons and Smith himself as the target odor. In a photo taken that day, he is dressed in a light-colored shirt and bell bottoms, his arms folded across his chest as he stands behind barrier number 5. He had been tipped off by his lawyer about the forensic strategy, and reportedly scrubbed himself vigorously in the shower that morning, to no avail. Crow, Foley writes, "went directly to the incredulous Lemuel Smith." The police conducted the experiment three times in all, and the dog trotted over to Smith each time, regardless of which plywood barrier the suspect was hidden behind.

Forensic entomology, remarkably enough, also factored heavily in an early version of a police lineup. In its first known use, described in a thirteenth-century textbook titled *The Washing Away of Wrongs*, a Chinese lawyer and investigator named Sung Tz'u described how he used blowflies to help solve a stabbing murder near a rice field. After all the workers in the field set down their sickles, as Sung instructed, traces of blood on one of the harvesting blades began to attract the flies. Confronted with the evidence, the tool's owner confessed to the murder.

Now widely acclaimed as one of Europe's foremost forensic entomologists, Mark Benecke's "childlike curiosity" first attracted him to the field when he was studying genetics at the Institute of Legal Medicine at the University of Cologne, where he received his doctorate in 1997. He worked in the building's tiled basement, where corpses were brought in for forensic exams. "Since I was a biologist, I went over to the corpses and the most interesting thing was everything that was creepy-crawly stuff. These were the insects," he told me. Benecke continued his own informal study of insects while working as a forensic biologist at the Chief Medical Examiner's Office in Manhattan in the late 1990s. He and a colleague would linger in the Decomp Room, where frozen corpses were stored, and retrieve "spare parts" from the toxicology portion of autopsies to conduct their own experiments in an unused room next to the lab's shower. For one experiment, they added maggots to liver and muscle tissue samples from people who had died of drug overdoses to see whether drugs altered the insects' growth. The ad hoc experiment ended abruptly when someone else in the office—they never found out who—threw out the tissue and maggots. Benecke supposes that the smell and disgust probably factored into the decision.

He has since taught and spoken around the world and become a prolific author on topics ranging from forensic entomology and snails to body modification and youth subculture. The freelance forensics expert is heavily tattooed and sometimes called Maggot Man as a nod to his avid interest and expertise in their development. This unusual skill set would prove pivotal in investigating the German toddler's death.

Forensic pathologists often sift through poop at crime scenes to extract DNA or RNA and glean clues about what the victims or suspects ingested. The presence of toxins or other unexpected substances may help explain a suspicious illness or an unnatural death. As in life, our bodies also host a wide range of colonizing species after we die. Hundreds of kinds of arthropods, especially flies and beetles, use human corpses for shelter, food, or breeding grounds as part of nature's decomposition process. Some scavengers arrive immediately after death while others subsequently take over in overlapping periods of colonization known as faunal succession. From the identities and growth rates of different colonists, forensic entomologists can work backward to estimate when they arrived and ultimately, when a person died.

Many countries have the same general progression of insect colonists. But differences exist, Benecke said, especially in less industrialized places. When he travels for work, he often uses a bottle of rum or whatever alcohol is on hand to preserve insects plucked from corpses until he can get back to his dissecting microscope to more carefully examine them. For the German toddler's case, Benecke received a shipment of maggots preserved in a more typical ethanol solution. Of those retrieved from the girl's face, he identified larvae from the bluebottle fly, *Calliphora vomitoria*. This species commonly inhabits corpses early on, he noted, and based on the estimated larval growth rates during that unusually cool July in Germany, he estimated that adult flies had arrived and laid eggs six to eight days before her body was discovered. The finding helped to establish her time of death as sometime between July 3 and July 5, 2000. But that detective work still left open the question of how long the young girl had been left alone.

● ● ●

Reconstructing a sequence of events is always a tricky science, and the difficulty grows over time. Witnesses' recollections grow fuzzy, records are lost, evidence is buried. Sometimes the gaps blur the circumstances of a person's death. At Cahokia, a pre-Columbian city along the Mississippi River famous for its dozens of earthen mounds, time has obscured the fate of an entire metropolis. "America's first city," as it's sometimes known, boasted a population of at least 10,000 and possibly up to 20,000 people at its height in 1100 CE. At that size, it would have rivaled medieval London. Cahokia, in what is now a suburb of St. Louis, had desirable neighborhoods in its center and more pedestrian suburbs on its outskirts. There were open courtyards for playing an ancient game with stone discs— called chunkey—and markets that sold exotic products like copper and shells from the Great Lakes and Gulf of Mexico to the more well-to-do. Farmers supplied the city residents with maize and other crops, while hunters and fishers contributed to the meat supply. And then, according to prevailing storylines, it all disappeared, leaving behind a ghost town full of enigmatic mounds.

To fill in the blanks in the city's timeline and more objectively estimate the ebbs and flows in the region, anthropologist A. J. White and colleagues turned to what's known as a fecal stanol population reconstruction. Stanols, molecules that come from the plants and animals we eat, can act as remarkably persistent markers of where we've been. As omnivores, we tend to have more of a stanol called coprostanol in our poop than strict plant-eaters do. Coprostanol forms when gut bacteria partially digest the lipid cholesterol (cholesterol is notoriously hard to break down, which is why it can clog arteries).

Based on the ratios of eleven telltale stanols, an international group of researchers created a kind of chemical library that can be used to match fecal signatures to humans and nine other animals. Reindeer, for example, favor lichen and leave behind more lichen-derived compounds in their droppings than other herbivores like sheep and cows. The signature for human poop, they found, is closer to that of fellow coprostanol-producing omnivores like pigs and dogs. Even so, the ratios of multiple stanols can distinguish us from both animals; despite often eating the same food as our canine companions, we don't process it in quite the same way.

White's analysis was further eased by the fact that large, domesticated animals didn't live in Cahokia until they arrived with the first Europeans, meaning that a coprostanol-heavy signature could be reasonably attributed to humans. The researchers collected two sediment cores from Horseshoe Lake, a freshwater basin that collected runoff from nearby fields and other places by the city where people defecated. The fecal compounds glommed onto particles of silt or sand and fell to the bottom and accumulated in layer after layer of the lake's sediment. Back in the lab, the scientists used a sensitive method for measuring chemicals to detect the faint traces embedded in the lake cores.

Carbon molecules trapped in organic matter like twigs and leaves helped the researchers date each layer of lake-bottom sediment. The fecal stanol measurements suggested that the Cahokia area's human population was already relatively high in the 800s, and then spiked around 1100. By 1400, both archaeological and stanol evidence suggests that the city's population had plummeted. After that date, White said, the prevailing narrative focuses on abandonment of the metropolis. "So it would be very easy to think that that was the end of Indigenous people in this area because that's where the story stops." Only it doesn't, at least not according to evidence like his team's measurements.

Around 1500, according to the fecal stanol data, the population partially rebounded and then plateaued in the 1600s before European missionaries and colonists began arriving. That uptick suggests the apparent abandonment of Cahokia was only temporary, and that a subsequent wave of Indigenous people, perhaps migrating from the north and east, repopulated the area. By 1699, when French missionaries began documenting the region, it was already populated by subgroups of the Indigenous Illinois Confederation. Grass pollen trapped in the sediment of Horseshoe Lake suggests that by then, southern Illinois had transitioned from woodlands to more open prairie land, attracting both bison and bison hunters. A separate spike in bits of sedimentary charcoal during the 1600s, White said, might have come from wildfires. But the confluence with the area's apparent population growth might alternatively signify an increase in cooking fires or controlled burns as part of a grassland management strategy.

The same kind of lake sediment analysis provided researchers with evidence of another ancient settlement's shifting fortunes on what is now Vestvågøy Island in Norway's Lofoten archipelago north of the Arctic circle. From matching the fecal evidence with chemicals linked to environmental conditions, the study suggested how the climate and other changes impacted the settlement over time. About 2,300 years ago, the researchers found, levels of human and animal feces rose significantly, agreeing with other signs of settlement on the island's lakeshore. The human and grazing animal populations both peaked around 500 CE, coinciding with a transition from forests to grasslands. After a subsequent decline and partial rebound in the early Middle Ages, emigration to Iceland and the arrival of the Black Death may have contributed to another population dip between 1170 and 1425. Around 1650, in the midst of a colder period sometimes called the Little Ice Age, a spike in grasses and shrubs suggests that the remaining inhabitants may have burned peat and cleared some of the surrounding forest for firewood.

The reasons for Cahokia's initial depopulation are less clear and remain hotly debated; White and colleagues believe that a Mississippi River flood and periods of drought around 1150 could have generated "significant stress" for the city's inhabitants and contributed to a partial exodus. A separate research group found no evidence of increased local flooding of creeks caused by rampant deforestation and upstream erosion, though—an argument against another common narrative of a self-inflicted "ecocide," or widespread destruction of the environment. White backs the idea that such willful degradation didn't cause Cahokia's contraction.

White's interpretation that the city subsequently rebounded in the 1500s and 1600s has attracted controversy as well due to the lack of significant archaeological finds dating to that period. "This is where I think the poop comes in and fills in gaps in the archaeological record because archaeology isn't perfect," he said. Cahokia's earlier inhabitants, after all, had built 100-foot-tall earthen mounds that tended to stand out, while the traces of later residents with different cultures may be far subtler. The physical evidence of an Indigenous village may have been destroyed or paved over, or perhaps researchers just haven't found it yet.

"The collapse narrative is unequally applied to Indigenous people throughout the world including Native Americans," White told me. "We talk about the collapse of the Anasazi or the collapse of the people of Cahokia." But a physical site that remained unoccupied for a time, he said, is not the same thing as the collapse of an entire civilization. And as his research suggests, the physical site of Cahokia was then repopulated. "For whatever reason, I think people find the collapse more exciting and interesting and that's maybe why we focus so heavily on it," White said. He did as well when he began his research.

But history doesn't end when we stop looking. "We can't just stop there. We have to tell the whole story," he said. The fecal stanol population reconstruction corrects the previous "collapse" narrative by pointing toward a more continuous human presence in the area. Today, multiple Indigenous groups claim strong ties to Cahokia, including the Osage Nation and the

Peoria Tribe. Both were forcibly relocated to Oklahoma reservations in the nineteenth century. In other words, White said, it wasn't environmental changes that pushed them out. The US government did.

• • •

As a storyteller that has helped to reshape our understanding of the past, human poop can be as illuminating as a runestone. But until recently, scientists often treated it as more of a nuisance. In a written reminiscence of his early career, Texas anthropologist Vaughn Bryant recalled the casual disdain with which researchers greeted discoveries of fossilized feces, or coprolites:

> During my undergraduate studies in the early 1960s, I visited my first archaeological site: a dusty rock shelter perched in the side of a canyon wall in west Texas near the Rio Grande. I noticed that each morning during the screening process the workers found dozens of flat, cow patty–shaped human coprolites (dried human feces), which they would carefully remove and pile up at the foot of the screens. These were considered worthless junk and a nuisance because the smaller pieces clogged the screens and delayed the process of looking for what they considered to be far more important artifacts. Later, we were treated to after-lunch entertainment when the screeners gathered at the edge of the shelter for their daily game: "Frisbee throwing." As each coprolite sailed out over the canyon the crowd would cheer or laugh, depending on how far the thermal updrafts carried each coprolite. It was great sport and I even tried my luck at throwing along with the rest. I did not know it then, but we were discarding some of the most valuable data being excavated from that site.

Around the same time, a researcher at McGill University in Montreal named Eric Callen was pioneering the emerging field of coprolite analysis, which was proving especially useful in detecting eggs from human parasites and grains of pollen from past plant life. But Callen lamented that his university colleagues ridiculed him and considered his work a "waste of time"—a tart dismissal given his specialty. When Callen died of a heart attack in 1970 at an archaeological site in the Peruvian Andes, Bryant managed to save his coprolite collection and move it to Texas A&M University.

As Bryant and archaeologist Glenna Dean later noted in their tribute to Callen, he would have been pleased to see how the field had expanded and gained respect. "The desire of the public to know and to touch their ancestors, anybody's ancestors, whether dead for one century or for millennia, is almost insatiable. Re-creating the very being of individual men and women from DNA and hormonal evidence left in coprolites, those most personal of artifacts, resonates with the public in a way that even facial reconstruction of skulls cannot," they wrote.

In Copenhagen, researchers have captivated the public by re-creating the daily life of a forgotten Renaissance neighborhood in large part through some very personal artifacts deposited in two wine barrels used as latrines. Sometime in the late 1680s, during the construction of a new thoroughfare to Copenhagen's northern city gate, workers demolished and paved over a house and what may have been a small outbuilding with a sunken space that contained two Rhineland wine barrels. Roughly forty years later on October 20, 1728, a catastrophic fire, aided by "strong winds, empty water conduits, drunken firefighters and narrow streets," destroyed that district and almost half of the city's older medieval section. Copenhagen again built over the rubble—this time putting in wider streets and a grand new square named Kultorvet (Danish for "Coal Market").

During renovations of Kultorvet square in 2011, archaeologists unearthed the wine barrels, which had lain undisturbed for more than three centuries. Researchers quickly determined that the exceedingly well-preserved barrels, each about three feet wide, had been repurposed as latrines and rubbish bins. Based on the grains, seeds, fruits, pollen, animal bones, and parasitic worm eggs found within them, the scientists launched an ambitious effort to reconstruct the diet, lifestyle, and general health of the barrels' users.

Mette Marie Hald, one of those researchers and an archaeologist at Denmark's National Museum, has developed something of a reputation as the go-to person when someone unearths another latrine in the country. You might think that this is an uncommon occurrence, but you would be wrong. In 2020, Hald and her colleagues published a study of twelve historic latrines found around Denmark, including the one at Kultorvet, Copenhagen's public square, and another that dates back to the Viking Age in the ninth century. She figures there might be forty or fifty similar discoveries scattered around the country, which to me suggested a novel travel itinerary, but they're not always analyzed because even archaeologists are sometimes dismissive: "Well, one more. Do we need to know? It's a latrine."

Hald's work, however, has provided some rather revealing looks at old neighborhoods and ancient settlements. Her specialty is identifying plant remains such as seeds and grains. One of her colleagues examines plant pollen, another looks at animal bones, and yet another analyzes parasite eggs. A fourth is extracting DNA. So far, the pollen evidence alone has provided the oldest archaeological observations of cucumber, rhubarb, citrus, and cloves in Denmark.

Viking-era pit latrines aren't always so clear-cut. "It's a bit like a compost heap, basically," Hald said, with food scraps and other waste sometimes intermingled with the poop. Prehistoric latrines are even harder to discern and often rely on the extraction of DNA or isolation of compounds characteristic of human feces. In 2008, scientists announced that they had extracted DNA from 14,300-year-old human coprolites in the Paisley Caves of southern Oregon. At the time, the find pushed back the date of the earliest-known human arrivals in North America (the 2021 discovery of fossilized human footprints in New Mexico dramatically extended that date to at least 21,000 years ago). A competing fecal stanol analysis, though, suggested that the Paisley Cave DNA could have come from an herbivorous animal instead. In 2014, one of the researchers from that opposing study led a separate chemical analysis of 50,000-year-old Neanderthal coprolites from an ancient hearth in southern Spain. That study, again relying on chemical markers in the fossilized poop, deduced that our prehistoric relatives weren't entirely carnivorous, contrary to many previous hypotheses. Some early critics questioned whether the coprolites could have come from bears, but a follow-up study isolated human gut-associated microbes from the Spanish poop and found evidence for what may have been a core microbiome shared by both Neanderthals and modern humans.

Medieval and Renaissance latrines tend to be more obvious, Hald said, and often include the remains of surrounding structures like the benches that people sat on while doing their business. To ease the excavation work at Kultorvet square, a worker sliced lengthwise through one of the wine barrel–latrines, and Hald immediately recognized plant seeds in the newly exposed layers of organic matter. "I remember coming out there seeing cherrystones and fig seeds and stuff like popping out of the layers there," she said. "Yeah, it was really, really clear from the side. It was quite amazing." Unlike the compost-like material from older latrines, the dense layers of Renaissance shit had the consistency of butter, she said, and could be cut out in cubes. And from those cubes, Hald and her colleagues retrieved some fascinating vestiges of daily life in the seventeenth century:

Bits of brick
Sand and gravel
Moss, straw, and hay
Whipworm, roundworm, tapeworm, and mite eggs
Rib of a kitten
Bone fragments from a small bird
Tooth of a piglet
Fish including herring, cod, and eel
Apples
Raspberries
Wild cherries
Wild strawberries
Elderberries
Dried figs
Bitter lemon or orange peel
Grapes or raisins
Turnips
Lettuce
Mustard
Coriander

Flush

Hops
Cloves
Buckwheat
Rye
Barley
Oats

The unfortunate kitten and bird may have been found dead and disposed of in the latrine. Hald thinks the straw, hay, and moss could have been used as toilet paper while she believes the buckwheat likely wasn't food but instead a packing material for clay tobacco pipes imported from the Netherlands. The barrel's mix, in essence, offers an unvarnished view of what the residents pooped out or threw out. "It's really interesting to see whatever else appears in this rubbish context because it's real life," she said. "It's like completely unfiltered. This is what they had. This is what they ate. This is what they disposed of, for whatever reason, rather than the facade that they would have wanted in front of other people."

After an early focus on pots and "the shiny gold stuff" like coins and jewelry, British paleopathologist Piers Mitchell told me, archaeology matured, and scientists eventually considered examining someone's intestinal contents to be a perfectly reasonable field of study. "If you don't study the people that made your coins and your pots and your jewelry, you are only looking at half of that population's important information," he said. In Herculaneum, Italy, entombed by the eruption of Mount Vesuvius in 79 CE, ash helped to preserve such a window into the lives of residents from an entire block. Excavations in one branch of the city sewer beneath a residential and commercial complex that included a bakery, wine shop, and two stories of apartments unearthed more than 700 sacks of human excrement and other organic matter. In the sewer, which probably served as both a cesspool and rubbish bin for kitchen scraps, archaeologist Erica Rowan and colleagues identified the remains of 194 kinds of plants and animals—114 of which could be considered food. A nutritional analysis of those foods suggested that varied, nutrient-dense, and healthy diets weren't limited to the city's wealthy residents but allowed those of more moderate means to "achieve modern day stature as well as survive and recover from illness."

Some finds have provided even more detailed views of what might have been happening within residents' guts. In a fourteenth-century latrine discovered under a city square in Namur, Belgium, researchers isolated genes that encode antibiotic resistant proteins from bacteriophages, or viruses that can infect bacteria like those in our feces. The discovery suggests that phages, which normally kill the microbes they infect, aren't always destructive; in this case, their genes may have helped protect gut bacteria from naturally occurring antibiotic compounds. The finding also shows how microbes were finding ways to acquire resistance long before the widespread use of medicinal antibiotics.

Paleoparasitology has likewise revealed the accumulation of helminths, the intestinal worms that have lived with us over the millennia. Mitchell, who studies a wide range of ancient diseases, has identified potential "heirloom" parasites (like roundworm, whipworm, pinworm, threadworm, and beef and pork tapeworms) that have infested humans throughout our evolution in Africa and our migration across the globe. With a ready source of food and lodging deep within the human gut, the parasites often do little to betray their presence. In large numbers or in more vulnerable hosts, though, they can cause abdominal pain, diarrhea, bleeding, weight loss, anemia, and other complications. Even England's King Richard III, who died in battle in 1485, was colonized by roundworms. Mitchell and colleagues found the royal heirlooms by examining soil samples from where the king's intestines would have spilled the parasite eggs into his grave. That grave, incidentally, wasn't discovered until 2012, when archaeologists found it in the remains of a church beneath a parking lot in the city of Leicester.

Mitchell distinguishes the old guard of heirloom parasites from the dozen or more unwanted "souvenir" species we've picked up along the way—examples of zoonotic diseases that can be transmitted from other animals to humans. Along China's famous Silk Road, relay stations built in the Han dynasty offered travelers a convenient place to rest, recharge, and relieve themselves. Here, too, researchers have found ancient parasites. Travelers at the time used sticks wrapped with cloth to wipe themselves (similar to the use of sponges attached to sticks in ancient Rome). From the well-preserved equivalents of toilet paper at a 2,000-year-old relay station in China's arid northwest, Mitchell and collaborator Hui-Yuan Yeh identified the eggs of a Chinese liver fluke, which would have lived in fish inhabiting a freshwater marsh more than 900 miles away. The fluke, in other words, likely found a new home after someone ate a raw-fish dinner and then accompanied the traveler west as a unique souvenir for several months or more.

The ubiquity of heirloom and souvenir parasite eggs in Egyptian mummy guts, Roman and Chinese latrines, medieval graves, and toilets throughout history has suggested to Mitchell, Hald, and other scientists that neither wealth nor good diets were enough to ward off infectious disease amid the lack of health-promoting sanitation. "There was no evidence that the Romans or the medieval people had toilets to make the place healthier, or to reduce your chance of getting sick or getting diarrhea," Mitchell told me. "The main reasons seem to be the convenience of, you've got to go to the toilet somewhere and if you don't want people to see your backside, then you can't do it in the street." So the Romans invented public latrines to keep the streets clear of smelly feces and urine. Sanitation as initially conceived, he said, was more about decreasing unpleasant odors and increasing public accommodation.

From the archaeological record, we can seldom say precisely who let loose into any of these latrines or containers. Then there was the "Latrine of the bishops" in Aalborg, Denmark. Researchers there reconstructed the diet of Bishop Jens Bircherod (or possibly of his wife, the only other person who used the same latrine) through what Danes have dubbed "the bishop clump." The holy shit, originally found in a broken bottle in a latrine in what had been the bishop's palace from 1694 to 1708, had been kept in storage since the palace's demolition in 1937 in what must have been a truly dedicated act of reverence. When researchers finally analyzed the lump nearly eighty years later, they learned that the bishop had been partial to blackcurrants, buckwheat likely sourced from his home island of Funen, cloudberries presumably brought over from Norway, peppercorns imported from India, and figs and grapes that were probably imported, too, but possibly grown in a palace garden.

In York, UK, a roughly seven-and-a-half-inch-long turd from the Viking era is even more celebrated. "I've seen it; it looks like it would have been uncomfortable to produce," Mitchell told me. The famous Lloyds Bank coprolite, found beneath the future site of a bank in 1972, dates from the ninth century and has revealed that its creator subsisted mainly on meat and bread. Unfortunately for the unnamed Viking, the poop was also infested with the eggs of parasitic whipworms and giant roundworms, the latter of which can reach a length of nearly fourteen inches and cause intestinal blockages in some cases. Roundworm larvae, meanwhile, can migrate to the lungs and be coughed up, swallowed, and returned to the small intestine to fully mature. Anyway, one of the coprolite's discoverers deemed the relic "as precious as the crown jewels," though comparable security was perhaps lacking when a visitor to the Jorvik Viking Centre dropped the hefty poo in 2003 and broke it into three pieces that had to be glued back together.

Hald said it's not clear whether the Kultorvet latrine in Copenhagen would have been used by only family members or by their servants as well. But the contents and nearby artifacts such as china, coins, and imported tobacco pipes suggest that some of the neighborhood residents may have been Dutch merchants or heavily influenced by Dutch culture. The figs and bitter lemon or orange peel could have come from the Mediterranean, and cloves would have been imported from Indonesia. Despite the family's cosmopolitan and surprisingly healthy diet, archaeologists again unearthed abundant evidence of parasites in their feces. Whipworm and roundworm eggs suggested a lack of hygiene (little changed from earlier eras) while tapeworm eggs—the oldest yet found in Denmark—were likely spread through undercooked meat.

The discoveries made such a splash, Hald said, that several museums got into the spirit. Copenhagen Museum re-created a kitchen garden on its grounds based on plants found in the city's ancient latrines. The National Museum got even more creative and held an outreach event for children on four Saturday mornings "so they could come in and hear about, well, ancient shit," Hald said. They called it *Lortemorgen*. Crap Morning. The hugely popular event used brown Play-Doh as a touchable stand-in for the real thing and taught the kids how spices found in the country's latrines were traded across the world. "We had some good fun," Hald said.

• • •

For my dog, Piper, few games are more engrossing than racing through the house to sniff out a hidden treat. At the Working Dog Center at the University of Pennsylvania, playtime with the incoming students includes a somewhat more sophisticated game called Puppy Runaways, which is basically hide-and-seek with a rotating cast of human volunteers. Cynthia Otto, who had long worked with search and rescue teams, opened the training center in 2012 as a legacy of the 9/11 terrorist attacks and the detection dogs that toiled in the ruins. All of the training center's canines are good at scent detection, she told me, but they vary in their focus, independence, confidence, and other behavioral traits. In 2020, the program graduated its one hundredth trainee, with a 93 percent success rate based on overall aptitude for detection work. "One of the reasons we have a high success rate is because we let the dogs choose their career path based on what they're good at," Otto said. Many of the graduates have been trained for disaster or wilderness search and rescue missions, while some have been trained to detect human remains, drugs, or explosives for law enforcement agencies. More recently, the center and other groups have expanded into training the dogs for disease detection.

Even as puppies, detection dogs are good at finding humans by our telltale odor. Most trainees also excel in picking out individual scents, including a synthetic blend called the universal detection calibrant that is a kind of generic odor for training purposes that doesn't commit them to any specific career path. Their options, though, are continuing to grow. As pest control assistants, dogs can identify the presence of bedbugs and termites. In conservation work, they can direct their handlers to sea turtle eggs and invasive plants. In Washington's Puget Sound, a Labrador retriever mix named Tucker helped his human colleagues, including one dubbed "the guru of doo-doo," detect the floating poop of orca whales. Samuel Wasser, director of the University of Washington's Center for Conservation Biology, told me that Tucker wasn't so unique.

Detection dogs could distinguish among the feces of eighteen separate species, he said. In Canada's Jasper National Park, for example, they could tell the difference between grizzly and black bear scat. From their perch on the bows of research boats, dogs have guided their handlers to whale poop floating up to one nautical mile away. Even more impressively, Wasser suggested that some dogs have matched one sample of scat to all others from the same individual animal, including specific weasel-like fishers in California and maned wolves in Brazil.

One review of canine scent detection studies put the limit of their sniffing abilities at roughly one part per trillion, or three orders of magnitude better than available lab instruments. The authors suggested that the feat is like detecting a drop (of bleach, for example) in a volume of water equal to twenty Olympic-sized swimming pools. That super-sensitivity is more than adequate to sniff out the changing bouquet of chemicals released from our bodies, whether in life or in death. One famous search and rescue dog named Falco was reportedly able to locate a blood-dipped Q-tip hidden within a five-acre parcel. In 2017, other cadaver dogs detected the twelve-and-a-half-foot-deep grave of three murder victims on a farm in Bucks County, Pennsylvania. In that case, twenty-year-old Cosmo DiNardo confessed to murdering four young men and burying them in two separate graves on his parents' property. DiNardo's cousin, Sean Kratz, was subsequently convicted for his role in three of the killings.

But can dogs distinguish one person's unique scent from another's? Otto, at the Working Dog Center, said the prevailing evidence suggests they can. After decades of conflicting results, researchers in the Czech Republic reported that highly trained German shepherds could even distinguish between the scents of identical twins. In the Albany double-murder case, we still have only the faintest whiff of an idea of what Crow homed in on when he consistently matched the soiled priest's alb to Lemuel Smith. We're learning, though, that the human body emits thousands of small chemicals that easily evaporate in the air. A 2021 tally of these volatile organic compounds or VOCs, as they're known, suggested that we collectively release at least 2,746 of them through our breath, sweat, feces, urine, saliva, blood, and breast milk or semen.

There are undoubtedly many more. We create these small chemicals as by-products of cell metabolism and bacterial processes, and through the breakdown of what we ingest, inhale, or absorb through the skin. Scientists call this library of compounds the human volatilome and it may act like an odor-based signature. Even twins vary in the kinds and amounts of chemicals that they release, which may help explain the Czech research results. An odorprint, in other words, may mark us as individuals in much the same way that a fingerprint does, and dogs seem to be uniquely good at detecting its constituents.

Some of the VOCs found in human feces match those in foods like cumin, truffle fungus, citrus fruits, and cold-water fish, meaning that the compounds themselves probably originate in the environment. Researchers suspect that the gut microbiome produces many others. Bacterial fermentation of the amino acids tyrosine and tryptophan in feces, for example, yields the VOCs phenol and indole, respectively (recall that indole has been variously described as flowery, musty, and "the scent of a healthy 'inner soil'"). Gut bacteria also produce hydrogen sulfide and methanethiol, thought to be the main contributors to "pungent flatus." VOCs can exit with poop or change their composition as they disperse through the blood to other organs like the liver and bladder.

The Albany investigators didn't know any of this and concluded that their scent evidence against Lemuel Smith was unlikely to be admissible in court. Nevertheless, the positive identification and a separate bite-mark pattern analysis linking him to yet another murder may have contributed to Smith's willingness to tape a confession as part of a convoluted insanity defense strategy devised by his lawyer. In his confession, Smith suggested that he had gone to the Albany store to sell his religious art—depictions of crucifixions on glass. Maybe his residual anger from a court-mandated psychiatry session that morning bubbled over when Margaret Byron, the store clerk, sounded doubtful about whether Robert Hedderman, the owner, would be interested in buying any of Smith's pieces. Maybe the store triggered a complicated religious fanaticism that had been stoked by Smith's antipathy toward his fire-and-brimstone father, John, and his bizarre feelings about his older brother John Jr., who had died from encephalitis before Lemuel was ever born. Lemuel nevertheless described John Jr. as a volatile force in his life who whipsawed between protectiveness and antagonism and as an instrument of death called "The Great Punisher" of women.

During his confession, Smith suggested that it was John Jr.—one of his three distinct personalities as described by a social worker—who had taken off his clothes and put on the alb before killing Hedderman, and then Byron. Smith had cut his right pinkie during the murders and used the alb and a white rag to stanch the flow of blood, but he didn't explain why he had defecated in the robe. One psychiatrist theorized that it "symbolized a desecration of God"; the FBI's Behavioral Science Unit suggested that it betrayed "the nervousness of the perpetrator"; and the police crudely speculated that prison sex might have loosened Smith's sphincter. Albert B. Friedmaan's 1968 essay, "The Scatological Rites of Burglars," suggests a different rationale: leaving a ritualistic *grumus merdae* (pile of feces) was a long-standing compulsion among thieves who considered it a good luck token—if an ultimately self-betraying one.

Regardless of the reasons for Smith's behavior, the tape-recorded confession proved to be a critical piece of evidence in his murder trial. An Albany County jury deliberated for only three and a half hours on February 2, 1979, before finding him guilty of four counts of second-degree murder and one count of robbery. A judge sentenced Smith to two consecutive life sentences. By that point, Smith was the chief suspect in five murders but was never tried for the other three. He killed again in prison when he raped and murdered a guard, Donna Payant, and his subsequent conviction in that case sparked an emotional debate over the death penalty in New York.

No one can say with any certainty whether Crow's ability to match Smith to the soiled alb changed the outcome of the case—or what might have happened had Crow chosen a different man. Like the sight of human eyewitnesses, a dog's sense of smell isn't infallible. But as one among many potential forensic tools, scent detection has consistently upended our expectations about what other animals can sense about our whereabouts, identities, and maladies. I found myself wondering whether the German shepherd's verdict weighed on Smith just as heavily as Harras von der Polizei's did on Duwe, the murderous German farmhand, and others throughout history.

Forensic entomologists like Benecke, through careful observation of how insects behave and respond to subtle environmental cues, have likewise helped to crack multiple cases. Bug bites have linked some suspects to crime scenes, while insects that live in one area but are found on a corpse in another have provided evidence that the body was moved after death. And in a type of forensic analysis that Benecke has pioneered, the presence and size of certain insect larvae can help investigators estimate how long a child or adult may have been neglected prior to their death.

From the German toddler's genitals and anus, Benecke identified two species: larvae from the false stable fly, *Muscina stabulans*, and the little house fly, *Fannia canicularis*. Both are drawn to decaying organic matter but vary in their specific appetites. The little house fly is strongly attracted to urine and feces; in their absence, it will still colonize a human body but generally not until the decomposition a few days after death draws it in. The false stable fly is strongly attracted to human feces as well but less so to corpses and can prey upon other fly larvae once it reaches a certain size. Based on the development of the false stable fly larvae, Benecke conservatively estimated that they had been living on the girl's body between seven and twenty-one days but most likely around two weeks. That meant the girl's diapers probably hadn't been changed or her skin beneath them cleaned for about a week prior to her death, offering evidence of neglect by her mother.

A judge sentenced the young girl's mother to five years in prison without probation for manslaughter, and Benecke's estimate of when the child had died allowed the family to inscribe a date on her tombstone. In effect, forensic evidence that many might find disgusting brought a small measure of justice after a needless tragedy. But it didn't have to end that way, he and Lessig noted pointedly in a write-up of the case and critique of the city's welfare office. "The entomological evidence recovered from the body made it highly likely that the neglect set in earlier, maybe even much earlier, than the actual death. This would mean that the child may have been saved by legal action that was, in fact, not carried out."

The case is far from unique, as Benecke has found. After an older German woman was found dead on her couch in 2002, he documented a lack of fly eggs or larvae inhabiting her eyes, ears, and nose—the preferred spots for initial colonization after death. Instead, he found evidence that false stable fly and little house fly larvae along with adult larder beetles had colonized her body while she was still alive. The presence of the little house fly larvae in particular suggested that egg-laying female flies had been attracted by her feces and urine, and the woman's son was subsequently prosecuted for neglecting her. More recently, Benecke documented a similar case involving an older man with a psychiatric illness and incontinence who had been left to die alone in southern Italy.

Unconventional detective work is nothing new, but forensics is littered with failed ideas. At the turn of the twentieth century, a pseudoscientific idea called "optography"—promoted by crime writers and a handful of physicians—seized the public imagination. The quacks who popularized the method argued that the retinas in our eyes functioned as a kind of photographic plate for the outside world, and as such, they likely recorded the last thing a person saw at death. Removing and examining a murder victim's retinas might therefore yield a negative image of the killer's face.

Optography was, of course, utter hooey, and a testament to how long we've spent reading the wrong signs in seeking out the records of someone's life and death. Phrenology, or measuring someone's skull to reveal their mental traits, has likewise been discredited as bunk. Bitemark analysis, which figured in the investigation of Lemuel Smith, has faced growing skepticism in the ensuing years and detractors have since labeled it a "forensic pseudoscience." Other critics have questioned the reliability of lie detector tests and hair analysis, both of which were also used in the Albany double-murder case.

We have sought the truth in the parts of ourselves that seemed the most complicated, mysterious, or elegant. The eyes. The brain. The mouth. The hair. Little did we know that many of the clues we sought were bubbling away just a few feet to the south, noticed not by us but by other creatures or stashed away for centuries in humble containers. It shouldn't be surprising that the same dogs we rely on to find survivors and retrieve our dead in the aftermath of tragedies or sniff out bombs to prevent new ones can also sense our shit. Then again, it shouldn't be surprising that the same chemical cues that help insects find food and shelter can act like clocks to mark our final moments. History doesn't end when we stop looking.

When I interviewed Benecke, I couldn't help but ask him about the meaning behind two of his many tattoos. Both are designs found on maintenance covers: one from Warsaw, Poland, on his right bicep, and another from Bogota, Colombia, on his left. They're souvenirs of the varied places where he has worked, and a kind of ode to sewers and observation. "If you just look, you will find all the evidence that fell down; everything that fell down and nobody takes care of anymore is down in the drainpipe under the manhole cover," he said, waving his hands for emphasis. Benecke's wife likes to say that it's all "hidden in plain sight," but he wouldn't even call it hidden. "It's just there and you just have to lift the fucking cover. And then it's all collected there, like with the insects," he said. "If you just feel disgusted or have any stereotype about insects then you're just not looking. But it is there and the information, it's all there."

Some of the richest stories are embedded in the basest forms. And if we understand that smelly compounds and coprolites and parasites and latrines and maggots can resolve historical mysteries and untimely departures, maybe we can do a bit better at reading our own filth to prevent sickness and promote health while we're still alive and kicking.

CHAPTER FIVE

Portent

I T WAS A CONDITION ONCE considered so gruesome, so fearsome, and so universal in Western countries that doctors dubbed it the "disease of diseases" and "the cause of all the hideous sequence of maladies peculiar to civilisation." Without the proper precautions, medics warned, people's innards could be poisoned, "depraved," or "deranged and corrupted." Behold the awesome power of constipation.

As the medical historian James Whorton found, the horrors of irregularity can be traced as far back as a description in an Egyptian papyrus book of pharmaceuticals from the sixteenth century BCE. The ancient Egyptians believed that rotting intestinal waste could poison the rest of the body—a concept of "autointoxication" that persisted for 3,500 years (and still figures in some ads). In the late nineteenth century, a French physician declared that a constipated person "is always working toward his own destruction; he makes continual attempts at suicide by intoxication."

The fearmongering did wonders for sales of constipation cure-alls such as abdominal massagers, enemas, chocolate-coated laxatives, and bran cereals with names like DinaMite. Whorton's historical research even unearthed a 1938 ad for Dr. Young's Ideal Rectal Dilators, conveniently sold as a set of four progressively larger flanged torpedoes. They may have indeed stretched or dilated the inner and outer sphincter muscles and caused "some momentary pain or thrill" but were basically indistinguishable from sex toys known as butt plugs.

These cure-alls and contraptions had a darker side. After its introduction in 1900, a compound called phenolphthalein skyrocketed in popularity and became the top-selling laxative in the United States "on the strength of its marketing campaign to rescue innocent children from the clutches of autointoxication," according to Whorton. Overuse of laxatives in general can decrease the absorption of nutrients and other medications, cause electrolyte imbalances, create dependency, and actually worsen the constipation by interfering with the colon's natural contractions. Some consumers have even used phenolphthalein and other laxatives as dangerously ill-advised weight loss pills. Nearly a hundred years after its debut, studies linked phenolphthalein to multiple carcinogenic effects in rats and the US FDA reclassified the drug as "not generally recognized as safe and effective," prompting manufacturers to remove it from their laxative formulations. Although the carcinogenic risk hasn't been confirmed in humans, separate studies have suggested that the chemical can damage DNA. Multiple countries have banned its use in over-the-counter products, but phenolphthalein is still one of the most common drugs found in adulterated dietary supplements.

Our desire to read the undigested tea leaves, to seek clues about our own mortality floating in a toilet bowl, is something that charlatans have taken advantage of since the dawn of human history. Part of the problem is that we don't have a stellar record when it comes to interpreting true signs of trouble. As we've learned, for instance, the ubiquity of intestinal parasites in the poop of our predecessors likely signaled poor public sanitation and personal hygiene. But at the time, as paleopathologist Piers Mitchell recounts in his history of human parasites in the Roman Empire, many medical practitioners were heavily influenced by the philosophy of Greek physician Hippocrates. The good doctor asserted that disease, including intestinal worms, could be triggered by an imbalance among the four "humors" of black bile, yellow bile, blood, and phlegm, or a corruption of any one of them. Hippocrates at least intuited the importance of maintaining an inner equilibrium if not the identities of what required balancing.

Aelius Galenus, the Greek physician for three Roman emperors, built upon Hippocrates' humoral ideas in crafting his own explanation and cures for intestinal helminths. The parasitic worms, he believed, formed through spontaneous generation in putrefied matter as it heated up. To return the humors to their proper balance and presumably deworm his patients, Galenus recommended that they undergo treatments such as diet modification, bloodletting, and medicines thought to have a cooling and drying effect on the body. All told, Mitchell said, the beliefs of Hippocrates and Galenus on the origin and treatment of intestinal parasites remained accepted wisdom in Europe and the Middle East for about 2,000 years, until the time of the Renaissance and the Enlightenment.

Dubious ideas about disease and bodily functions hardly ended there. As Steven Johnson's book *The Ghost Map* documented, Victorian London in the mid-1800s was rife with ads for quack cholera remedies—some touting disinfectants that promised to lift the foul miasma widely assumed to cause the deadly diarrhea, others hawking opium to ease upset stomachs, or mixtures with linseed oil or castor oil or brandy, when simple rehydration with clean water and electrolytes would have saved thousands of lives. During the COVID-19 epidemic, remarkably similar and dangerous claims and hoaxes proliferated on the Internet, touting everything from bleach, industrial methanol, and hydrogen peroxide to the horse dewormer ivermectin, urine, and cocaine as miracle cures or preventatives.

Social media may have replaced the sandwich boards and newspaper ads, but our feces have fueled a similar cottage industry of products that promise easy answers about how we're doing or clear signs that an at-home analysis or cleanser or "detox" is aiding our health. For every sensible and reasonably informative fecal test or chart or word of advice, there are countless extravagant claims reminiscent of the Victorian era's constipation cure-alls. "Quickly drop that extra 5+ lbs of toxic poop in your colon," an ad for the $29.95 zuPOO cleanser promised me. For $195, The 10-Day Reset by Sakara would send me a beautifully packaged DIY detox box of tea, powder, bars, probiotics, recipes, and separate "Beauty Water Drops" and "Detox Water Drops" that could do everything from clear my skin to heal my gut and restore "digestive harmony." Companies like Goop have made big businesses out of products that promise to restore inner body parts with words like scrub and cleanse and detoxify and rejuvenate. Nor has Silicon Valley been immune from the hype. Once-lauded gut microbiome–testing company uBiome filed for bankruptcy in 2019 and federal prosecutors charged its two cofounders with more than forty counts of health care, securities, and wire fraud in 2021. Among the allegations, prosecutors charged that the company deceived investors and health care and insurance providers about poop-based tests that were "not validated and not medically necessary."

The seductive siren songs of quick fixes may ring loudly in the ears of anxious and sleep-deprived parents. Baby poop, after all, can be a visceral and frightening force of nature, beginning with the greenish-black meconium. When a baby jumps the gun and poops while still *in utero* and then inhales the sticky fecal matter, meconium aspiration syndrome can cause serious respiratory distress. And just like that, a dark green staining of amniotic fluid or of a baby's vocal cords as seen through a laryngoscope can provide one of the first warning signs in a newborn's life. A bewildering palette of fecal consistencies and colors often follows.

"Sleep training? Vaccines? Forget it. The most important parenting decisions and signals involve poo," write Anish Sheth and Josh Richman in *What's Your Baby's Poo Telling You?*—the cheeky follow-up to their original owner's manual, *What's Your Poo Telling You?* "Poop got a lot more interesting when I had kids," one mother told me. Before children, she said, she never envisioned the animated daily discussions she and her husband now have about it (my husband, Geoff, and I have similar in-depth conversations about our dog's output: Did she poop on her walk? Really, that much? Did it look OK?). Nor would she have thought herself capable of fretting over the fresh contents of a training potty—wondering if she was looking at evidence of functional constipation—and never once pausing to dry heave.

As we've seen with disgust, love can truly transform our senses— if not always our sense. To protect ourselves from manipulation, we'll need a better understanding of what poop can really tell us, and what we can realistically do about it. From a scientific standpoint, we're in an exciting moment of renaissance and reassessment: what we thought was a pretty simple substance is anything but. Every stool can deliver a blizzard of signals from birth to death. The catch, of course, is decoding the true messages amid the noise and knowing whether and when they demand action. Some of the signs we can clearly sense on our own.

Flush

Some of them require more careful scrutiny, like a doctor's eye, a dog's nose, or a machine's measurements. And some of them are just emerging as intriguing possibilities of what scientific sleuths might predict and prevent by carefully piecing together complicated patterns. As I discovered about my own output, we still have plenty to learn about the lifelong companion we thought we knew.

• • •

In a toilet or diaper, even basic colors can be telling. Poop that is frequently yellow and greasy could suggest that you're not digesting fat well and that an evaluation for celiac disease or another cause of the malabsorption is warranted. A green stool could indicate that food is passing through your colon too quickly and not giving your bile a chance to break down into its more familiar brown coloring (unless you happened to eat a black-bun Halloween Whopper during Burger King's 2015 promotion fiasco, in which case everything likely came out the same color as the Chicago River on St. Patrick's Day). A black stool could indicate bleeding in the upper gastrointestinal tract or a recent snack of black licorice, while a red stool might mean you ate beets or are bleeding in the lower tract—often from hemorrhoids but sometimes due to a more serious problem such as colon cancer. A white or clay-colored stool, devoid of the coloring that normally comes from bile, may suggest a blockage in a duct that ferries bile from the liver to the gallbladder (in infants, this blockage is known as biliary atresia and can be life threatening if not corrected). Mercifully, my poop is a normal if boring variation on the color brown.

Given the notoriously colorful hues of baby poop, it may comfort parents to know that many pediatric gastroenterologists have a simple rule: after the meconium, basically every shade is normal except for black, red, or white. Pediatricians have even created their own color scales and at least one smartphone app to help assess whether a photographed diaper deposit looks abnormal and warrants a follow-up. A small pilot study suggested that the snap-and-share method was accurate, though the closeup photos made me thankful that I wasn't there for the real thing.

Then there's the matter of form. On one of his many *Oprah* appearances before going on to host his own show, cardiac surgeon and purveyor of often questionable medical advice Mehmet Oz told the studio audience that paying attention to our bowel movements could provide a good indication of proper diet and digestion. While gastroenterologists say this is certainly true, Oz had a *very* specific kind of movement in mind. "You want to hear what the stool, the poop, sounds like when it hits the water. If it sounds like a bombardier, you know, 'Plop, plop, plop,' that's not right because it means you're constipated. It means the food is too hard by the time it comes out. It should hit the water like a diver from Acapulco hits the water: swoosh." Also, the perfect poop should be shaped like an S and be brown with just a hint of gold, Oz said. (So maybe bronze?) Not in pieces, he said, which suggests that "you don't have enough of it left to poop out in the right way and probably it's hurt the colon that has to process it." Also, we should fart more, he asserted.

I'm a big believer in having goals, though I confess that delivering the perfectly bronzed and fully intact swooshing S and letting loose with more regularity were not things that I had aspired to as the paradigms of intestinal health. Nor are they terribly realistic, for that matter, given that the shapes and sizes of our output can vary from day to day. It is true, however, that the first toot and plop after a medical procedure can herald a return to normal business, as I found after my gallbladder surgery.

Oz may envision stools as graceful divers or bombing runs, but one of the easiest and most common ways to track poop over time is a visual chart that interprets a range of seven fecal forms. Devised by the UK's Bristol Royal Infirmary in 1997, the Bristol stool scale uses shapes and consistencies as proxies for how long it takes food to pass through our guts. The chart is now a mainstay of surgical offices, many with their own flourishes to help patients identify their stool types. Much like imagining shapes in the puffy clouds on a summer day, for instance, Stanford Medicine's Pediatric General Surgery Department has helpfully likened each type to a familiar form. Under the banner "choose your POO!" the shapes range from rabbit droppings and a bunch of grapes (suggestive of constipation) to corn on the cob and sausage ("ideal"), to chicken nuggets (trending toward diarrhea), to porridge and gravy (diarrhea). The idea is to aim for the corn-and-sausage middle and avoid the extremes.

At the constipation end of the scale, the colon can absorb too much water, resulting in infrequent or uncomfortable bowel movements despite considerable effort. The common condition, which impacts an estimated 16 percent of American adults, has so many potential causes that researchers are still hotly debating how to define and classify it. Fifteen major classes of drugs can cause constipation, according to one comprehensive tally. Even Tylenol can stop things up.

At the opposite end of the scale, multiple drugs can cause diarrhea by increasing the intestinal muscle movements that push everything onward. Allergies, infections, and contaminated or overly spicy food can likewise trigger a defense mechanism in which the body speeds the transit through the intestinal tract and tries to rid itself of the perceived danger as fast as possible. Diarrhea can be a consequence of conditions like irritable bowel syndrome and lactose intolerance. Diseases, drugs, or damage can inflame or tear the intestinal lining, interfering with water absorption and provoking a massive immune response that can make matters worse. In some cases, the colon doesn't have enough time to absorb water from undigested food and by-products. In others, like during a cholera infection, more water pours into the intestinal canal from surrounding cells. Either way, the colon can essentially turn into a flume and flush out the poop.

Given the multiple constipation-inducing pain and antinausea drugs stacked against my colon's post-surgery conveyor belt, it's remarkable that the first clump of grapes found its way out after little more than two days. Soon thereafter, I proudly produced a smooth sausage on the Goldilocks scale of just-right poops (though not, sadly, a perfectly swooshing diver), before sailing well into the realm of chicken nuggets and definitely not-right porridge. The overshoot may have reflected the residual effects of my multiple other diarrhea-promoting medications.

Although the Bristol stool scale is the go-to standard for distinguishing among ideal and less-ideal forms, some scientists have questioned its reliability and argued that clocking the gut transit time provides a more useful indicator of health and microbiome function. Remember my sesame seed and corn kernel transit tests? Scientists funded by a health science company called ZOE used blue-dyed muffins as their own visual marker to measure the gut transit times of 863 healthy volunteers. Their study, "Blue Poo: Impact of Gut Transit Time on the Gut Microbiome Using a Novel Marker," suggested that royal blue dye works just as well as expensive tracking options like wireless smart pills. For the participants, they calculated a median transit time of roughly twenty-nine hours but observed an impressive range of four hours to ten days. Although neither extreme is ideal, nutrition scientist Sarah Berry told me the range reflects the surprising variability that she and her colleagues have seen in a larger set of studies investigating how people respond to meals.

Most study participants had a gut transit time between fourteen and fifty-eight hours (my independent tests suggest that I fall into this group as well). For those with longer times, the study found a clearer association with higher levels of visceral fat and post-meal spikes in blood sugar and circulating fat, which are independent risk factors for cardiovascular disease. In a seemingly counterintuitive result, the study found that those longer transits were also associated with greater gut bacterial diversity. The potential exception to the rule that more diversity is better may be explained by the increased availability of bacterial food in this case: the more time it takes for a meal to pass through the colon, Berry said, the more time microbes have to feast on the remainders, allowing some to bloom at the expense of others. A particularly long journey through the gut, in other words, may add nuance to the desirability of microbial diversity and another variable to the list of microbiome influences. "If the diversity includes a greater diversity of unfavorable microbiome microbes, then that's not necessarily a good thing," Berry said.

For the diaper-wearing set, scales developed in Amsterdam and Brussels have provided other alternatives for judging gut transit times based on stool forms. After studying the representative photos of baby poop, I again sympathize wholeheartedly with new parents. As the developers of the Brussels Infant and Toddler Stool Scale point out, however,

Flush

reliably assessing a young child's stool can be hugely important in evaluating recurring patterns and diagnosing gastrointestinal disorders such as functional constipation, diarrhea, and irritable bowel syndrome.

Newer methods may be more refined, but diaper divinations have been around for centuries. In his nineteenth-century Norwegian folktale, "The King of Ekeberg," Peter Christen Asbjørnsen tells readers that "the enlightenment" has come to Oslo because the residents there no longer blame trolls for their youngsters' afflictions. Instead, according to an English translation by Simon Roy Hughes, they let an insightful woman "cast metal for the child suffering from rickets, enchantments, and devilry. Or they send one of the child's nappies to Stine Bredvolden, who is so wise that she reads the child's sickness and fate, and thereafter determines how it will fare." Molybdomancy, or fortune telling by dropping molten metal into water and interpreting the resulting shapes, is no more valid than measuring skulls or reading the retinas of murder victims. Bredvolden, on the other hand, may have had her own early stool scale. And as I found out, reading the contents of a toilet bowl every day for a year did indeed provide a surprising wealth of information. The big question was whether it would make any difference in improving my health and well-being.

● ● ●

Some people might spend the day after Christmas nursing hangovers and puzzling over a new smartphone or a thousand pieces of the Eiffel Tower on the dining room table. I spent the day nursing a hangover and admiring scenes of torture in my new Hieronymus Bosch art book. Then I got to work upgrading two poop-tracking apps and assembling a bamboo Squatty Potty. The stool wasn't a Christmas gift, strictly speaking, but it seemed appropriate to give it a whirl after a night of beef Wellington, champagne, and steamed cranberry pudding. I had already been using three tracking apps for much of December, beginning my journey of personal revelations, and felt it was time to take my daily routine to the next level: exactly seven inches off the bathroom floor.

The Squatty Potty, Internet famous for its memorable commercial of a squatting unicorn filling cones with soft-serve rainbow ice cream ("the creamy poop of a mystic unicorn"), is a glorified footstool designed to help people achieve a more natural squatting angle: technically, the anorectal angle. Squatting, the idea goes, allows for a fuller evacuation of the bowels than sitting at chair height because it relaxes a muscle called the puborectalis that loops around and constricts the lower intestinal tract like a kinked garden hose. Relaxing the muscle and increasing the anorectal angle helps unkink the rectum and produce a more natural, well, stool.

This is important because working too hard to produce a few rabbit droppings or grapes can be bad for our health. Toilet seats and benches have been around since antiquity, but most people in Western countries still squatted to relieve themselves until the nineteenth century's seated water closets began prioritizing comfort, privacy, and convenience over functionality. That design flaw, critics contend, may have contributed to several unintentional side effects related to the strain of defecating. Forceful exertion during a bowel movement, for instance, can tear the main intestinal lining and cause anal fissures. Diverticula, or marble-size pouches where the gut tissue bulges outward due to intense pressure on the colon, can likewise tear and cause diverticulitis if the damaged tissue becomes inflamed or infected. And hemorrhoids, which can pop up around the lower rectum and anus and are akin to varicose veins, can be caused by excessive straining or prolonged sitting on the pot. As we've heard, frequently bearing down to get everything out, one hallmark of constipation, can lead to more serious conditions like defecation syncope—fainting on the john—or stroke. Poor Elvis.

Squatting isn't necessarily a cure-all; one study in India suggested that it, too, can increase blood pressure, and considerable disagreement persists over whether it can prevent or alleviate constipation and hemorrhoids. But the few investigations of its merits suggest that outspoken advocates like the German writer and physician Giulia Enders may be onto something. In one small but detailed Japanese study, researchers filled the anus and rectum of six volunteers with a liquid contrast dye. Then they measured the abdominal, rectal, and anal sphincter pressures and filmed the volunteers as they passed the poop substitute in three positions (yes, the volunteers had to be refilled each time). Greater hip flexing in the squatting position, the scientists found, widened the anorectal angle and led to less straining while defecating. In 2019, gastroenterologists at the Ohio State University recruited fifty-two volunteers to test out what they termed a defecation posture modification device (clearly missing an opportunity to call it a defecation modifying-posture device, or DMPD). Anyway, the study found that the device, basically a stool, did indeed speed up self-reported bowel movements, decrease strain, and encourage more complete evacuation.

The toilet seats in our house reach a regal height of 17½ inches—great for sitting on the throne but apparently less great for efficiently conducting one's business from on high. My art books or old science textbooks would have done in a pinch for adjusting my angle, but I decided the time was right to see whether I could improve my form with the Squatty Potty. It might spoil the suspense to confess that I didn't notice a big change the first few times I used it, with neither a creamier nor a swooshier delivery. But I had already learned that I wasn't having much trouble depositing the goods, as my smartphone apps helpfully charted.

Tracking the frequency, speed, quantity, consistency, and color of every bowel movement outside of a research investigation or art installation might seem like a kind of torture that would have delighted Bosch. But the three competing apps I downloaded operate under the guiding principle that poop knowledge is power, at least for understanding normal patterns and identifying potentially worrisome trends. So I used all three to track my daily bowel movements. Let the Danes have their Crap Morning. I had a full Crap Year.

The default screen of one app is a monthly calendar with tiny icons based on the representative poop of the day. For my start date of December 9, a single vertical "smooth and soft sausage" icon appeared above the numeral 9, forming a sort of exclamation point that suggested every stool would be special. Two apps let me snap photos of each output and even send them to friends—though perhaps they had doctors more in mind. Based on my subjective Bristol stool scale classifications, one of the apps then rated each deposit as GOOD in green or BAD in red—a sort of Tinder for ordure. The third app has a running timer since the

"Last Plop," helpful background information, a "Plop Analyzer," and a full page of statistics. Under a "What can I do?" header on the analyzer page, it had already reached a verdict after my first entry: "You are doing great. Please keep up the good work." At the age of fifty, I had received a gold star for pooping.

Amusing, yes, but also surprisingly revealing. Over twelve months and 996 plops, I learned that I typically go two to three times a day— on the high end of average. I never missed a day and went more in the spring, less in the summer and fall, mainly in the morning, and almost always between 8:00 a.m. and 6:00 p.m. when I wasn't traveling or staying up late to write. My tracking confirmed that the first coffee of the morning was nearly guaranteed to get things moving. That makes sense in light of studies like one in which University of Iowa researchers somehow convinced a dozen healthy volunteers to undergo a tap water enema and then be outfitted with a flexible probe taped to their buttocks and inserted about two feet into the colon for more than eighteen hours. The probe, equipped with six pressure sensors, was designed to objectively measure the strength and coordination of the volunteers' colon muscle contractions. The wave of contractions and relaxations, or peristalsis, provides the gastric motility that moves food through the intestinal tract. After an overnight stay, the volunteers drank a cup of black coffee, decaffeinated coffee, and hot water in random order, arranged around a large (and relatively unhealthy) lunch. Black coffee not only stimulated the colon's motor activity but also nearly equaled the lunch's effect in propagating muscle contractions toward the exit doors, though for only about half as long. The coffee, in turn, revved the colon's motor 23 percent more than its decaffeinated counterpart and 60 percent more than the water.

It turns out that a cup of java can kindle "a compelling need to defecate" in roughly three out of every ten drinkers, at least according to one small study. Caffeine may not prompt that call of nature, though, given the multiple lines of evidence that decaffeinated coffee can also propagate intestinal contractions. Instead, researchers suspect that the highly complex brew may act indirectly through an increase in gut-brain signals or motility-promoting hormones like gastrin, motilin, and cholecystokinin.

My daily tracking showed that my peristaltic motor revved even more during a trip to Mexico and less when I visited my parents in Minnesota, and that a week of looser stools coincided with a trip to the Oregon Coast. Each time, though, the patterns reverted to normal when I returned home. Thankfully, I used the gravy icon only once and the porridge one sparingly. But the relative abundance of chicken nuggets populating my apps' monthly calendars suggested that I needed more fiber to firm up my output.

I already had a bottle of Metamucil on hand that a friend had given me for my fiftieth birthday. Because of course he did. As its fiber of choice, the old standby uses seed husks from the blond psyllium plant, *Plantago ovata*. The plant, incidentally, has been used medicinally for centuries; today's brands have merely repackaged that traditional knowledge in a modern form. Because psyllium husk is highly water-absorbent, it can add more bulk to stools and help regulate their transit times. The increased mass can aid constipation by stimulating more movement while *also* aiding diarrhea by slowing down the race through the colon and reducing the overall number of bowel movements. My initial week on Metamucil did seem to regulate my frequency and consistency. Taking two pills per day, unfortunately, was apparently too much too soon for my delicate plumbing. I was downright uncomfortable, with gas, bloating, and intestinal cramps, so I temporarily shelved the gut hack until I could chart a friendlier course.

The science suggests that my symptoms may have derived in part from the absence of existing gut microbes that could digest the fiber's complex carbohydrates. For those who lack the right fiber-munching microbes, fermentation expert Robert Hutkins told me, easing into a higher-fiber diet may help their bodies acclimate by recruiting more of the necessary bacterial colonies. In essence, the slow but steady increase can "train" the microbiota into adapting accordingly. Recruiting enough microbes can take a while, he said, and may require adding more fiber-digesting specialists like those in yogurts and other fermented foods.

Computational biologist Lawrence David said the necessary bacteria could be there all along, but just in low numbers that need time to build up. Some of his newer data suggest another possibility: the right bacteria just haven't had a reason to activate the right fiber-processing genes or metabolic pathways until it's there for the taking. "When you give people fiber, or you do this in an artificial gut, you see that the bacteria there don't really degrade it as much as if you give it to them the next day," he said. The second time around, the microbes seem to adapt to the new food and break it down much more avidly.

Beyond feeding our curiosity about how the body works and providing lively anecdotes for cocktail parties, experiments that test how the intestinal tract responds to inputs like coffee and fiber can be critical for understanding and controlling conditions that alter our output, such as too-slow constipation and too-fast diarrhea. Coffee, after all, is one of the most popular beverages in the world. Although it may aid the constipated, studies like the University of Iowa experiments suggest that even decaf may be less advisable for people with chronic diarrhea or fecal incontinence.

● ● ●

Just as coffee's complex brew of chemical compounds give it an immediately recognizable aroma, odor is another unmistakable attribute of poop. It's also a useful indicator: particularly foul emanations can signal poor digestion or nutrient absorption and warn of motility disorders like irritable bowel syndrome or other conditions like celiac disease, Crohn's disease, and pancreatitis. Because gut microbes play such a big role in dictating which volatile organic compounds we release, it makes sense that changes in the microbiome might alter the chemical signature of someone's feces. That shift, in fact, may explain the ability of researchers to use fecal VOCs to differentiate among infectious diseases, and the ability of detection dogs to sniff out signs of intestinal C. diff infections in patients, stool samples, and surfaces in some hospitals.

In Amsterdam, a beagle named Cliff—billed in 2012 as the first bacteria-detecting dog in the world—accurately identified C. diff cases by sniffing patients in the wards of two hospitals. When he detected a

Flush

case of diarrhea caused by the pathogen, Cliff would quietly sit down by the patient's bed. (The researchers considered using detection rats as well but concluded that dogs would be easier to train and more acceptable to patients and staff.) Teresa Zurberg, the Canadian dog handler honored at the Peggy Lillis Foundation gala, trained her dog Angus to identify the bacterial pathogen's odor from pure cultures and C. diff–positive fecal samples. A 2017 study suggested that Angus likewise did exceedingly well in detecting known positives and in avoiding false alarms during a trial of his detection abilities. In a clinical unit of Vancouver General Hospital, the dog alerted staff to potential C. diff contamination of equipment and surfaces more than eighty times.

A more recent trial at Toronto's Michael Garron Hospital comparing the C. diff–sniffing abilities of two detection dogs—a German shepherd named Piper and a border collie-pointer mix named Chase—yielded more disappointing results. The variability between Chase's and Piper's detection capabilities and a lack of agreement on some positive and negative identifications, the study authors suggested, raised questions about the reliability of detection dogs as "point of care" diagnosticians. That doesn't necessarily mean the dogs weren't able to pick out the distinctive odor of a C. diff infection, though. Variable performance in detection work can be influenced by distractions or a lack of motivation: in this case, Chase seemed to have "moved on," lead author Maureen Taylor told *STAT*, and was easily distracted by breakfast trays on patients' beds. "The dogs found it hard to pass a toilet without drinking out of it," Taylor also noted.

Detection dogs may be imperfect sleuths for medical forensics, but their potential is vast. As medical aides, Cynthia Otto's team at the University of Pennsylvania has shown that they can identify ovarian cancer through a unique odor in the blood plasma of patients. Similar detection work has focused on bacterial biofilm infections of prosthetic implants. And in 2020, Otto's center launched a new project to determine whether dogs could detect the presence of prions—the infectious proteins linked to chronic wasting disease—in deer feces. In a subsequent study, she and her colleagues reported that the center's dogs could detect signs of COVID-19 in patients' urine, sweat, and saliva samples.

"Diseases cause definite changes in our odor and it's just a matter of training on the proper odor through the samples," she told me. In fact, Otto views the progress in dog detection research as a re-visitation of an age-old concept. "I mean, if you look back in ancient Greece, they talked about the smells associated with different diseases," she said. Hippocrates, for one, suggested the diagnostic utility of body odors and wrote about disease-specific smells in urine and saliva. Modern humans, by contrast, have largely left that sense behind as we rely on other cues. If we slow down and refocus, "we *might* start to retrain our noses," Otto said. In the meantime, we can rely on our canine companions or try to extract the knowledge embedded in the odors they detect.

Though dogs can still clobber machines in the sniffing department, most hospital and clinical labs can clue in on an impressive range of other fecal constituents. For some patients with diarrhea, white blood cells in the stool can warn of bacterial infections or inflammatory bowel disease. Another common test, a gastrointestinal pathogen panel, can detect the DNA or RNA of more than twenty bacterial, viral, and parasitic disease-causing organisms to help narrow down a diagnosis.

Because it begins forming in the second trimester, even a baby's meconium can act as a kind of medical archive. Once expelled by a newborn, the substance can be tested for the breakdown products of drugs to determine whether the mother used them during the final four or five months of pregnancy. Salt Lake City–based ARUP Laboratories, for instance, offers the service for opioids, cocaine, marijuana, and six other drug types. There are clear medical advantages to knowing what a baby might have been exposed to *in utero*. But perhaps sensing the creepy law enforcement implications of using babies to drug test their mothers, the lab states that its method is intended for medical purposes only and isn't valid for forensic use.

Feces can chronicle the body's attempt to rid itself of heavy metals as well. In India, scientists measured the concentration of toxic metals in the fecal pellets of blue rock pigeons to monitor the pollution at six industrial areas in the city of Jaipur. And in Zambia, scientists reported detecting "extremely high" levels of lead and elevated levels of cadmium in the feces and urine of infants and toddlers living in polluted townships near a lead and zinc mine. Although blood and urine tests are more often used for such biomonitoring, the research suggested that fecal samples can also work well for public health surveillance of poisoning from lead and other metals.

Multiple commercial labs, in fact, offer stool tests that can measure the levels of lead and more than a dozen other metals. But there's a downside, particularly with direct-to-consumer testing services that are charging worried families hundreds of dollars to test their poop. "This fecal metals test reveals how many metals are moving through you and out of you," an option by a company called Life Extension claims, though it comes with the disclaimer that the laboratory services are "for informational purposes only" since they haven't been licensed by the FDA to provide medical advice. Sidestepping the involvement of a doctor puts the onus on concerned consumers who may overestimate their true risk, receive unreliable results, or not know what to do with them.

Most at-home tests, including ones mailed back to a lab for processing, aren't regulated by the FDA. Some of the ones that are, though, have become integral to public health strategies for HIV and other sexually transmitted infections, hepatitis C, colon cancer, and COVID-19, among other diseases. Driven by privacy and convenience—and in 2020, by limited access to other options during the pandemic—they have soared in popularity. From mid-March to mid-April in 2020, testing company LetsGetChecked reported a 477 percent increase in demand for its at-home colon cancer screening test. It's just one of many that look for faint signs of blood in the stool—technically, "fecal occult blood"—as a potential indicator of tumors or polyps. "Occult blood" may sound like it belongs in a mystical pagan ritual that may or may not involve a goat, but in reality, gathering my own stool sample for a $49 kit sold by health testing company Everlywell was more like a preschool art project. Using a long-handled blue brush, I dabbed at my poop for five seconds, shook off the excess, and then painted a small square on the collection card. Then I repeated the process with a second brush and a second square, creating two miniature abstract watercolors. This fecal immunochemical test, or FIT, detects blood through the presence of hemoglobin, the major oxygen-carrying protein in red blood cells.

Everlywell, dubbed "the Uber of lab testing" by public health reporter Kim Krisberg, sells convenience in exchange for the chance at repeat business. On that promise, at least, it delivered: I returned the test on a Friday and had my result the following Tuesday: negative (a positive result would have prompted a discussion with my doctor about whether another colonoscopy was necessary to determine the source of the blood). For people who live in remote communities or who have been poorly served by the deeply dysfunctional US health care system, timely lab results can be a godsend.

More sophisticated at-home tests that also look for altered DNA as a potential sign of cancer, like a prescription-only option by Cologuard, are more accurate but also an order of magnitude more expensive without insurance. Because I had already received a colonoscopy and a clean bill of health a few years prior, I chose the simpler option. The collection process was easy enough, though it could be confounded by bleeding hemorrhoids, blood in the urine, or some other gastrointestinal injury like a stomach ulcer.

There were other drawbacks. On the "My Everly" online dashboard, the company suggested that I could become a member and get a lab test every month (for $24.99), like a bouquet or book-of-the-month club but focused more on things like cholesterol and sexually transmitted infections. Nothing says Valentine's Day quite like a chlamydia and gonorrhea test. The site also asked for personal medical history and family history information that I declined to provide, even though it would have helped to "personalize" suggestions and my experience. Er, no thanks.

More testing isn't necessarily better, especially given that some of Everlywell's services, like two versions of a food sensitivity panel, have been panned by some experts as gimmicks with no established medical benefits. Other hormone and vitamin deficiency tests haven't yet proven their utility to independent assessors like the US Preventive Services Task Force. Everlywell has asserted that it never sells customer data and uses state-of-the-art security safeguards. Even so, it was a bit disconcerting to discover that a colon cancer test could be the gateway to an expensive health journey of questionable value.

My own poop, at least, likely wouldn't trigger a national security incident if it ever fell into the wrong hands. Number two from number ones, however, *has* figured prominently in accusations of "espionage via excrement," in the words of a 2016 BBC report. During the 1949 visit of Chinese communist leader Mao Zedong, or Chairman Mao, to Moscow, Soviet spies reportedly diverted the toilets in his remote guesthouse to secret containers and brought the poop to a lab for analysis. The spying plot, reportedly ordered by Soviet leader and fellow communist Joseph Stalin, may have been aimed at constructing a psychological portrait of Mao. North Korean dictator Kim Jong Un might have had that subterfuge in mind when he reportedly brought a portable toilet to a 2018 summit in Singapore. A stool analysis could have revealed a medical condition or other secrets about Kim's health, but his elaborate security precautions "will deny determined sewer divers" insights into the supreme leader's stools, according to *The Chosun Ilbo* newspaper in South Korea.

Scatological spies would no doubt be impressed by Stanford University's memorable addition to The Internet of Things: a "smart" toilet that operates autonomously thanks to pressure and motion sensors, four cameras, and a computer interface. In addition to a color-based urinalysis, the toilet uses "deep learning" to classify all incoming deposits according to the Bristol stool scale. For bathrooms with multiple users, the toilet's flush lever reads each person's fingerprint and a camera that would make even airport security personnel blush matches each user to "the distinctive features of their anoderm." That's science-speak for "butthole print."

Researchers reported that their automated system—far more complicated than my tracking apps—rivaled trained medical personnel in identifying disease markers. But their ideal scenario of toilet-based "precision medicine" and an app that would alert the user's health care team to signs of trouble begs the question of whether the added complexity could contribute to false assurances or over-diagnoses. It also left me wondering whether the potential benefit really warrants the technology's elaborate privacy and security features needed to shield someone's anoderm or diarrhea from prying eyes. "The smart toilet is the perfect way to harness a source of data that's typically ignored—and the user doesn't have to do anything differently," said one of its creators in a news release. But maybe we *should* do things a bit differently to avoid the potential pitfalls of outsourcing our own attentiveness to a computer algorithm.

David, the computational biologist, has adopted a more wait-and-see attitude about smart toilets. He views the inclusion of multiple measurements as all part of the experimental process and akin to what the more established field of human genomics went through during its early years. "There's only one way to find out which of these is actually useful to people," he said.

German engineers, ironically, created a much simpler toilet decades ago that encourages careful observation—to the horror of many US travelers. Older German toilets feature a rear "poop shelf" above the waterline upon which the deposits rest, allowing each producer to inspect the goods before they're whisked through a drain toward the front of the bowl (the high-and-dry platform also cuts down on excessive splashing). Benecke, the German forensic entomologist, told me it was commonly used to check for parasitic worms in feces, though that particular need has waned over time. A toilet brush is usually necessary to help clear the stools, but the Flachspüler, or flat-flush toilet, is particularly useful for first examining their shape, color, and consistency. Science comedian Vince Ebert calls it "the German platform of contemplation."

• • •

If nothing else, being more aware of my poop set a baseline from which I more readily noticed modest deviations. My tracking adventure, though, paled in comparison with David's odyssey when he was a graduate student at the Massachusetts Institute of Technology. He and his graduate advisor, Eric Alm, not only characterized their daily bowel movements for a full year, but also collected them along with saliva samples to conduct a more extensive analysis. Using a reconfigured cell phone app, they tracked 349 variables ranging from the poop's weight and odor to their diet and mood; recording all of it took about an hour every day. When it was all over, David told me, he was so sick of the routine that he swore off smartphones for years.

Combined, the two researchers collected more than ten thousand measurements of how they and their inner microbes lived. The study suggested that amid ongoing competition within the gut, most of their microbial communities remained remarkably stable for months, with a few notable exceptions. Specific disruptions like food poisoning, international travel, and dietary changes were all associated with clear shifts in gut flora. After Alm came down with a nasty bout of *Salmonella* food poisoning accompanied by diarrhea, for instance, his gut flora changed dramatically. Many of the stalwarts vanished. Minor constituents suddenly flourished. And some entirely new species appeared. But when the researchers took a closer look, they saw that the replacements in his gut largely replicated the functions of the former communities, like the ability to digest certain carbohydrates. Although the infection wiped out many of Alm's bacterial species, the ecosystem seemed to rebalance itself and conserve its workflow if not its workers.

In the midst of the study, David accompanied his wife to Bangkok for fifty-one days. He went out of his way to try new foods and to this day considers his time there a culinary highlight, despite contending with diarrhea for almost one-third of his stay. To his surprise, however, his gut hadn't adopted a new slate of microbial inhabitants. "I'd always assumed that you would pick up other bacteria from all over the world, and I didn't really see a lot of evidence of that," he said. Rather, some species that had been there all along but in relatively low numbers took off while other previously abundant ones receded into the background. When he returned home to Cambridge, Massachusetts, the pattern reversed itself in about two weeks.

After following David's journey, I reasoned that if I really wanted to get to the heart of my own gut health, I should have my microbiome sequenced too. I decided to go with the Floré testing service from Sun Genomics (normally $249 but on sale for $169), in large part because it promised to tell me which microbial species were cohabitating with me instead of just providing hazier "gut health" indicators. "Get to Know Your Gut," its ads suggested. Once I had sampled what I thought was enough poop from a collector pad and sealed the samplers in biohazard bags, registered the kit, and dropped off my well-secured bundle to FedEx, Floré promptly offered me an "add-on" $99 test that measures a biomarker for inflammatory bowel disease. But there was a catch: screening the sample already in transit wouldn't be considered a valid diagnostic test; that would require a new sampling kit for $129. Er, no thanks.

I had initially hoped to contribute my microbiome sequence data to the American Gut Project, a crowdsourced effort launched in 2012 that is comparing "in the wild" gut microbiomes from thousands of participants, predominantly in the US, the UK, and Australia. In the midst of the pandemic, though, the project's scientists pivoted to COVID-19 research and stopped sending out regular collection kits to would-be collaborators. The project's database isn't all that representative of the planet's gastrointestinal microbes: the data have been stripped of identifying details, but the researchers concede that the participating citizen-scientists are nearly all from industrialized countries and are wealthier, healthier, and more highly educated than average. Despite its limitations, the public database has amassed a diverse collection of bacterial species and gene sequences by virtue of its sheer volume, and researchers mining the data have uncovered some tantalizing trends. In 2018, a data analysis suggested that self-reported categories like "vegan," "vegetarian," and "omnivore" weren't terribly predictive of participants' gut microbiome diversity. Rather, a small subset of volunteers who reported eating more than thirty types of plants every week harbored more diversity than a comparison group who ate ten or fewer. In particular, the plant-lovers hosted more fiber-fermenting specialists such as *Oscillospira* and *Faecalibacterium prausnitzii*.

David's own analysis showed that whenever he ate fiber-rich foods, his gut microbiome shifted the next day to accommodate more fiber-digesting bacteria such as *Bifidobacterium*, *Roseburia*, and *Eubacterium rectale*. And whenever he ate yogurt, his microbiome harbored a greater abundance of Bifidobacteriales, the bacterial order that includes *Bifidobacterium* and other close relatives commonly added to yogurts as live cultures. David, now a self-described professor of "poop research" at Duke University, had already shown that putting a group of ten Americans on an entirely meat-, egg-, and cheese-based diet for five days quickly shifted their microbiome in a different direction. Their gut communities transitioned toward bacteria that were more bile-tolerant (high-fat diets release more bile acids) but less adept at metabolizing complex plant carbohydrates. Their poop samples, in turn, contained higher concentrations of deoxycholic acid, a bile acid by-product of bacterial metabolism known to promote DNA damage and liver cancer. Among its findings, the study added to evidence that high-fat diets can promote the growth of microbes that aid the development of inflammatory bowel disease.

When the same ten people ate a plant-based diet, their gut microbes shifted less dramatically but toward more fiber-degrading abilities, as David saw glimpses of from his own diet. The pronounced differences prompted by the volunteers' temporary diets, in fact, "mirrored differences between herbivorous and carnivorous mammals." The rapid ability of our gut microbiome to switch between the two states, David and his coauthors suggested, may reflect a key evolutionary adaptation for our ancestors, given the fluctuating availability of meat.

Today, people in the US eat about 30 to 35 percent of the fiber we should be taking in; Hutkins and other researchers routinely refer to this shortfall as the "fiber gap." Because it has an outsized effect on poop's weight and volume, the average deposit in Western toilets is a fraction of what it is elsewhere—another example of how "normal" isn't necessarily the same thing as "healthy." In 2021, microbiome researchers Erica and Justin Sonnenburg and colleagues published what they called a "stunning" new observation that suggested we might be missing something just as important. Healthy volunteers who ate a diet rich in fermented foods like yogurt, kimchi, and kombucha tea for ten weeks had significantly improved gut microbiome diversity and reduced signs of inflammation. Counterparts who instead ate a high-fiber diet didn't show the same increase in microbial diversity and shed more carbohydrates in their stool samples. One potential explanation, exemplified by my less-than-fulfilling fiber supplement experience and suggested by both David and Hutkins, is that people in more developed countries also lack sufficient fiber-digesting specialists. The bacterial food alone isn't enough; we need the bacteria that can eat it.

I decided to try eating yogurt every day. For good measure, I experimented with a daily probiotic called Probulin that supplied twelve bacterial species plus a prebiotic—a favored food for microbes—called inulin. The new pill gave me indigestion and made me burp more frequently, though it also seemed to work well as a preventative against pathogens during my own week-long culinary adventure in Mexico. Having increased my intake of bacterial fermenters, perhaps I could also diversify my fiber, I reasoned.

And that's how I came upon Pure for Men, which is a proprietary dietary supplement made from psyllium husk, aloe vera, black chia seed, and flaxseed. The variety of fibers intrigued me more than Metamucil's boring psyllium—maybe I could open up more niches for fiber-loving intestinal flora! Pure for Men had other advantages in mind, however, which pertain to that "for Men" part of the brand and may be further explained by the product's slogan, "Stay Clean. Stay Ready." Yes, the ability of fiber to efficiently sweep everything toward the exit doors might help me, as a gay man, upgrade my sexual wellness routine and "bottom with confidence," the website informed me. "Enjoy playtime! Clean & worry-free," the bottle said under the helpful depiction of a bed. It reminded me of the enterprising company that sold butt plugs as constipation remedies.

I will freely admit that having a company pitch me its fiber supplement as a sexual aid was not on my 2021 Bingo card. Then again, for many people, turning a work zone into a recreational area requires a change of scenery (leaving aside the sexual fetish of coprophagy). Remember the "poophoria" sensation some people apparently feel on the toilet that may be due in part to an abundance of nerve endings in the region? One friend suggested with a straight face that the right fiber would make pooping "orgasmic." I, alas, did not get high on the pot. Nor did I feel especially clean and worry-free. I wasn't as uncomfortable as I was during my first fiber-palooza; I had learned to gradually increase my dose over time. But I was still missing something.

● ● ●

Just as diet can shape the intestinal tract and its complex stew of resident microbes, those microbes hold sway over complicated determinants of health and disease. While disgust and its mechanism of behavioral immunity might help some people avoid communicable diseases like cholera, for instance, bacterial signatures within poop itself might indicate whether certain resident microbes offer another form of protection. One project in David's lab is examining how cholera susceptibility may be influenced by interactions between preexisting gut microbes. Similar to how bacteria in a fecal transplant can out-compete C. diff for the gut's limited growing niches, people who don't get as sick from cholera may have endemic bacteria that are better at battling the invasive species for space and resources. Alternatively, David mused, their microbiome patterns could act more like biomarkers of an underlying biological or immunological mechanism that confers more protection. Either way, examining the microbial patterns of someone's stool might indicate whether they are more or less susceptible to diseases like cholera.

There's yet a deeper level to decipher. David said researchers are putting a growing emphasis on identifying which bacterial proteins or metabolites, made by bacteria as part of their daily living, are being produced and what they do rather than which microbes are producing them. He and Alm, his former thesis advisor, were surprised to find relatively little overlap in their gut species when they compared their microbiomes, making direct comparisons more difficult. Specific bacterial functions or products made by multiple species, on the other hand, may be broadly shared among people even if the microbial producers aren't.

We may be awash in far more of these bacterial products than previously thought. In 2019, another team led by researchers at Stanford University discovered that bacteria residing in the human mouth, gut, skin, and vagina make thousands of small proteins, most of which had been overlooked by scientists. "It's probably not the case that the species of bacteria itself is affecting you," David said. Rather, a secreted compound or surface protein or chemical reaction may be the thing that impacts your sickness or health. If a scientist's goal is to develop disease biomarkers or therapeutics, then, it makes sense to shift the focus from the microbes to the soup of proteins, carbohydrates, fats, and smaller metabolites they churn out. David's lab, for instance, is studying the gut microbiome in children with obesity to better understand how the presence or absence of specific molecules might indicate a metabolic shift that helps drive the weight gain.

Microbiome studies, in fact, are increasingly turning toward the art of prediction based on poop: a kind of twenty-first-century prophecy that is delving deeply into the complex relationships between us and our resident microbes. Finnish researchers found that the microbiomes of newborns, sequenced from DNA in their soiled diapers, could predict whether the infants would be overweight by the age of three. Microbiome researcher Susan Lynch told me that her own group had found similar results and was beginning to clarify some of the mechanisms by which microbial molecules in the infant gut might change the physiology of cells lining the intestinal tract, making them look like the cells of patients with established obesity. "I firmly believe that those studies are leading us towards the developmental origins of obesity," she said.

Microbiologist David Mills and colleagues have successfully predicted whether infants were nursed or ate formula by the presence of particular molecules in their diapers. Nursed infants receive oligosaccharides in human milk that act as a natural prebiotic food source for *Bifidobacteria*. The Mills lab developed a dipstick method to measure the amount of unconsumed oligosaccharides in an infant's diaper, which might be extendable to adult stool samples as an easy measure of how well people are digesting fiber. His group also found that children who have higher levels of *Bifidobacteria* in their intestinal tract have lower levels of genes linked to antimicrobial drug resistance, perhaps because the milk-loving colonists produce acids that can lower the pH and keep drug-resistant microbes at bay.

Berry and her fellow "Blue poo" researchers have grouped certain bacterial species with longer or shorter gut transit times, suggesting that the relative abundance of specific microbes might help predict when the blue dye reappears. Poop could provide predictive clues about the development of human cognition and immunity as well. In the gut microbiomes of one-year-olds, separate research groups in the US and Canada both found that a relatively higher abundance of certain bacteria, and from the genus *Bacteroides* in particular, was associated with improved neurodevelopment one year later. In the Canadian study, one-year-old

Flush

boys who had more *Bacteroides* in their microbiome showed subsequent signs of better brain development, particularly in their cognition and language skills. If the relationship bears out, the bacterial benefit could be conferred by Bacteroidetes-produced molecules called sphingolipids, which play a key structural role in the cell membranes of neurons as well as a regulatory role in supporting brain development. But why only boys? It's not yet clear, although some research has suggested that communication between the gut and brain may be more sensitive to early-life gut microbiome disruptions in boys than in girls.

From examining the gut microbiome in stool samples of one-month-old infants, Lynch and her colleagues uncovered a signature of allergic risk within baby poop and successfully predicted which of the children would develop allergies by the age of two and asthma by the age of four. Just as David's work has suggested, the markers have gone beyond simply implicating certain bacterial species. Lynch's group has begun to identify some of the individual, bacteria-made molecules that seem to be driving the immune dysfunction that predisposes the children to allergies and asthma.

Antigens, or proteins stuck to the outer surface of cells and viral particles, act like identification tags for each microbe that we're exposed to in life. When we don't encounter them early on, one hypothesis suggests, we're less able to tell friend from foe and the immune system can end up inadvertently overreacting to harmless cells or antigens down the road and attacking the things it shouldn't. That overreaction, in turn, may help predispose us to autoimmune diseases like rheumatoid arthritis and ulcerative colitis. Several studies have suggested that bacterial metabolites can likewise influence how our immune cells function. That means what we eat also matters, Lynch said, since diet is a big factor in determining which microbes colonize the gut and how they behave while there.

The environmental influences, as we've seen, can begin *in utero*. As one example of their long-term impact, tobacco and antibiotic use during pregnancy can alter the mother's microbiome and shift the bacterial production of molecules that influence the risk of allergies and asthma in her infant. In one study, Lynch and colleagues showed that the microbiome in the meconium of newborns at high risk for allergies and asthma was significantly different from the microbiome in babies born to healthy, non-asthmatic and non-allergic parents. "So from the outset, those seeds, those first species colonizing that system and educating the immune response and dictating the physiology of the human cells that are present, are different—vastly different in fact—in those children at high risk versus low risk," she said. Lynch believes that the environment of infants—everything they're exposed to early in life— further determines which microbes will colonize their guts and which ones will stay away.

For instance, research suggests that a caesarean section instead of a vaginal birth and infant formula instead of breast milk alter the microbiome in ways that increase the risk of asthma and allergies. At-risk babies seem to be delayed in their ability to accumulate microbes from the environment. This altered microbiome, Lynch and her colleagues think, can drive inflammation that strongly selects against other microbes attempting to colonize the gut niche. "The microbes that can colonize can withstand inflammatory responses and they typically tend to be pathogens," she said. The initial colonizers can then shape the early conditions of that gut ecosystem by dictating which other species will be kept at bay or permitted to join them as the microbiome develops. The critical window of development, when early microbial colonists and their products collectively educate the child's immune system, may set the trajectory for lifelong health outcomes. Babies who are exposed to a broader diversity of microbes, Lynch found, seem to be more protected and at lower risk.

Her research has convinced her that our microbial residents are the true drivers of our complicated relationship. "I think we're just biological entities surviving in a microbial environment, essentially," she told me. But her work also has raised the intriguing possibility that we might be able to manipulate the manipulators by reengineering the developing gut environment of high-risk infants. Encouraging the colonization of highly influential keystone species, for example, might modify the rest of the early microbiome and properly train the immune system.

Certain species might work together to produce small molecules that shape immune function. "We see that as the rheostat for human health. So if you can figure out the rules of engagement between microbes and host immunity, you now have a pretty incredible lever on how to shift the system into specific states," Lynch said. In other words, we might be able to hack the code for how our microbes regulate our immune system and readjust the dial or lever. "So it's really thinking very differently," she said: a more holistic view of our ecosystem that considers how something that happens in the gut can influence the response of immune cells in the lungs, and how something that happens very early in life can shape the risk of disease years later.

Based on their research findings, Lynch and colleague Nikole Kimes cofounded a biotech start-up called Siolta Therapeutics (siolta is Gaelic for "seeds"). The start-up, Lynch said, is conducting clinical trials on microbiome-based drugs, including an oral therapeutic consisting of three bacterial species that might help reseed the gut of infants at high risk for developing asthma and allergies. Tinkering with the gut microbiome, especially in early life, carries risks and Lynch acknowledged that rigorous trials and follow-ups will be necessary to make sure that Siolta's interventions are scientifically and medically sound. "You may have a short-term benefit, but perhaps a long-term deficit," she said. "We need to think about what the long-term impact of that is. That's not something that you can play with."

● ● ●

Tinkering with my own gut microbiome carried less risk, though it was humbling to realize how little I understood about my intestinal inhabitants. At the tail end of summer in 2021, Geoff and I decided to go on a pandemic diet together and started to increase our daily consumption of fruits and vegetables while cutting back on alcohol, bread, and refined sugar. After four weeks of encouraging success, regular bowel movements, and near-daily cups of yogurt, I decided it would be a great time to try the sexy fiber again along with a somewhat toned-down version of a daily probiotic pill that contained ten bacterial strains but no added inulin. I would take the probiotic in the morning and the fiber in the evening, and surely the third time would do the trick and elevate my diet game to another level.

What actually happened was that although the leafy greens and other healthy foods were successfully decreasing my body's bulk, my colon's engines were taking a bit longer to rev in the morning and I was depositing less bulk. The daily probiotics and yogurt may have been diversifying my gut, but they weren't doing much to ease the expulsions. Nor was the sexy fiber, for that matter. So back came the Squatty Potty, which seemed to help with the initial delivery. I wasn't having consistently diver-swooshy, unicorn-creamy, or clean-as-a-whistle movements. But maybe I didn't need to. I felt better because I was eating better, and my output was still arriving in reasonably predictable patterns. Like me, David had been wondering where to start with all of the gut probiotics and supplements on the market, so his lab recruited volunteers for a clinical trial to study six off-the-shelf fiber supplements. The researchers assessed the gut microbiome's response to each by measuring the relative production of butyrate, a metabolite made by fiber-fermenting bacteria and a key source of energy for cells lining the colon. "Our finding is that what matters more than the supplement that individuals pick is their prior diet," he said. "Regardless of which supplement you pick, what's going to be more informative for how your microbiome responds is whether or not you've already been eating a diet that's rich in fiber."

David said he's noticed that many of the take-home messages from microbiome studies have agreed with decades of sensible and straightforward health recommendations like, "Eat more vegetables." The Sonnenburgs' study on a diet rich in fermented foods similarly echoed long-established wisdom that kimchi, kefir, and yogurt are good for you. Research, though, can attach a scientific rationale to the intuitive advice. "At the end of the day, if healthy eating is the goal, maybe this is one way that just makes the difference for some people," David said.

In my own search for better health, the most useful apps and tests and supplements all did the same thing: made me more mindful of what I was putting into my body and what was coming out of it. Common sense isn't sexy. You can't bottle it and sell it at a premium. But I found some optimism and relief in the intuitive findings that being healthier, reflected in part by better poop, relies more on sensibility than a magical alignment of microbiome and diet or a daily dose of the perfect supplement.

Two weeks after sending in my Floré kit, I received an email that my microbiome test results were ready. I found them fascinating, puzzling, and ultimately sobering. My score of 74 for overall gut balance seemed somewhat arbitrary, though I noted with a twinge of pride that the needle was in the green part of the scale (excellent) and higher than 90 percent of other Sun Genomics customers. But what did that really mean? A chart listing all of my microbes and their relative abundances provided some clues but also suggested how complicated and variable the human gut microbiome really is. Just two bacterial species made up a quarter of my gut flora and eight made up half of the total, though the sequencing found 189 species in all. The top one, the fiber-fermenting *Faecalibacterium prausnitzii*, contributed more than 13 percent to my total, a bit beyond the fairly wide range of 1 to 13 percent for most customers. I had other fermenting specialists in the mix as well, like *Eubacterium rectale*.

Under a section on health and nutrition recommendations, the report nevertheless suggested that the relative abundance of *Bacteroides* species in my gut indicated that my diet was too high in saturated fats. Six weeks into a successful diet in which I had minimized my fat intake, I was a bit skeptical of the suggestion and wary of what the company would offer to improve my "healthy gut ratios," given that Sun Genomics also offers personalized probiotics. The report did find that several probiotic species had already made themselves at home ("Keep up the good work," it said, echoing one of my poop-tracking apps). One, *Streptococcus thermophilus*, likely came from the yogurt I was regularly eating for breakfast. More surprising was the complete absence of many others, including all ten of the species in the probiotic supplement I was taking. Either the species weren't viable or didn't take to my gut, even as temporary residents.

The vast majority of my gut species were commensal microbes thought to have a benign effect on the microbiome, though most of them remain poorly characterized. I harbored a small amount of

E. coli bacteria, which I found oddly pleasing despite the variable nature of the potential pathogen, given that I had studied the species for so long. I also hosted a small amount of the archaeal species *Methanobrevibacter smithii*, which digests some leftovers of bacterial fermentation and produces methane. And then the report took a sharp turn. On the small list of unfavorable microbes, I saw a familiar and disturbing name: *Clostridioides difficile*. After writing so much about the horrors of C. diff, I was shocked to see the pathogen listed as a resident in my own gut, making up about 0.6 percent of my microbiome. I was apparently an asymptomatic carrier who didn't suffer ill effects from the microbe after being colonized by it. Mystified about how I might have been exposed, I was unsettled to learn that I could remain colonized for months. When I talked with two microbiome researchers about it, though, they weren't surprised at all; a scientist from Floré told me it was rare for her to see a report that *didn't* include C. diff, given the ubiquity of the bacterial spores in hospitals, clinics, and doctors' offices. Healthy people commonly carry the pathogen in low abundance, agreed microbial ecologist Sean Gibbons (although published estimates have varied widely; one Japanese study reported a colonization rate of 17.5 percent among 120 asymptomatic volunteers). My lack of symptoms, Gibbons explained, meant that the ecology of my commensal microbial community was keeping it in check: a planted garden crowding out the weeds.

In one of her talks, Lynch likened algae overgrowth in Lake Erie to disturbances in the human gut. She told me she's always viewed human health and disease and our relationship to our microbiome through the framework of ecology: How do complex ecosystems develop and respond to disruptions, whether in a lake or a gut, and what's their capacity for resilience? It's a far more nuanced and intricate view of health and development than the quick fixes we're often sold by marketers.

In Whorton's historical treatise on constipation, he writes that doctors of the late 1800s and early 1900s were generous with their prevention advice. Even so, "recommendations to eat more fruits, vegetables, and whole grains; to be more active physically; and always to respond promptly to nature's morning call to evacuate seemed to many people to require more self-discipline and sacrifice than they cared to exercise. The public, anxious about autointoxication, thus fell easy prey to all manner of marketers of anticonstipation foods, drugs, and devices." More than a century later, a proliferation of gut- and poop-based tests can similarly project a mirage of easy answers and oversimplified value judgments on what's good or bad when the reality is far messier. Instead, David said, it may be more useful to take a conservationist approach to the microbiome and think about how to manage our own inner ecosystem.

I was reminded of environmental writer Emma Marris, who argues in *Rambunctious Garden* that as the most influential creatures on the planet, we've fundamentally changed even the most remote landscapes. We have damaged or destroyed much of the environment, true. But we've also distanced ourselves from what we imagine to be the remaining bits of untouched wilderness, not understanding that our fates are already intertwined. "We have hidden nature from ourselves," she writes. Preserving what remains, she argues, will require a hybrid strategy in which we accept our role as change agents in the world around us and embrace our responsibility as caretakers for a vast, half-wild garden.

The gut envisioned by hucksters and schemers has long been a dark, dangerous, and disgusting place full of poison and corruption. A sewer or tube full of toxins to be cleansed and expelled. But if it's more like a half-wild garden, maybe it's time we all become committed gardeners. Apps and tests and supplements, if backed by science and used wisely, can help point out imbalances and problem spots but they can't replace the actual grunt work of tending to our inner flora. Some of the complicated patterns in poop clearly require more research to decipher the signs of trouble. For the rest, we already have our sense and senses to heed the burst of signals floating in the toilet bowl. Best of all? Our own built-in technology will never be obsolete.

CHAPTER SIX

Monitor

WE BEGAN OUR HUNT FOR the killer on a cloudy morning in July near the corner of East Wright Avenue and East T Street in Tacoma, Washington. Past a parked Winnebago with a blue and green Seattle Seahawks football blanket as a privacy curtain for a side window. Then down to the bottom of a large catchment system in one of Tacoma's poorest neighborhoods.

I stopped with the others in front of a maintenance hole cover under a highway overpass. The waste stream from an estimated 15,000 to 18,000 city residents flows past this point in the sewer line, and we were hoping that the assassin hiding within might betray its presence. Steven George, an environmental technician with the city, lifted the cover with a metal hook and then shone a flashlight into the murk below. His work colleague, Haley Abbruscato, was already busy taping a folded industrial paper towel around the end of an extension pole called Mr. LongArm to form a kind of oversized swab.

Casey Starke, an undergraduate researcher overseeing the reconnaissance mission, peered down the hole in search of some suitable buildup for Abbruscato to sample and singled out a small concrete bench just above the flow. Problem spots in the sewer system like overhangs and ninety-degree turns, he said, tend to concentrate solid matter and make for prime sampling locations. With the target identified, Abbruscato lowered the pole until it made contact and then slowly turned it to get her sample. Up again, and she had clearly hit the mark.

Starke opened a clear plastic tackle box that doubled as a portable test kit, with blue gloves, sterile Q-tips, and a tray of carefully labeled polypropylene tubes filled with a buffer that both inactivates the

SARS-CoV-2 virus and preserves its genetic material. The stench was unmistakable through our face masks as he dabbed at the pole's soiled swab with a Q-tip before transferring the sample to a tube. "This is a very ripe site, isn't it," he said matter-of-factly.

"It is," Abbruscato agreed.

"But unfortunately, that correlates to good sample types," he said.

"So the riper the better?" I asked.

"Yeah, that's unfortunately what I've got to work with."

"Usually, you go for the greasy, nasty, gunky stuff," Abbruscato explained. "So if it smells *real* bad, it's probably in there." By *it*, she meant the killer in the sewer. Not the horrifying It of Stephen King's wicked imagination but an even more murderous villain that had wiped out more than 138,000 Americans by that point.

Starke, an aspiring doctor with an avid interest in epidemiology, was volunteering with a nonprofit biotech start-up called RAIN Incubator. The pilot project would help determine whether the team could accurately detect genetic material from the SARS-CoV-2 virus at two wastewater treatment plants and five other strategic points across Tacoma and correlate the signal trends to the county's heat maps of COVID-19 cases. The goal was to combine the test results with socioeconomic data to point out the most vulnerable hotspots and help public health officials determine where to direct more resources. The researchers had their work cut out for them, but if the pilot project proved itself, the city could be further subdivided into areas that drain into some three dozen pump stations where the sewage is pulled to higher ground and from which sludge could be easily collected. "It's ready made for epidemiology," Starke said.

Just as your poop can reveal plenty about what's happening within your gut, a community stool sample can illuminate what may be lurking within an entire population. The growth of wastewater-based epidemiology, turbocharged by the desperate need to get a better handle on a devastating pandemic, could establish the methods and infrastructure for tracking other deadly pathogens and dangerous drugs like the highly addictive opioid painkillers that have fueled a parallel epidemic. Cities across the globe geared up to read these stories in the sewer in the spring and summer of 2020, as hundreds of researchers converged upon the realization that a relatively low-tech surveillance method might give them a critical early warning.

The precedent, in fact, had been set more than eighty years earlier. In the summer of 1939, polio raged around the world and a team of investigators from Yale University set out to conduct tests in three cities with large epidemics: Charleston, South Carolina; Detroit, Michigan; and Buffalo, New York. They had tried before, unsuccessfully, to confirm the presence of the poliovirus in sewage during a 1932 epidemic in Philadelphia and again in 1937 in New Haven, Connecticut.

But this time, they struck gold, first in July when they sampled from a Charleston pumping station that collected sewage from a hard-hit district as well as a nearby isolation hospital. In the crude confirmatory experiments of the time, the researchers inoculated two rhesus macaques with a sample of the sewage, upon which both monkeys developed polio. To confirm their findings, the researchers used tissue from the central nervous system of the stricken animals to inoculate additional monkeys (and occasionally other lab animals), which subsequently developed the clinical signs of polio as well. As in humans, the virus signaled its presence in the monkeys through a fever, spinal cord lesions, and acute flaccid paralysis—a fast-progressing weakness or paralysis in the arms, legs, and even lungs. Tests conducted later in the summer when the epidemic had waned and again in the fall came back negative.

In Detroit, the researchers conducted the first successful wastewater test from a single building when they sampled sewage from a basement trap where the sewer pipe left an isolation hospital. On three occasions in August and September, their inoculation tests in monkeys confirmed the presence of infectious poliovirus. The team also made the first comparison between positive sewage test results and the concurrent burden of polio cases in the hospital's isolation wards. The investigators had less luck in Buffalo despite the city's ongoing epidemic. But they inadvertently demonstrated the blunt cruelty of such early experiments and the danger posed by other toxins and pathogens in wastewater when they inoculated monkeys with sludge collected from a treatment plant.

"The sludge material proved unusually toxic in that both of the monkeys inoculated with relatively small doses of it, promptly died," the authors noted. Another ten monkeys died from bacterial infections contracted during the series of experiments.

The scientists had, however, demonstrated that a lethal virus could be tracked through a community's sewers, and other researchers quickly took note. "As soon as we were informed of this discovery, it seemed to us important to get it verified," wrote a team of Swedish researchers who began testing the sewage of Stockholm during an outbreak in the city that same year. The scientists not only verified the earlier results by detecting polio in a sample collected that October, but also determined that the virus could retain its virulence for weeks. After storing the sewage sediment at roughly thirty-nine degrees Fahrenheit for two months, the scientists used it to successfully infect macaques. The authors argued that sewage that hadn't been disinfected was posing an imminent public health threat, adding to the growing consensus that polio was a water-borne disease. "Hence the serious consequence that during those periods when infantile paralysis is epidemical, we have to reckon with *the sewage as an important source of infection, from which the disease can spread over vast areas*," they wrote.

But wait, a French researcher responded. What about an animal vector like, say, sewer rats? In a droll bit of Scandinavian shade, the Swedish scientists roundly rejected the suggestion: "This idea may, of course, appear plausible if one considers the extraordinary abundance of rats prevailing in certain parts of the Paris sewers." But in the closed drains of Stockholm's sewer system, they asserted, "the rat has certainly no prospects of existence, still less of propagating." That would be welcome news indeed to the residents of what some have more recently called "the rat capital of Scandinavia," and to the listeners of Radio Sweden, which warned that a record invasion in Stockholm was being abetted in part by enterprising rats that were entering homes through the "protected environment" of sewer pipes.

The Swedish researchers similarly—and wrongly—dismissed the idea that insects could be carriers: "These arthropodes certainly very unwillingly, if ever, look for such a vehicle as sewage when they are about

Flush

to lay their eggs." Au contraire, as we saw from Mark Benecke's forensic entomology. Amid the missteps, the authors introduced an important concept: "One possibility presupposes the presence of a considerable number of healthy virus carriers living within the drainage area whence the infected sewage came." The idea of a silent outbreak was born, if deemed much less likely than the possibility that a living being, perhaps a single-celled protozoa, was somehow enabling the virus to multiply in the sewer.

Joseph Melnick, a pioneer of virology, environmental surveillance, and polio vaccine research, poured cold water on the idea that sewage was playing a direct role in the spread of the disease in a 1947 paper. But he championed the idea that determining the existence or nonexistence of virulent poliovirus in the sewer—a binary yes-no signal—could provide critical epidemiological information about whether it was present only during epidemics or *all the time* in urban environments. As scientists now know, only about one in every 200 polio infections leads to paralysis. Even so, the virus can efficiently replicate in the intestinal tract of both symptomatic and asymptomatic carriers and spread to others who ingest virus-laden fecal particles or are indirectly exposed through contaminated food or water. In the sewer, signs of the virus can track the rise and retreat of local outbreaks even in the absence of clear cases.

In 2021, polio was still endemic in Pakistan and Afghanistan, the last holdouts of a disease that has stubbornly resisted decades of effort to fully eradicate it from the planet. But periodic outbreaks of both the wild virus and a form that can occasionally escape from oral vaccines made with live but weakened versions of the virus have hit other countries. Israel has conducted sewage-based environmental surveillance for polio since 1989, with monthly samples collected and tested at sentinel sites across the country. In May 2013, wastewater monitors picked up signs of wild poliovirus type 1, the first time it had been detected in Israel since 1988. Investigators soon traced the silent outbreak to the southern city of Rahat, the largest predominantly Bedouin community in the country. Public health officials launched a vaccination campaign and the outbreak subsided in 2014. In a little more than six months, however, an estimated 60 percent of susceptible individuals, mainly children under the age of ten, had been infected in the community.

Surveillance systems in Israel and elsewhere have also had to account for chronic shedding of poliovirus particles by immunocompromised individuals who cannot fully clear the intestinal infection. Through the end of 2019, the World Health Organization had tallied nearly 150 cases of prolonged or chronic shedding from individuals who had been infected with polio through an attenuated vaccine and who represented potential reservoirs of disease. In one exceedingly rare but remarkable example, researchers were tracking a man in the UK who, at last count, had been continually shedding virulent poliovirus particles through his poop for more than thirty years.

The ability of poop-based surveillance to sketch out the contours of diet, disease, and even drug habits clearly raises ethical questions about who deserves to know what we're carrying around with us and when the public good outweighs personal privacy. In aggregate, though, testing wastewater may afford a measure of anonymity that monitoring cell phone use and personal health data doesn't. David Hirschberg, RAIN Incubator's founder, told me that because poop is so widely devalued, people are far more likely to agree to surveillance of wastewater samples than blood samples. As he put it, "I think people are done with their shit."

Roughly 40 percent of people infected with the SARS-CoV-2 virus shed it in their feces, potentially allowing sensitive tests of pooled sewage samples to detect even a handful of cases that might otherwise go undetected. Results can be available within a few hours of when a sample arrives at a lab for testing. Compared to clinically confirmed COVID-19 cases, multiple researchers have agreed, daily sewage tests can provide a head start of about a week in detecting the virus in a community. This monitoring requires a public sewer system, of course, and more than one-fifth of US households use private septic systems instead. Similar pooled tests, though, have also worked on wastewater samples from cruise ships and commercial airplanes.

A research group in Denmark, for example, detected multiple antibiotic-resistant genes in the toilet waste of eighteen international flights arriving in Copenhagen. The researchers found more evidence of *Salmonella enterica* and norovirus on flights originating in South Asia, and more evidence of C. diff on flights originating in North America. Beyond the clear implications for monitoring the spread of emerging diseases and antibiotic-resistant pathogens, the surveillance could help scientists estimate the prevalence of specific microbes in the city of origin. Testing poop from planes, ships, and buildings, in turn, could prompt follow-up tests and the identification of infected individuals. In sounding the alarm on silent outbreaks, we'll need to carefully consider how to strike the right balance between promoting public health and protecting the right to privacy.

Wastewater-based epidemiology is, at its core, a means to an end. A tool. A potent one, certainly, but the beginning point on a pathway of decisions that rely more on what we're willing to do than on what we're able to see. The signal is useless if we fail to heed the warning or fail to use it wisely to prevent the preventable. Will we invest in the long-term planning and infrastructure needed to identify global threats and in the ethical discussions needed to gain the public's trust about what happens next? Will we inconvenience ourselves to protect the most vulnerable, with the understanding that a lingering threat to one is a threat to all? Yet again, our own shit is suggesting how we might live *in* the world instead of apart from it. Whether we do so is entirely up to us.

● ● ●

On January 21, 2020, the CDC announced the first confirmed US case of what would eventually be called COVID-19, in a man in his thirties who had recently returned to western Washington from a trip to Wuhan, China. That afternoon, I raced around Seattle, on assignment for the Daily Beast, and looked for one of the first signs of public worry: Were people snapping up face masks? They were. Several drugstores were already sold out and I talked with a woman from China who was buying one of the last boxes in a Walgreens for her girlfriend, who would soon travel back to Chengdu. We know what happened next: terrifying scenes from Wuhan and then northern Italy contrasted with an apparent lull in the United States as officials confirmed a trickle of seemingly isolated cases from people with direct ties to Asia.

And then at the end of February, the epidemic exploded into public view in the Seattle region. A high school student with no known links to other cases. A woman who had returned home from South Korea. An outbreak at a nursing home called Life Care Center in suburban Kirkland that would be linked to 167 confirmed cases and thirty-five deaths by mid-March. Several of those early diagnoses were serendipitous: the student's case was caught by researchers at the Seattle Flu Study who had shifted course to test collected samples for the SARS-CoV-2 virus, while others were flagged by a suspicious infection control team at EvergreenHealth Hospital in suburban Kirkland.

But as shocked researchers discovered later, the virus may have been silently swirling around western Washington for three to six weeks until it reemerged with a vengeance when it reached a highly vulnerable population. Computational biologist Trevor Bedford and colleagues used genomic sequencing data to suggest that the outbreak could have sprung from yet another silent introduction of the virus in late January or early February. By March 1, Bedford estimated, the Seattle region already had 1,000 to 2,000 active cases; state officials had confirmed thirteen.

Multiple factors contributed to the stunning lag. Only individuals who received a test—most often prompted by their symptoms—and who tested positive were officially diagnosed with a case of COVID-19, assuming that the results were accurate. Early missteps by the CDC delayed the rollout of widespread PCR-based diagnostics, and stringent case definitions meant that only symptomatic people who had visited China's Wuhan region were assumed to be at high risk and in need of testing for a virus that had long since gone global. Epidemiologists initially thought that infected people wouldn't transmit the coronavirus until they started displaying symptoms—typically within a few days of exposure but sometimes up to two weeks later. We've since learned that presymptomatic carriers are often still infectious and that an estimated 35 to 40 percent of people infected by the virus have no symptoms at all but are likewise capable of spreading it on to others. Once testing became more widely available, US labs were swamped by demand and had turnaround times of up to two weeks before improving to about two days.

Flush

Around the world, parallel efforts were finding other ways to track the virus. In the Netherlands, scientists reported finding SARS-CoV-2 gene fragments in the sewage of three cities less than a week after the country's first confirmed case. In Amersfoort, the early March discovery predated the first diagnosed COVID-19 cases in that city by six days. Beyond the early confirmation, the study raised the prospect that the viral concentration might be correlated to the prevalence of cases.

As with polio detection decades earlier, other countries soon followed suit. In Italy, environmental virologists Giuseppina La Rosa and Elisabetta Suffredini at the National Institute of Health in Rome had worked with colleagues for more than a decade to track a range of pathogenic viruses in wastewater. Hepatitis A and hepatitis E. Norovirus. Adenovirus. A menagerie of lesser-known intestinal viruses that may or may not cause gastroenteritis. Until SARS-CoV-2, La Rosa told me, doctors had doubted the utility of their environmental surveillance. She and Suffredini had even published a paper detailing how they and colleagues used archival wastewater samples to detect a group of norovirus variants in Italy before doctors reported finding them in patients. "The clinicians are not so interested in the environment," she said. And now? "SARS-CoV-2 changed everything."

Like other virus hunters, they quickly pivoted and confirmed the country's first signs of the coronavirus in wastewater collected from Milan on February 24, 2020, three days after Italy's first confirmed COVID-19 case. The pandemic rampaged through northern Italy soon thereafter, and the researchers again turned to archival wastewater samples, this time from Milan, Turin, and Bologna starting in October 2019. Might there be earlier signs of SARS-CoV-2's arrival in the region? La Rosa and Suffredini each led their own analyses in separate labs to provide independent checks on the tests. Their findings agreed on a shocking conclusion: the virus had been circulating in Milan and Turin since mid-December 2019, and in Bologna since the end of January 2020. The results spurred a pilot project to set up a national surveillance system and help detect other silent outbreaks throughout Italy.

In Spain, a water utility called Global Omnium launched its own sewer surveillance system after hearing about the success in the Netherlands.

The Spanish system soon expanded to more than twenty cities representing more than ten million people. Unlike most other strategies at the time, the surveillance embraced the same kind of detailed sampling that Tacoma's RAIN Incubator was doing. Pablo Calabuig, the CEO of GoAigua North America, a Global Omnium subsidiary, led the effort to construct a digital dashboard that provided, almost in real time, results from about 800 sampling points. A team of ten researchers and technicians tested the samples for signs of the virus every two to three days. In one of the more successful interventions, the surveillance team subdivided the city of Valencia into thirty sewer-sheds; based in part on the wastewater surveillance, officials imposed stricter prevention measures in specific parts of the city.

Many of the first cases had arrived in the city when more than one-third of the Valencia soccer club and many of its fans contracted the virus during an ill-fated trip for a February 19 Champions League soccer match in Milan with Bergamo's Atalanta club. The "Game Zero" match, as Italian media later dubbed it, was attended by throngs of supporters from Bergamo and has been widely blamed for accelerating the northern Italian city's devastating COVID-19 outbreak. A pulmonologist at a Bergamo hospital later called the match "a biological bomb." Some of the shrapnel landed in Valencia, worsening that city's outbreak as well.

After spending several weeks adapting and refining a testing methodology that Global Omnium's researchers had earlier used to detect norovirus, the company began testing Spain's wastewater for the SARS-CoV-2 virus in early May 2020. Carina González Taboas, an environmental microbiologist with the company in Valencia, told me how everything moved so fast. She didn't sleep much, and time became a blur. After the launch, she became a familiar face on television as she explained the rationale for the testing strategy.

At the time, Calabuig said, the testing didn't reveal the number of cases but could accurately track, and foretell, trends over time. Maps that have tried to quantify COVID-19 risk levels for states and counties typically rely on the concentration and growth trajectory of confirmed cases. GoAigua's surveillance platform does essentially the same thing but combines the wastewater test results with demographic and income data to more accurately estimate a community's vulnerability. As one example, Calabuig said Global Omnium's testing arm in Spain sampled the wastewater of nursing homes in suspected hotspots; a positive signal prompted pooled saliva-based tests to help identify infected individuals.

Much of Spain was hit hard by the first wave of COVID-19 and went into lockdown. With Valencia's proactive approach that combined aggressive contract tracing with targeted PCR-based diagnostic testing based on the suspected hotspots, Calabuig said the city was able to avoid the worst of the second wave, with fewer hospitalizations and a lower prevalence rate than many other cities in the country. The experience convinced him that health officials should adopt separate surveillance strategies for subsequent phases of an epidemic. In a city with few or no known cases, frequent testing at the level of a wastewater treatment plant may detect an outbreak as soon as possible in the most cost-effective way. After an outbreak has been confirmed, testing less often but at more places in the city might be better for understanding where the virus is concentrating or dispersing.

The burgeoning research effort benefited from drug detection efforts as well. In Cambridge, Massachusetts, Mariana Matus and Newsha Ghaeli had initially focused their research on measuring bacteria, viruses, and chemical compounds in sewers. Their research team at the Massachusetts Institute of Technology, known as Underworlds, created a succession of portable collection robots with names like Mario, Luigi, and Yoshi—and eventually led to a spinoff company called Biobot Analytics. By discreetly sampling the "urban gut" of defined neighborhoods, they hoped to detect telltale compounds in the sewage that could estimate the local burden of health concerns like obesity or opioid addiction. The city of Cary, North Carolina, was among the first to use Biobot's technology to track its opioid epidemic at the neighborhood level after an alarming spike in overdose deaths. Critically, the company's ability to identify drug breakdown products, or metabolites, allowed the researchers to distinguish between opioids that had been flushed down a toilet and those that had been consumed. City officials then used the data to fine-tune their funding and policy initiatives.

Drug detectives, in fact, were honing their surveillance tactics in multiple cities. In Tacoma, researchers tested wastewater from two treatment plants for a major breakdown product of marijuana (mostly expelled through urine). Over a three-year period, most of it following the start of legal retail sales in Washington State, the scientists reported that marijuana consumption had doubled—leading to this classic headline in the *Seattle Times*: "Gee whiz." In 2016, researchers at the European Monitoring Centre for Drugs and Drug Addiction similarly revealed the drug proclivities of major cities based on drug metabolites detected in their wastewater. "Cocaine use appears higher in western and southern European countries, while amphetamines are more prominent in northern and eastern Europe," the center's European Drug Report asserted. The extensive report, perhaps predictably enough, spurred concerns about Big Brother snooping in the sewers and prompted jokes about the partying habits of northern cities such as Antwerp and Amsterdam, where Ecstasy (technically MDMA) seemed to be present in significantly higher concentrations than elsewhere.

Scientists in Australia took their Gladys Kravitz community snooping to the next level with their suggestion that they could map out social, demographic, and economic differences by testing the influent from twenty-two wastewater treatment plants representing six states and territories. The study concluded that biomarkers for vitamin, coffee, citrus, and fiber consumption correlated with more socioeconomic advantages in a given community, as determined by dozens of factors tabulated during Australia's national census. Conversely, biomarkers for the opioid tramadol, several antidepressants, an anticonvulsant, and the high blood pressure medication atenolol correlated with more socioeconomic disadvantages. In communities with older residents, they found higher concentrations of morphine, two medications for high blood pressure, and an antidepressant. Multiple "garbologists" have shown how easy it is to glean personal details about family households based on what they throw away. Sewer-based surveillance projects have demonstrated that the literal data dump swirling around in our toilets (sorry) can offer a surprisingly revealing look at neighborhoods and cities as well.

The onslaught of COVID-19 provided a critical testing ground for the evolving surveillance technology to prove itself in a new arena: a global pandemic unfolding in real time. To be useful, the wastewater signals would need to be reliable and give health officials enough early warning to act upon them. And then, as the strategy took off around the world, a larger question loomed: *Would those warnings do any good?*

From samples collected at a treatment plant in the Boston suburbs on March 18, 2020, Biobot reported the first detection of SARS-CoV-2 in North American wastewater. Matus said their results suggested more than 95 percent of all virus particles shed during the course of an infection were released within the first three days, on average. The clear signal in the sampled wastewater supported the idea that the surveillance was offering a leading indicator of new cases. The discovery then begged the question of whether the method could quantify *how much* virus was being released into the sewage and whether that amount could reveal anything about the number of people infected. For a region with 446 confirmed cases at the time, Biobot's initial "back of the envelope" calculations estimated that the true tally could have been anywhere from 2,300 to 115,000.

Wastewater plants can provide the flow rates of sewage and viral gene-detecting PCR tests can help estimate the virus concentration in a sample. But the average concentration of virus in the poop of infected people was driving much of the uncertainty in case estimates, Matus acknowledged, and wasn't well known at the start of the pandemic. Measuring it experimentally in a representative group of people might reduce the uncertainty, but that would require knowing quite a bit about the individuals, including who was infected and who wasn't. Refining the signal by collecting from smaller areas also required trade-offs. Workers at wastewater treatment plants knew how to collect samples of incoming influent or sludge and could do the same for wastewater-based surveillance. Sampling at specific sites along the sewer line might provide more granularity, Matus said, but also required more resources.

Based on its demonstration that the virus was at least detectable,

Biobot began fielding inquiries from across the US. The company accepted about 400 wastewater-treatment plants from forty-two states into a pro bono testing campaign, which confirmed that it could reliably measure the virus from a variety of sites and provide useful data to the respective communities. In June 2020, Biobot launched a commercial wastewater testing service for SARS-CoV-2. "The response has been amazing," Matus told me in mid-July, with 150 communities already on board. The experience had been highly motivating if somewhat surreal, given that the outside world had slowed while her company had never been busier and already quadrupled in size. Suddenly, she said, "people understand the power of waste and wastewater and how we're tapping into this very rich source of data that is produced naturally by everyone, every minute of the day, on a daily basis."

• • •

At the headquarters for RAIN Incubator, a former homeless shelter that now houses multiple labs and workspaces, I met Hirschberg and his friendly dog Moby in a second-floor conference room. Hirschberg told Starke that he had just been reading up on maritime law after receiving an inquiry from the owners of an Alaskan fishing boat. The company was interested in sewage testing for its international crew, and maritime law suggested that although individual workers couldn't be tested at sea, their collective waste could. Any positive signals could then prompt follow-up testing when they reached shore.

Compared to the mass testing of individuals, Hirschberg said wastewater surveillance could provide a less biased look at transmission trends: because everyone poops, the surveillance can capture data from a significant fraction of those who are asymptomatic or who lack access to health care and would be otherwise hidden from view. Hirschberg said that advantage could be key for some of Tacoma's largely segregated neighborhoods. Across the US, COVID-19 was disproportionately killing African American, Indigenous, and Latino residents. An international study later suggested that the cumulative toll had reduced the life expectancy of US men by more than two years, the largest drop among the twenty-nine countries studied and a stunning decline largely attributed to higher mortality rates among working-age men. From an equity standpoint, Hirschberg said, testing the sewage in underserved neighborhoods might provide timelier warnings that bypassed the lack of access.

As Moby padded over to let me pet him, Hirschberg said he hadn't yet sold public officials on the value of his nonprofit lab's environmental surveillance. But he had at least secured the city's help in collecting samples and hoped the pilot project would help change minds. Hirschberg, a veteran of multiple molecular biology and diagnostic labs, had worked alongside other virus hunters on tests for HIV and other major pathogens. He knew from his previous work on severe acute respiratory syndrome, or SARS, which first appeared in China in 2002, that SARS-CoV-2 would also likely be shed in sewage. Both he and Stanley Langevin, RAIN's head of scientific development at the time, cited a famous investigation in which scientists traced a major SARS outbreak in Hong Kong's Amoy Gardens apartment complex to faulty plumbing. Aerosolized bits of fecal matter, the researchers determined, had been drawn back into a resident's bathroom through the floor drain and then into a nearby air shaft, from which the virus spread to other units. Studies so far have suggested that SARS-CoV-2 viral particles in poop and sewage are far less dangerous. Although the viral RNA can persist in wastewater and transmission through contaminated feces hasn't been ruled out, researchers haven't yet found any clear evidence that the virus can be spread from person to person through wastewater.

Hirschberg was proud of RAIN's scrappy, DIY ethos of building equipment from scratch, renovating used machines, and training students to be similarly resourceful. "We do blue collar biotech," he said. When the nonprofit pivoted to COVID-19, his team developed its own viral RNA and antibody-detection tests. The former relies on an ultrasensitive, machine-driven copying process called polymerase chain reaction, or the now ubiquitous PCR. It's the same test that labs use to diagnose COVID-19 in people. Once researchers have sequenced the genetic material of a virus like SARS-CoV-2, they can select representative genes or genetic regions and design small pieces of DNA that stick to them like bits of Velcro. These primers, as they're known, allow the copier to churn out multiple copies of the selected sequences. If they're present in sufficient numbers, the copying process can confirm the presence of the virus, or at least of its genetic material. Because SARS-CoV-2 uses single-stranded RNA instead of double-stranded DNA as its genetic code, an additional step converts the RNA to DNA so the PCR machine can properly read its sequence data.

Since the RAIN team had begun testing at Tacoma's T Street site in mid-May, it had returned a positive result every week (over a five-month span ending in early October, it would deliver only a single negative result). At the wastewater treatment plants, accommodating workers provided fecal samples from the sludge that precipitates to the bottom of large settling tanks, while Starke and city employees initially tested at other spots around the city by sampling the flowing wastewater in the sewers. Then a discussion with Langevin about the samples' fatty consistency led to a pivotal aha moment. The researchers realized that the virus, with its fatty outer envelope, likely stuck to clumps of fats, oils, and greases—the notorious FOGs that can form sewer-clogging fatbergs and are normally the bane of sanitation engineers. The smelly, hated globs might act like snares to capture and concentrate viral fragments, and Langevin figured he could get a stronger signal from a swab of that sludge than from a liter of wastewater.

Other researchers were reaching similar conclusions. Civil and environmental engineer Krista Wigginton and colleagues had previously found that other viruses covered by fat-filled envelopes stick to the solids in raw sewage. A team led by Wigginton and collaborator Alexandria Boehm then confirmed that SARS-CoV-2 behaved the same way in wastewater samples from Palo Alto and San Jose, California. The settled solids in wastewater treatment plants, they reported, contained dramatically more viral particles than samples captured from the liquid influent.

I joined Langevin at a table on the first floor of the RAIN Incubator to hear more about the group's work. He was likewise a veteran of virology and disease surveillance labs and had been chasing signals for twenty years, he said. Langevin pointed out the obvious flaw in the main signals being used to track COVID-19. Hospitalization and death, he said, are the two *worst* ways to conduct disease surveillance: they're endpoints and do nothing to give you advance warning. In explaining their work to me, the Italian scientists La Rosa and Suffredini had made the same point in a slide show that included a surveillance pyramid: reported hospitalizations were at the tip, representing only a tiny fraction of cases, while environmental detective work was at the broad base. Science writer Ed Yong used a different analogy to explain why the US had been repeatedly behind the curve with the coronavirus. "Pandemic data are like the light of distant stars, recording past events instead of present ones. This lag separates actions from their consequences by enough time to break our intuition for cause and effect. Policy makers end up acting only when it's too late. Predictable surges get falsely cast as unexpected surprises."

As a research fellow at the CDC's lab in Fort Collins, Colorado, in 1999, Langevin played a key role in trying to find more proactive surveillance methods for the mosquito-borne West Nile virus that emerged in New York City that summer. He and his colleagues exposed twenty-five bird species to virus-infected mosquitoes to show that passerines like blue jays, common grackles, and American crows were among the biggest contributors to the viral transmission cycle that was keeping the virus in circulation. They were falling victim to West Nile from mosquito bites too. But before dying, their blood was acting as a viral reservoir from which more mosquitoes could become infected. "We tracked mosquitoes to be ahead of the human curve. Then, miraculously, this new signal came out when birds started falling out of the sky," Langevin recalled. His group was one of the first to push for "dead bird" surveillance as a leading indicator of subsequent outbreaks. "It was amazing how predictive it was," he said.

As a science reporter for *Newsday* at the time, I wrote extensively about the epidemic and remember officials enlisting the public's help in reporting the stricken birds; many readers called us about them as well. In 2001, the first four dead crows signaled the start of West Nile season on Long Island. Environmental signals like these are noninvasive, easily collected, and generally subject to fewer regulations and politics, Langevin said. "And if they're done in the right way and interpreted in the right way, they can be extremely powerful."

Langevin and a growing number of researchers were convinced that using environmental signals to track COVID-19 could be just as effective, but the sewer-based strategy still had limitations. Although testing our poop for what's passed through us makes sense, it's not infallible given all the other things that pass through our sewers. Detergents or bleach flowing from commercial and industrial sites can dampen signals by degrading the viral particles. Rainwater in combined sewer systems that collect runoff, domestic sewage, and industrial wastewater will dilute the sample. So will showers, washing machines, and dishwashers during certain times of the day. Higher temperatures can increase the decay rate of the viral RNA. Some patients will shed considerably more of the virus than others. And the yes-no binary signals that might point out the initial appearance or reappearance of an outbreak or lingering reservoirs during periods of relatively limited transmission can be overwhelmed during epidemic peaks when everything lights up.

As COVID-19 surged in December 2020, Richard Danila, the deputy state epidemiologist for Minnesota, wasn't yet convinced of the need for wastewater-based epidemiology. "This is a waste of money, a waste of resources. I don't need to look at wastewater when I got cases everywhere," he said. And he did: "We got cases in every single county, eighty-seven counties in Minnesota. You don't need to tell me about wastewater." Maybe in a place where there were no cases. Fine. But otherwise, it was just meaningless information to him. And then he zeroed in on the crucial question for the field: Even if scientists found it, what then? "I mean, it's the human that's important, not the wastewater." The surveillance based on wastewater epidemiology might be a nice academic exercise, he told me. But he needed proof that it would do any good. In principle, the concentration of SARS-CoV-2 viral particles in a sewage sample should tell you something about the relative burden of COVID-19 in that community. But whether the virus could be accurately quantified was hotly debated. Langevin and some other researchers were adamant that there were just far too many variables.

Flush

The presence of potential "super spreaders" who release vast numbers of viral particles meant that efforts to estimate the number of cases based on viral concentrations in the sewers were doomed to fail, he said. The wildly variable early estimates in the Boston region and elsewhere seemed to bolster his point. "We're not all equal. We don't poop out the same amount of virus," he said.

Drilling down to smaller sub-sewer-sheds that drained the waste from defined parts of the city, Langevin said, offered the resolution necessary to find out where most of the transmission was happening in the absence of mass testing. The yes-no signals pointing out consistent COVID-19 hotspots could then guide the distribution of masks and other resources necessary to help bend the curve of new cases and empower local communities to protect themselves. The RAIN team had already started dividing the T Street Gulch site, which had lit up all summer, into smaller sections to see if they could collect and test sewage samples from as few as 150 households. "I really, truly believe that the sewer system is our signal," he said. "It's the best one we got."

By attacking the same problem from multiple angles, a remarkable convergence of research efforts was rapidly pushing the science forward and chipping away at the remaining variables. In Michigan, epidemiologist Kevin Bakker teamed up with engineer Krista Wigginton to collect daily samples at wastewater treatment plants in Ann Arbor and Ypsilanti toward the goal of setting up a COVID-19 early warning system in the state. Bakker was initially interested in tracking viruses like respiratory syncytial virus, norovirus, poliovirus, and Enterovirus D68, which can cause polio-like paralysis in children. Wigginton had been focusing on intestinal pathogens and coronaviruses like the ones that cause SARS and Middle East respiratory syndrome, or MERS. Along with scores of other researchers, they both turned to tracking the SARS-CoV-2 coronavirus.

As a mathematical modeler, Bakker said he was excited by the potential that data from multiple test sites might help him and his colleagues hash out the main factors working against a reasonable estimate of a community's COVID-19 burden. Because a primary mandate of wastewater treatment plants is to clean the incoming wastewater, the plants already collect reams of data about chemical and physical conditions like water temperature and turbidity. "If we have a bunch of readings from different locations, it gives us a better grasp on what the unknown parameters are," Bakker said. His group was also working with researchers to better understand the starting concentration of coronavirus in the stools of infected people. But how might they account for poop diluted by other wastewater or rainwater, especially in a combined sewer system?

It turns out that the pepper mild mottle virus we heard about back in chapter 1 makes an excellent marker. Although the virus is best known as a plant pathogen that infects hot, bell, and ornamental peppers around the world, it's nearly ubiquitous in our poop, providing a very useful indicator of our presence. We ingest the virus when we eat peppers or hot sauce; it passes through the digestive system along with the food but can be easily detected in our poop (someone who has never eaten a pepper or pepper product may not shed the virus but would be vastly outnumbered by pepper fans in most communities). The pepper mild mottle virus is relatively rare in animal feces but the most abundant RNA virus yet found in human feces. Its relative concentration in the sewer or in a treatment plant's influent, Bakker said, can help researchers estimate the human contribution to the overall flow. As such, the virus has been used as a proxy to point out fecal pollution in global waterways and warn of potential food or water contamination by harmful intestinal counterparts: wherever the pepper mild mottle virus is, other gut inhabitants are likely there too.

But that still left the unresolved question of how to account for our different SARS-CoV-2 shedding rates. Environmental microbiologist Ian Pepper has looked for faint signals in wastewater and biosolids, the organic matter recycled from wastewater treatment plants, for more than forty years. As director of the University of Arizona's Water & Energy Sustainable Technology Center, Pepper has led efforts to reclaim and purify water and to detect viruses, bacteria, and other pathogens in the environment. In late February 2020, the university center advertised that it was accepting wastewater samples from utilities to test for signs of SARS-CoV-2. Like Biobot, the center received hundreds of responses and began processing samples from as far away as New York, Florida, and Canada.

Soon afterward, university officials began devising a complicated fall reentry plan for students. Pepper led the wastewater-based epidemiology team. His team's task: monitor all of the campus dorms. Like Langevin, he reasoned that maintenance holes were the logical choice. Fortunately, campus planners had engineered the complicated sewer system so that each dorm drained into a separate line. Pepper's team would sample sewage from twenty campus buildings, including dorms and the student union center, three times a week at eight thirty in the morning.

Based on his experience interpreting the data from samples sent by other wastewater treatment plants, Pepper came up with five levels of concern for the university, from Level 0 indicating a lack of detectable virus to Level 4 reflecting "sky-high" virus concentrations. On August 24, 2020, students returned to campus and Pepper's team began sampling the sewage. Nothing. Then the next day, a dorm lit up. "Even though we had been preparing for that event all summer long, when it actually happened all hell broke loose," Pepper recalled. Based on the viral concentration in the wastewater, the team decided to test every resident in the dorm. Those tests uncovered two asymptomatic cases and the students were moved to a quarantine location to reduce the risk to others.

The researchers also decided to retest the wastewater and collected samples from the dorm's outflow every five minutes over a half-hour period. The virus concentrations were virtually identical at each time-point. "That kind of validates the theory that when viruses are in feces, which is then in the sewer, the viruses disperse," Pepper said. In other words, a simple flush wouldn't send them hurtling on toward the wastewater treatment plant. Like scientists had previously found with SARS, the particles tended to bounce around in the pipes and stick around for a while.

Cases at the University of Arizona peaked in mid-September and then dropped and remained relatively flat through the end of November even as counts were rising elsewhere in Arizona. The success of the surveillance, Pepper said, was changing the dynamic between the university and the surrounding community. "At the beginning of the semester, the community was concerned that students would be spreading the disease to the community. Well, I think it's the other way around now," he said.

Sewage-based tests had lit up consistently since that first positive result, and Pepper estimated that the university had averted another eighty outbreaks through its follow-up testing based on the location of each signal. "And so the university has been successful in remaining open, which many universities have not been able to," he said. The targeted testing, he added, was far more economical than continually testing every student.

At first, Pepper said, local epidemiologists viewed his environmental surveillance efforts with polite skepticism. No longer. "Poop doesn't lie. If the virus is in the wastewater, it came from someone." The team rarely encountered false positive results, and unlike clinical tests, weekly sewage tests could predict whether the local burden of cases was increasing, plateauing, or decreasing. When we talked in early December 2020, Pepper said the clinical case rate had been fairly low, but he worried about a reading from the Agua Nueva Wastewater Treatment Plant in Tucson, which had just registered one of the highest viral concentrations yet seen. "Memorial Day, Independence Day, Labor Day, and now Thanksgiving: each time, one week after that holiday, we see a spike in the virus concentration. Two weeks later, we see a spike in the number of cases."

What about the contentious issue of whether wastewater surveillance could estimate the true number of cases? Pepper said the dorms provided a gold mine of epidemiological data: they were defined communities with known numbers and identities of residents and known numbers of symptomatic and asymptomatic cases. An equation allowed his team to predict the number of infectious cases based on the virus concentration in the wastewater, number of residents, average amount of feces, and viral shedding rate. The latter variable was still a concern. But Pepper's team already knew the number of cases in each dorm from the follow-up testing triggered by substantial concentrations of SARS-CoV-2 in the wastewater. Some infected students would shed more than others, of course. But within a population of 300 or so, the highs and lows would average out into what's known as the mean shedding rate. "So we back-calculated the shedding rate and found it to be virtually identical in every dorm," Pepper said.

The sleuthing based on the fortuitous existence of a well-characterized population at the university allowed the team to use the mean shedding rate to correlate the viral concentration to the overall number of cases in other wastewater utility districts. Geographic information and discreet sampling within particular zones or zip codes could yield heat maps that predict upcoming hotspots based on the expected number of cases. From the number of reported cases, the analysis could then estimate the number of unreported cases, the vast majority of which would be asymptomatic.

Though they differed in their approach, both Langevin and Pepper saw enormous implications for redistributing public health resources. The proof of principle in the dorms suggested that the same process could be used to test high-risk populations in nursing homes, prisons, food processing facilities, and other buildings or complexes served by individual sewer lines, just like the polio hunters had done in the Stockholm hospital.

• • •

The Yuma Center of Excellence for Desert Agriculture, affiliated with the University of Arizona and funded primarily by the agricultural industry, has focused on an increasingly vexing problem. How can a region blessed by abundant sunshine but constrained by just over three inches of rain every year continue to grow much of the continent's fruits and vegetables? The southwestern Arizona region produces about 80 to 90 percent of North America's leafy greens like lettuce, spinach, and kale in the winter months. But COVID-19 was hitting agricultural workers hard and posing another major threat to the multi-billion-dollar industry. When Paul Brierley, the center's executive director, heard about the success on the University of Arizona campus, he began working with Pepper to bring wastewater-based surveillance to Yuma County too.

Initially, Brierley and his team hoped to test wastewater in the portable outhouses used by workers in the fields. But the combination of heat and deodorizing chemicals quickly degraded the virus and compromised the ability of tests to accurately identify it. Undeterred, they figured the surveillance might be useful elsewhere in the county. With Pepper's help, they made their case in a webinar attended by representatives from the city and county, agricultural community, local schools, public health agencies, Yuma Regional Medical Center, Marine Corps Air Station Yuma, and US Army Yuma Proving Ground. Minutes after the webinar ended, a county supervisor called to express interest, and the board subsequently committed $220,000 toward developing, equipping, and staffing a testing lab.

Brierley and his team were ready by early November 2020. Instead of handing the lab's data off to individual clients or the health department, they formed a steering committee that included representatives from many of the same agencies that had heard their initial pitch, along with each municipality. Brierley said the coordination meant that if his team happened upon any unexpected findings, the committee could immediately gather. The early warning, they knew, would be useful only if they could act quickly.

The seasonal workforce at Yuma's DatePac processing plant, the largest handler of Medjool dates in the world, normally swells to a temporary peak of more than 1,500 for the end-of-summer harvest. In 2020, COVID-19 limited that peak to 450 before the workforce fell back to its core crew of about 200 workers. Juan Guzman, the plant's senior vice president of operations, told me that he and the shift supervisors had tried to do what they could to protect their employees, who are primarily Latina women. But for COVID-19, they were flying blind. Then they heard about the potential of wastewater-based surveillance. "When somebody is telling me here, 'I can give you a data point, which is there's somebody sick in your facility and you can avoid an outbreak,' well, it's a no-brainer, right?"

Brierley's team began sampling the plant's wastewater twice a week, right after break time for the first and second shifts. Nothing. And then the week after Thanksgiving, the first positive. "And all hell breaks loose," Guzman recalled. The Yuma County Public Health Services District arranged for the Regional Center for Border Health to set up a mobile testing unit from a van in the company's parking lot. From the initial and confirmatory tests, four employees tested positive. None had any symptoms. Guzman sent them home with pay and asked them to self-isolate.

Because the remaining workers were already spaced and wearing masks, the health department didn't advise additional quarantines, only close monitoring for any symptoms. "And sure enough, nobody else tested positive. So it was very exciting," Guzman said. The ripples of fear that had emanated from the positive tests dissipated as the workers realized that the precautions had worked. No one can say for sure what would have happened had the four cases not been identified, but Brierley likes to think that the averted outbreak saved a lot of Christmases. The employees responded by increasing their compliance with the mask rules, and the cleaning crew noted that the dispensers of soap and hand sanitizer had to be refilled more often.

When the wastewater-based surveillance revealed another asymptomatic case in February 2021, workers clamored to be first in line for testing, Guzman said. A month later, the plant distributed voluntary sign-up sheets for the newly developed COVID-19 vaccines. Of the 200 employees, only ten didn't immediately sign up. By the time the second dose arrived, all but five workers had agreed to be vaccinated.

Keeping your employees healthy, of course, is good for business. For an agricultural county often saddled with the reputation of having one of the nation's highest unemployment rates, though, Guzman told me he was proud that Yuma County might be known for using science to protect its workforce. Throughout the pandemic, the US struggled to do likewise. Mounting a more effective defense the next time around may depend in large part upon the willingness of state and federal governments to reinvest in risk reduction efforts.

In September 2020, the CDC raised hopes for renewed prioritization of disease surveillance when it launched the nation's first wastewater-based monitoring network, called the National Wastewater Surveillance System. The system's dashboard includes an interactive map of hundreds of sampling sites across the country, color-coded by trends in the relative levels of SARS-CoV-2 RNA at each site. Environmental microbiologist Amy Kirby, the agency's program leader for the surveillance system, said calculating the tally of cases from those viral levels was still a difficult undertaking. For a broad surveillance network to be useful, though, absolute numbers aren't critical. Trends are. And in the case of emerging or reemerging pathogens, flagging their arrival in a rural or urban sewer system could trigger a cascade of additional testing.

Kirby said the CDC was helping rural utilities get the tools they needed to be part of the network; she hoped to expand the system to include some collaborating tribal nations as well. Surveillance at the level of facilities like colleges, nursing homes, businesses, and prisons may offer more details about viral hotspots, but she cautioned that the signal can also be noisier due to a smaller sample size that isn't as well mixed. Successful monitoring at the University of Arizona and other colleges, though, highlighted the potential of rapid collection and testing that could be acted upon quickly. The CDC, Kirby said, had launched an on-site testing project to see whether it could replicate the successful warnings at twenty jails and prisons across the country. If so, the project might help protect a particularly vulnerable population.

The dramatic mainstreaming of wastewater epidemiology over a two-year period was perhaps best illustrated by a database of global monitoring sites maintained by the COVIDPoops19 project. By January 2022, nearly sixty countries and more than 270 universities were using the method to track the SARS-CoV-2 virus. On a longer-term basis, wastewater treatment plants in key cities could be tapped as primary sentinels in large networks; a positive test in a city like Detroit or Chicago could prompt regional canvassing. "If we detect in some of those, we have the training and the logistics to be able to go out and sample in rural Michigan or Iowa," Kevin Bakker said. As a major agricultural center along a busy border with Mexico, Yuma might also be well-positioned as an important sentinel node. Signals from the highly contagious SARS-CoV-2 Omicron variant were unmistakable everywhere in early 2022: a tsunami of nearly vertical lines in graphs of wastewater data that closely tracked the exponentially rising case counts. But even there, sewage-based confirmation of the seemingly obvious helped foretell where and when the wave was cresting.

Beyond COVID-19, Kirby and other CDC scientists have developed a surveillance wish list; at the top, she said, are antibiotic-resistant pathogens. The network might also prove useful in estimating a community's burden of foodborne infections like *E. coli*, *Salmonella*, and norovirus. Patients can shed plenty of the microbes through diarrhea and vomit, but most tend to recover without needing to see a doctor, meaning that existing surveillance only captures a small fraction of cases. The network could help detect poorly understood emerging pathogens as well, like the fungus *Candida auris* that is fast becoming a global health threat and growing resistant to multiple antifungal therapies. So far, severe cases have been largely confined to hospitals and nursing homes, Kirby said, leaving researchers in the dark about the extent to which the fungus may be lurking in communities. The most-wanted list of dangerous pathogens could be included in a testing panel that Kirby said the agency was hoping to roll out within the next two years.

More threats will emerge from drugs and other synthetic chemicals and from spillover diseases that jump from birds or bats or primates or rats to us. Biosecurity expert Jeanne Fair explained that global warming can disrupt natural habitats and increase the risk of spillover events by bringing infected wild animals in closer contact with humans. By 2020, a pandemic early-warning project called PREDICT had detected about 950 new animal-borne viruses of concern, primarily in Africa and Asia. The program, launched in 2009 to better prepare the world for future outbreaks of infectious diseases, was shuttered by the Trump administration.

The COVID-19 pandemic, for all the suffering it brought, may provide an opening to help reinforce the value of sustained cooperation and vigilance. Maybe the outbreak could get the planet on the same page, "to understand that, yes, these are continuous problems," Langevin told me. Mariana Matus said she hoped Biobot's work with COVID-19 could be a stepping-stone to applications like an expansion of its monitoring work on opioid use. Bakker said he wanted the refined methods to be reapplied to polio as well, effectively circling back to the disease that helped launch the field. "We've been on the brink of eradicating it for going on thirty years now, and we need that final push," he said.

By proving itself, wastewater-based epidemiology has opened the door for more sophisticated methods like viral metagenomics, in which scientists try to identify all viral genes within a given sewage sample to detect new viruses and perhaps head off future pandemics. To be successful, though, this environmental surveillance will require engaging and building trust with the communities being monitored. People left in the dark about why the surveillance is needed, how it will be used, and how it might benefit them may be far less likely to cooperate and more liable to suspect bad intentions and believe misinformation.

Perhaps the biggest danger, then, is that we won't heed the lessons that are already in plain sight. We know that failing to invest in unsexy but necessary infrastructure and surveillance projects blinds us to future threats. We know that failing to counter the demonization of science and public health fuels a toxic swirl of fear, mistrust, and disinformation that increases the body count—poop may not lie, but people do. And we know that failing to confront and eliminate the threat to the most vulnerable among us inevitably prolongs the agony for everyone.

In *The Demon-Haunted World*, astronomer Carl Sagan and writer and producer Ann Druyan, his wife, mounted a full-throated defense of the scientific method and warned of a slide "back into superstition and darkness." The dumbing down of the US, they wrote, "is most evident in the slow decay of substantive content in the enormously influential media, the 30-second sound bites (now down to 10 seconds or less), lowest common denominator programming, credulous presentations on pseudoscience and superstition, but especially a kind of celebration of ignorance." Those words, written in 1995, presciently foretold our self-inflicted pandemic misery. It seems fitting that a way out of the superstition and darkness and ignorance and pseudoscience, at least in part, may run through the darkest recesses of our sewers, where the "greasy, nasty, gunky stuff" may help us see new threats in time to act upon them. *Will it do any good?* That, it seems clear, is entirely up to us.

CHAPTER SEVEN

Epitome

THE GROUP OF ABOUT THIRTY Jahai women, from a remote village in the Royal Belum State Park in northern Malaysia, could *not* stop laughing when they heard the request. In a photo that captures the moment, two of them are covering their mouths while others are smiling broadly, clearly amused. After Mathilde Poyet carefully explained the purpose of her team's research project, she saw the women's eyes change as it dawned on them. "At some point, all the women at the same time just realized what I was asking for: poop," Poyet told me. Uproarious laughter ensued. "Like it's thirty women just laughing for ten minutes. That was awesome." The women were amenable to donating; it's just that no one had ever before asked them for such a thing.

In another small village in the extreme western part of Rwanda, Poyet made a similar pitch to a group of men, women, and very curious children and told them she'd be available to answer any questions for as long as they needed. "And the leader of the village, she was a woman and she said, 'No, no, that's OK. You wait here and we'll come back,'" Poyet recalled. An hour later, she had forty stool samples: an equally memorable instance of poop on demand.

In more than a dozen countries around the world, Poyet and Mathieu Groussin, cofounders of the Global Microbiome Conservancy, have sought out fecal donations from both urban and rural populations to isolate and preserve gut bacterial lineages before they go the way of dodos and dinosaurs. Scientists once scoffed at the notion that bacteria could go extinct, and our distinctly uncharismatic micro-fauna aren't exactly poster children for conservation in the way that orcas or polar

163

bears are. But recent research has nonetheless raised the specter of microbial "extinction events" in which some strains begin to disappear from their ecological niches.

"Just as global warming, deforestation, and environmental pollution impoverish ecosystems of the planet, the consumption of processed food and the abusive use of antibiotics and sanitizers contribute to the decrease of human-associated bacterial diversity," Poyet and Groussin wrote in a 2020 editorial. "As a result, some commensal microbes that have coevolved with us for millennia—and which represent an integral facet of human health and history—may soon go extinct. We already find many gut bacterial species that now exist almost exclusively among members of non-industrialized, isolated human populations; yet these populations, along with their lifestyle and culture, are under threat from globalization and climate change."

The realization that the world's microbial diversity might be decreasing roughly coincided with the discovery of methods to grow bacteria previously considered unculturable beyond the finicky confines of the gut. I had initially imagined the Global Microbiome Conservancy as a kind of Svalbard Global Seed Vault for the human microbiome, but Poyet and Groussin stressed that their work involved far more than stocking a doomsday vault. They were actively growing the bacteria their team had isolated from forty-four communities in fifteen countries visited so far: more than 10,000 separate bacterial strains representing 450 species. Think of it more like a mini-sanctuary for endangered microbes.

We're finding that the gradual erosion of our inner ecosystem may have come at a significant cost. We've already seen how the indiscriminate killing of both commensal and pathogenic gut microbes from antibiotics can increase our susceptibility to C. diff infections, and how the absence of certain microbes at birth can predict which toddlers will be more disposed to asthma and allergies. Multiple studies have linked the whittling away of the gut microbiome in more industrialized countries to a higher prevalence of autoimmune diseases and linked gut dysbiosis to the progression of metabolic disorders like diabetes and obesity. If true, increasingly rare microbiomes with high bacterial diversity may provide clues and even medicinal compounds to help counter the negative impacts of our microbial depletion. The intestinal contents of some communities, in other words, may represent particularly valuable natural resources.

The expanding roster of potential health applications has raised increasingly provocative questions and dilemmas as it creates a new class of haves and have-nots. Can the right shit make you thin or happy? Can it slow the aging process? A *South Park* episode called "Turd Burglars" lampooned our celebrity culture by imagining that fecal transplants had become so desirable as a fountain of youth and health that some of the town's adults conscripted the boys into stealing poop. After DIY transplants went disastrously wrong, Kyle saved the day by finding the priceless "spice mélange" of famously fit NFL quarterback Tom Brady (a spot-on reference to *Dune*), who kept bottles and jars and barrels of his own crap in a hidden room behind a bookcase.

In real life, a London shop called the Viktor Wynd Museum of Curiosities, Fine Art & Natural History earned headline after headline over its curio-cabinet displays of the (unverified) poo of singers Kylie Minogue and Amy Winehouse. It turns out that beauty products made from the feces of the young, athletic, or beautiful have a long historical precedent. In *History of Shit*, Dominique Laporte cites two eighteenth-century documents that describe the use of poop from young and athletic men as a sort of age-defying beauty cream. Laporte himself seemed a bit incredulous and skeptical but dutifully relayed several anecdotes. "In some instances, custom went so far as to exact meconium, the 'discharge of just-born infants.' In other instances, an individual was retained for the specific purpose of keeping the lady supplied," he wrote. That next-level mud masking may seem a bit far-fetched until you consider the enduring popularity of a facial made from the poop of the Japanese bush warbler. Uguisu no fun, as it's called, seems decidedly unfun to me but has nonetheless garnered rave reviews on Amazon.

Given the commodification of what popular culture has proclaimed as good shit, it's more than a little ironic that a decided lack of microbial diversity in wealthy countries like the United States has made us the most impoverished and potentially the most in need of help. Poop, yet again, is upending our notions of what is normal and what has value.

Fecal microbiota transplants, the once-derided remedies of last resort, have arguably had an outsized effect in forcing this reevaluation. When Mark Smith, the cofounder of OpenBiome, and I talked again in 2021, he had become the CEO of Finch, a biotech start-up based in Somerville, Massachusetts. FMTs, beyond their transformational effects on patients, had challenged ideas about where humans end and our inner microbial residents begin. For Smith, the shift toward a more ecological approach to human health had even called into question what it means to be human. Maybe, he mused, we're a super-organism made up of many species that coexist within a single dwelling?

In *I Contain Multitudes*, Ed Yong likewise contemplates our relationship with our inner colonists, the vast majority of which are mostly harmless. "At worst, they are passengers or hitchhikers," he writes. "At best, they are invaluable parts of our bodies: not takers of life but its guardians. They behave like a hidden organ, as important as a stomach or an eye but made of trillions of swarming individual cells rather than a single unified mass." It's hard to draw a simple dividing line between "us" and "them" when many of those microbes help us digest our food, synthesize our vitamins, regulate our immune system, and keep killer pathogens at bay. Those consequential roles, in turn, have raised questions about how diverse microbiomes may have aided our ancestors in important ways—helping them digest a wider variety of foodstuffs, say, or preventing more misguided autoimmune attacks—that have faded with the apparent demise of certain bacterial species. In that light, maybe it's not surprising that many researchers have adopted a conservationist attitude toward our own inner jungle.

From the thousands of *potential* bacterial species inhabiting any person's gut, most people in the US and other industrialized countries now host maybe fifty to 200 species. Poyet said exact comparisons with nonindustrialized populations can be tricky because of the high proportion of previously unknown species in those microbiomes. From the samples she and Groussin have looked at, however, rural populations living in nonindustrialized countries have microbiomes that are often twofold more genetically diverse than those of industrialized populations. From the few re-creations of past microbiomes published to date, the estimated diversity is higher yet.

If an emerging goal of medicine is to support the robustness of entire microbial communities instead of merely focusing on single molecules that target individual threats, then our role may be something akin to "park rangers for the microbiome," as Smith put it. More than just tending our own inner gardens, we would become guardians of the entire supply of seeds. Patrolling, protecting, and occasionally rebalancing a powerful but complex and fragile ecosystem might help preserve the ability of its many parts to continue working together toward our common good.

● ● ●

To really grasp what many of us are missing, it might be helpful to understand the power of what some of us still have. In industrialized societies like the US, uncommon people like Joe Timm can still be called super donors given their ability to cure hundreds or even thousands of C. diff patients via an FMT. The same technique has shown promise against other bacterial infections, including another deadly scourge in hospitals known as vancomycin-resistant enterococci, or VRE. And as FMT trials continue to branch out in the hope of addressing even more conditions, the value of a healthy microbiome is continuing to expand.

Finch, like several similar companies, was set up to develop therapeutics based on the gut microbiome. In all, the start-up was testing four products for five indications. One of them, CP101, was essentially a single-donor FMT for curing recurrent C. diff infections but in a pill form that allowed Finch to standardize the dose based on the total number of viable bacteria delivered per treatment. CP101 had performed reasonably well in a large Phase II clinical trial, nearing the cure rate for FMTs delivered via colonoscopy, and was farthest along in the company's developmental pipeline. "I think it's all been consistent with our thesis that delivering an intact microbial community in an oral capsule should behave pretty similarly to delivering it in a colonoscopy and hopefully is a lot easier for everybody involved," Smith told me.

More surprisingly, Finch listed the same CP101 product as a potential therapeutic for curing chronic hepatitis B viral infections. Adults who are exposed to the virus rarely develop chronic disease, except for immunocompromised individuals who can't fully clear it from their infected liver cells. For reasons that aren't entirely clear, however, roughly 90 percent of infants exposed to the virus immediately before or after birth develop a chronic hepatitis B infection for life. An eye-opening 300 million people around the world are carriers of the virus and at higher risk for liver cancer and cirrhosis than the general population.

Some drugs can suppress a chronic infection as long as they're taken continually but can't root out hepatitis B from its stubborn reservoirs. Weekly interferon injections, which deliver potent proteins normally released by the immune system to fight off infections, are among the few therapies that can sweep it out. But in so doing, the interferons can cause a constellation of severe side effects. In 2015, an influential study by researchers in Taiwan and China suggested that a well-established microbiome might offer a third therapeutic option. As in humans, the scientists found that younger mice with an immature gut microbiome largely failed to clear the virus while their older counterparts did so fairly quickly. Tellingly, adult mice given broad-spectrum antibiotics to deplete their gut microbiome became newly susceptible to a chronic hepatitis B infection as if they were youngsters. The developing immune system, the researchers found, seems to depend upon a family of proteins called toll-like receptors, which help it gauge whether to bear down or ease up on potential threats. One of the toll-like receptors may make the immune system overly tolerant of hepatitis B until the gut microbiome kicks in and helps give the virus the boot.

For countering C. diff, Smith said there's plenty of functional redundancy in a healthy microbiome. "There are a lot of different bacteria that are able to out-compete C. diff," he said. That means individual strains are less important than the health and diversity of the entire community, as I discovered in my own gut. But for countering chronic hepatitis B, the new research points toward a more specific defense mechanism. Patients with a chronic infection have a disrupted microbiome that may be a consequence of the virus shutting down part of the immune response and dysregulating the gut microbiota. Three small clinical trials in China and India have backed the growing hypothesis that FMTs from healthy adult donors might restore a microbiome-immune system feedback loop that clears away the virus from chronically infected patients. "We said, 'Well hey, we actually have a product that can deliver these intact microbial communities,'" Smith recalled. And so Finch began testing CP101 against chronic hepatitis B as well.

In parallel with the strength-in-numbers approach aimed at displacing pathogenic weeds within a patient or restoring the immune system's ability to pull them out, researchers are drilling down in search of seeds for flora with more targeted powers against complex conditions such as colitis. It's little different than discovering that a compound in, say, periwinkle flowers or sea sponges can fight cancer. But in this case, the new frontier for drug discovery exists within our own bodies. In a remarkable 2015 study that served as a case in point, doctors and researchers in Canada reported the results of the first-ever double-blind randomized controlled trial of FMTs for ulcerative colitis. (It was also the largest FMT trial for *any* disease at the time.) The Canadian team recruited six donors and delivered the goods via a fecal enema to thirty-eight patients once a week for six weeks, and a water enema to a control group of thirty-seven patients for the same duration. Neither the research investigators nor the patients knew whether they were dealing with real FMTs or placebos.

At first, the treatment didn't seem to offer any clear advantage over the placebo. A review committee trashed the trial's chances of success partway through its planned recruitment period and stopped it early due to "futility." The committee allowed one last group of already enrolled patients to continue, though. And that's when one of the FMT donors started to make up for lost time. Initially, the trial had recruited two healthy volunteers to be the suppliers, but "Donor B" had to take a four-month hiatus (due to an antibiotic prescription) after sending two patients into remission. Four other donors filled in, with little success, until Donor B reentered the picture and took over as the sole provider just before the review committee's thumbs down. In the final batch of patients, though, Donor B sent another five into remission and achieved an overall success rate of 39 percent—four times higher than the rate for all other donors combined. The volunteer's gut microbiome was markedly different and more diverse than a frequently used counterpart, Donor A, who hadn't helped a single patient. Even after the hiatus due to an antibiotic prescription, Donor B seemed to have something the others didn't.

The revelation helped to crystallize a tantalizing new idea, once only vaguely imagined. How many other Donor Bs might be out there? Might other conditions have their own saviors? If researchers identified individuals or communities who possessed that certain something, perhaps their poop could be distilled into personalized curatives. More recently, after two double-blind, placebo-controlled trials yielded conflicting results about whether FMTs could improve the symptoms of irritable bowel syndrome, researchers in Norway relied exclusively on donations from a single donor, a healthy thirty-six-year-old man. The third time was clearly the charm: while fewer than one in four patients given their own feces improved, more than three in four of those given an ounce of the donor's poo did. And for those given two ounces of what might be rightly described as liquid gold, almost 90 percent improved.

A new quest was underway to find out whether the frontier within us—perhaps within only a few of us—might reveal medical marvels obscured by the modern trend toward homogeneity. Smith told me that the ulcerative colitis trial results inspired Finch and its collaborator, the Takeda Pharmaceutical Company in Japan, to try a new approach toward developing an experimental therapy by identifying and isolating the rare "superpowers" of super donors. Unlike the screening problem of traditional drug development that requires testing large libraries of molecules to see if anything works, the researchers already knew from the clinical trials that had used FMTs to treat ulcerative colitis that *something* worked. To narrow down their search for what it was, Finch and Takeda combed through a dozen studies and zeroed in on the bacterial strains from Donor B and others that had engrafted into patients who responded to the treatment but were missing in patients who failed to respond. Company scientists collected donor stool samples containing the bacteria, grew them in a lab, and put them through additional tests to see whether they possessed promising mechanisms and could be developed into an interventional drug. From the combined data set representing more than a thousand ulcerative colitis patients, Smith said, the companies selected a collection of promising bacterial strains. The result was FIN-524, later renamed TAK-524, each dose of which contains a set number of viable cells from each kind of bacteria. The companies used the same approach to create FIN-525, a related therapeutic targeting Crohn's disease.

If TAK-524 ultimately works, Smith said the biological drug's activity against ulcerative colitis will likely be due to a complex mix of metabolites from the combination of bacterial strains that all work together. Smith likens the therapeutic to an implantable device that can deliver the mixture of molecules right to the patient's gut. "Rather than having to deliver those metabolites and figure out how to get them to release right at the epithelia where they belong and not in other places where they don't belong, we just have bacteria that can actually do that for us," he explained. The strategy takes advantage of how the gut bacteria have coevolved with us over millions of years to release their cargo right where it's needed. If future work points to a particular group of metabolites that are driving the process, Smith said, researchers may be able to increase their concentrations simply by adjusting the dose of their bacterial manufacturers.

That approach, however, still doesn't solve the fundamental problem that the University of Minnesota's Alexander Khoruts articulated back in chapter 3: the vast majority of Western donors may not have the right stuff—or at least *no longer* have the right stuff—to cure the kinds of plagues most associated with Western lifestyles. If truly healthy people are rare, truly healthy people with fully intact gut microbiomes may be unicorns.

● ● ●

Of the world's existing populations, a village from the Yanomami tribe of northern Brazil and southern Venezuela harbors the most diverse microbiome reported to date. The Indigenous Yanomami people lead a seminomadic hunter-gatherer lifestyle and live in remote villages in the Amazon, most of which rarely interact with the outside world. In 2008, an army helicopter spotted an isolated and previously unmapped village in Venezuela's Amazonas state, and a medical team made contact the following year. The team, which included health workers from other Yanomami communities, collected microbiome samples from nearly two-thirds of the villagers, including mouth and forearm swabs from twenty-eight individuals and fecal samples from twelve. Health workers then vaccinated children against measles and influenza and administered antibiotics to treat infections.

DNA sequencing revealed that the villagers' fecal and skin microbiomes were much more diverse than any other human population yet studied by scientists. For microbial ecologist Maria Gloria Dominguez Bello and a team of international colleagues including physician and microbiologist Martin Blaser, her husband, the fecal samples provided a treasure trove of information about commensal microbes that have evolved with us for millennia and have supplied a bounty of useful genes. "In a sense, their microbes were living fossils. The fecal samples were absolutely unique—and priceless," Blaser recalled in his book *Missing Microbes*. The fecal microbial diversity, in particular, blew away a comparison group from the US and easily bested the diversity documented in two other groups transitioning toward more urban lifestyles: rural Malawians and two Guahibo villages in southern Venezuela.

Despite the ubiquity of antibiotics, processed food, and urban lifestyles around the world, the Yanomami had managed to preserve a way of life that had persisted for thousands of years. The villagers traded arrows for machetes, cans, and clothes from other Yanomami groups. They gathered up wild bananas and seasonal fruits, plantain, palm hearts, and cassava from the jungle. They hunted birds and frogs and small mammals, crabs, and fish, occasionally supplemented with meat from peccaries, monkeys, and tapirs.

And in their guts, the villagers had preserved a remarkable archive of microbes like *Helicobacter*, *Spirochaeta*, and *Prevotella* bacteria that have become increasingly rare in other populations. The diverse microbes, in turn, provided a host of encoded functions through their genes: metabolizing amino acids from proteins, for example, and synthesizing vitamins such as riboflavin. Surprisingly, they also supplied numerous genes for antibiotic resistance despite the village's prior lack of antibiotic use, raising questions about how such genes in the community's collective "resistome" might be dispersed and maintained. If the genes arrived independently of any antibiotics, Dominguez Bello and her colleagues noted, characterizing that resistome would be important for designing and deploying new antibiotics that don't run up against preexisting countermeasures.

We've already seen how eating fiber and fermented foods can quickly shift the microbiome. In 2016, though, Stanford's Erica and Justin Sonnenburg suggested that there may be limits to what we can regain in the Western gut, at least on our own. Their study in mice, "Diet-Induced Extinction in the Gut Microbiota Compounds over Generations," first demonstrated that a diet low in plant fibers that the rodents' gut bacteria could use as food gradually eroded the diversity of those microbes. That's not surprising given what we've learned about our ability to disrupt the microbiome. The Sonnenburgs and their colleagues could reverse that reduction in diversity over a single generation if they put the mice back on a fiber-rich diet that included a variety of plants. Again, not so surprising given the rapid shifts seen in people fed plant-based diets. But here's the catch: if the researchers kept successive generations of the mice on the same low-fiber diet, the rodents progressively lost more and more of their gut bacterial diversity until they reached a tipping point at which a better diet alone wasn't enough to recover the losses. Rather, the researchers had to both supply the diverse dietary fibers *and* reintroduce the missing microbes through fecal transplants. The bacterial species they had driven to the brink by withholding their food source, they wrote, "are inefficiently transferred to the next generation and are at increased risk of becoming extinct within an isolated population."

The study may help explain why hunter-gatherer and agrarian populations harbor such high gut microbiome diversity while more industrialized populations have lost many of their bacterial lineages over successive generations. Poyet said the phenomenon is no different than the ecological concept that a forest with a greater diversity of trees will have a higher diversity of animals. The complex food molecules recruit bacterial specialists that possess the enzymatic machinery needed to break them down. Those by-products of digestion, in turn, can be harvested by other types of bacterial specialists. And so on. The more complex your primary food items, Groussin said, "the more complex the web of species you can build."

In that light, it makes sense that the startling loss of diversity in the average Western diet and our disregard for traditional foods that have long been staples of other cultures' diets might bring negative consequences. Given that immune function can be influenced by blood sugar, blood pressure, and cholesterol levels, Brooklyn dietician Maya Feller told me that a public reappraisal of whole and minimally processed but previously devalued foods like high-fiber beans, lentils, and cassava is long overdue. "It's about the foods that we put into our body and those foods that were previously not valued actually being the things that are lower on the glycemic index—so better for our blood sugars, the things that help us to manage our cholesterol metabolism, the things that help us to improve our cardiovascular health," she said. Poop that's high in fiber can be one indication of a healthier diet.

So could fiber-rich Yanomami feces become a valuable resource for scientific research? What would that mean, given ongoing exploitation of the tribe and other Indigenous communities amid threats posed by malnutrition and disease and by illegal logging, farming, and mining to extract other resources? In June 2021, the Brazilian government finally authorized the use of its federal National Security Force to help protect Yanomami people in the northern state of Roraima amid a surge of more than 20,000 "wildcat" gold miners. The garimpeiros, as they're known in Portuguese, have been illegally prospecting for gold on the tribe's vast reservation, polluting its rivers with mercury used to bind and separate gold particles from the surrounding sediment, and attacking Yanomami communities. The miners have brought influenza, malaria, and COVID-19. In 2020, after three babies died from suspected COVID-19 in a Brazilian hospital, medical officials compounded the tragedy by burying the bodies in a nearby cemetery, thereby preventing the Yanomami families from carrying out lengthy funeral rites that require the bodies to be cremated.

The discovery of unique microbiomes in Indigenous groups has spurred a scientific gold rush that could soon be joined by a medical one. If we begin to envision ourselves as producers instead of merely consumers, the mental shift may help us better understand our role in nature and reconsider the value of our natural output. But bio-prospecting in the by-products of remote communities and viewing their microbial communities as potential medicine cabinets raise threats of further exploitation. In this unfamiliar territory, we will need to devise new ways of equitably valuing and sharing the wealth derived from Indigenous groups without retreading past and present patterns of colonial oppression.

Anthropologist Alyssa Bader, who is Tsimshian, has collaborated with Indigenous communities in exploring how traditional diets help shape the oral microbiome. She also has an avid interest in the ethics of paleogenomics research and emphasized the importance of informed consent and close collaboration with Indigenous groups who may be impacted by a scientific project. Bader pointed out that we still know relatively little about what the microbiome might ultimately reveal about someone's health or lifestyle or ancestry or other personal details. Clearly communicating what's known and unknown, she said, can help communities think through the potential risks and benefits.

An even more eye-opening example of genomic diversity and the ethical issues it can raise arrived with the recent characterization of actual fossils: preserved coprolites from rock shelters in Utah and Mexico. If some researchers blithely tossed ancient poop like Frisbees, others had the good sense to at least tuck them away in collections. Fifteen-hundred-year-old coprolites collected sometime between 1929 and 1931, likely from a site in east-central Utah known as Arid West Cave, sat for about ninety years in a forgotten archive until an international research group decided to take a closer look. For an extensive analysis published in 2021, three of those ancient droppings were joined by counterparts collected from two other excavation sites. Researchers tested

2,000-year-old coprolites from the Boomerang Shelter site in southeastern Utah, where previous studies of the poop suggested that the cave dwellers ate diets high in maize. And from a site near the town of El Zape in the state of Durango, Mexico, expeditions in 1957 and 1960 recovered coprolites dating from the eighth century to the early tenth century CE.

In all, the researchers examined eight desiccated coprolites (technically known as "paleofeces" by some scientists). Just like the pavement of European plazas has protected underlying medieval and Renaissance deposits, the shallow cliffside caves and alcoves in Utah and Mexico sheltered the prehistoric poop from rain and other moisture that would have hastened its decomposition. The coprolites were so well preserved, in fact, that the scientists were able to extract enough DNA to reconstruct a remarkable catalog: 498 separate microbial genomes. To exclude those that may have been contaminated with modern DNA, the researchers took advantage of the fact that DNA degrades over time and retained only the 209 genomes that showed evidence of time-related wear and tear. By comparing the remaining sequences to those from known microbial genomes, the scientists were able to label 203 as the likely genomes of human gut microbes. Of those, 181 were highly damaged, suggesting that they were truly ancient intestinal residents.

These gut microbes were highly diverse, and here's where modern comparisons can be particularly telling. Genomic researchers tend to divide the world into two main groups: People who have an "industrial lifestyle" are less physically active and commonly take antibiotics. They also eat a Western diet characterized by processed foods that are high in fat, refined sugar, and salt (like a cheeseburger, milk shake, and French fries) but low in fiber from fruits and vegetables. People with a "nonindustrial lifestyle," on the other hand, tend to be more active, have limited antibiotic exposure, and eat more unprocessed foods that they've grown or raised themselves.

As might be expected, the ancient gut microbiomes had more in common with modern nonindustrial samples from rural Fiji, Peru, Madagascar, Tanzania, and Mexico than with industrial samples from the US, Denmark, and Spain. The ancient and modern nonindustrial microbiomes, for example, had fewer genes for degrading mucins (proteins bound to carbohydrates in the mucus lining the gut) and alginates (food additives), both of which are common in industrial communities. But the nonindustrial microbiomes, both past and present, had more genes encoding enzymes that help digest starch and glycogen (basically a multi-branched chain of glucose molecules). Those compounds are more reflective of diets rich in complex plant-based carbohydrates. The ancient and modern nonindustrial groups also had more spirochetes, an ancient phylum of corkscrew-shaped bacteria that often live symbioti-cally in the guts of insects like wood-eating termites and cockroaches, where they help break down cellulose. In humans, the microbes are more associated with syphilis, lice-borne relapsing fever, and tick-borne Lyme disease. From the poop samples, though, the researchers found that the ancient and modern non-industrial communities alike hosted a harm-less spirochete called *Treponema succinifaciens*, thought to be passed on to humans via termites and pigs but almost universally absent in urban populations.

Nearly four out of every ten of the ancient microbial genomes iso-lated from the archaeological sites had never been seen before by sci-entists. Compared to their modern counterparts, the ancient genomes had far fewer genes for antibiotic resistance, especially those conferring resistance to tetracycline. They did, however, include more genes for chitin-digesting enzymes. From a microscopic analysis of the ancient poop, the researchers determined that the diets included some chitin-rich foods such as locusts and cicadas, mushrooms, and corn smut (the fungus, a delicacy known as *huitlacoche* in Mexico, is particularly tasty in soups and tamales).

As the variety of our diets has constricted, we have literally starved some of our inner residents. Regular antibiotic use—certainly lifesaving in many cases—has forced out others. In *Missing Microbes*, Blaser argues that the inappropriate use of antibiotics has become an existential cri-sis for our inner ecosystem and has fueled "modern plagues" such as asthma and ulcerative colitis but also obesity and childhood diabetes. Research by Blaser's lab and others suggests that the risk to long-term health from antibiotic overuse is greatest during early childhood, when cognition, immune function, the microbiome, and other key systems are rapidly developing. And yet, as he observes, the average child in the industrialized world receives seventeen courses of antibiotics before the age of twenty.

Poyet and Groussin said they weren't surprised by what the study of ancient coprolites from Utah and Mexico suggests about our collective loss of bacterial lineages. Examining other ancient samples from around the world could put that loss into better context. But the emerging field of study also raises a host of ethical questions. Indigenous peoples in the Southwest, for instance, viewed the newly identified gut bacterial species as a tangible link to their past. "Can we consider these microbes are part of the biological heritage of Indigenous populations?" Poyet and Groussin wondered. "Should populations have control over the usage and sharing of this microbial genetic material? Should we consult with Indigenous populations before using this ancient DNA for reconstructing co-migration routes of hosts and their microbes?"

The collected coprolites weren't subject to US regulations like the Native American Graves Protection and Repatriation Act, but the study's authors wrote that they explained and discussed their research with Southwestern tribes who maintain strong cultural ties to the artifacts. Even so, some tribal members were upset that the scientists hadn't consulted them earlier, and other researchers said the evolving field may have to consider more far-reaching ethical guidelines given its foray into a gray area.

"We have this ever-expanding technology and innovation, and it gives us access to all these new areas of research, like the human microbiome and then the ancient microbiome," Bader said. "And we know that because of the nature of scientific exploration and innovation that the research is always going to outpace the ethics." In this quickly changing landscape, she said, ethical considerations shouldn't be about ticking off boxes on a checklist but rather helping scientists think about the larger picture. "Researchers aren't working in isolation in labs. Our work has real impact beyond what we can even imagine. And we need to be thinking about the role of our research and the questions that we're asking and the methods that we're using in that bigger context." Collecting samples from diverse populations to increase representation, for example, isn't enough for equitable or ethical research. Bader pointed out that researchers don't always retain control of the downstream applications of their samples, like drug development and personalized medicine. Even efforts to preserve a "disappearing" pool of microbes could be ethically fraught, she said, without an effort to carefully consider the environmental and cultural context in which they're appearing, why they're changing or disappearing, and how that might impact the microbiome donors. Scientists need to consider the people connected to the microbes and understand that the ethics of human research are closely intertwined with the ethics of microbiome research. "They're not separable," she said.

Poyet and Groussin maintain that including historically underrepresented populations in their own research is a critical social justice issue. They have positioned their conservation efforts as an effort to save a community resource that is being threatened precisely *because* of exploitation: the indiscriminate use of antibiotics, the deliberate destruction of indigenous land, and the damaging effects of human-wrought climate change. Microbiome collections and studies probing the links to health and disease have overwhelmingly focused on the less diverse bacterial communities of industrialized populations, Poyet said, meaning that scientific and medical advances like microbiome-based C. diff therapies tend to be tailored to those same well-studied populations. "There is absolutely no evidence that they would work with people from nonindustrialized populations with C. difficile," Groussin added. Including underrepresented groups in the research, they've argued, is essential for taking their own microbial traits into account when developing biomedical interventions and for preventing the further amplification of health care inequities.

Working with a variety of communities has required taking differing values into account as well, and the conservancy has worked with local collaborators in each location to build trust and adapt the collection and consent process to local customs. Strong cultural taboos against poop, at least so far, haven't been an issue, but some communities have declined to share saliva samples over privacy concerns. For an *Undark* exploration of the ethics around collecting Indigenous feces, journalist

Katherine J. Wu talked to a Nepalese researcher who told her the country's Raute ethnic group was "adamantly opposed" to donating stool samples for a separate research project. Upon death, he said, they believe that their bodies, everything that emanates from them, and all of their belongings should return to the soil.

For anything donors are willing to donate, Groussin said, community-specific consent forms explicitly state that they remain the full owners of their biological material, including all of the extracted bacteria and metabolites. Donors can also request that their donations be returned or destroyed. The nonprofit was working on a framework for how benefits derived from the donations would be equitably shared with the original owners. And after isolating and biobanking the bacterial species, the conservancy had begun sending copies back to local collaborators so each country could maintain its own storehouse of microbial biodiversity.

From research conducted on the bacterial samples so far, Groussin and Poyet found that genes for fiber degradation were being exchanged at high rates within the guts of some nonindustrialized societies, especially hunter-gatherers. Antibiotic resistance genes, meanwhile, were being exchanged at high rates among pastoralists who treated their livestock with antibiotics. The free exchange of antibiotic resistance genes, even in nonindustrialized societies, hints at the magnitude of the problem around the world. The swapping of fiber-degrading enzymes makes sense since bacteria that can glean energy from new carbon sources may be able to colonize new niches, Groussin said. The free exchange of bacterial enzymes might benefit their human hosts as well by improving their ability to digest more complex plant fibers. One helps the other.

● ● ●

Because early microbiome and fecal analyses were dominated by people from wealthier countries and skewed more male than female, our definition of normal or ideal or even valuable was likewise biased. The expanding historical and geographical reach of such research has provided a more complete picture and essentially flipped "normal" on its

head. Diminished microbial diversity may be the default in most industrialized populations, but the increasing homogeneity is now widely
seen as a deficiency and a potential contributor to Blaser's "modern
plagues" like asthma and obesity and ulcerative colitis. If judged by
historical standards, signs of abundant fiber and parasitic worms in
poop might be normal—even desirable. Paleoparasitology studies have
reinforced the normality of parasite infections by demonstrating that
even the most affluent people in history had worms. "Now us not having
worms is actually weirder than the fact that they had them," biological
anthropologist Tara Cepon-Robins told me. "And so we're the odd ones
out for that. I think that's kind of cool." Spoken like a true helminth fan.

The worms add more complexity to the question of what constitutes
good shit. Helminths are parasitic, and as part of their survival strategy, they activate a bizarre branch of the immune system called the Th2
pathway. Th1 is the full-on, proinflammatory branch triggered by bacteria and viruses while Th2 is like a dimmer switch that tamps down the
immune response. Mounting a full-blown attack on the worms would
kill them but cause collateral damage in the process. The coevolutionary
solution has been an uneasy trade-off. The Th2 branch, Cepon-Robins
said, identifies that the worms are there, and then turns down the
immune response to them.

For common helminths spread through fecal contamination, like
whipworms and giant roundworms, mild to moderate infections are
unlikely to cause significant malnutrition in children who have relatively healthy diets. Poor diets, though, can compound the consequences
of chronic infections, like weight loss and stunted growth and development. Hookworms, Cepon-Robins said, can be nastier since the adult
worms attach to the small intestine and draw blood, increasing the
potential for severe anemia through blood loss. The helminth-activated
dimmer switch seems to work on the entire immune system, meaning
that people infected with parasitic worms are in turn more susceptible to bacterial and viral infections and less responsive to vaccines. The
lessons extend to farming and gardening too: the persistence of some
parasites in soil, paleopathologist Piers Mitchell said, underscores the

importance of properly composting biosolids to avoid increasing the burden of parasitic infections.

On the other hand, he told me, "We do need parasites and other microorganisms in our intestine in a nice balanced way to match the ones that we've evolved for. Otherwise, you end up with a thin, weedy, washed-out microbiome of organisms that doesn't reflect what we're used to." Intestinal parasites, in other words, may not be entirely bad. One benefit of tamping down the Th2 immune branch may be to lessen the kind of overwrought immune response associated with allergies and autoimmune conditions such as ulcerative colitis and Crohn's disease. With no intestinal parasites to contend with, one line of thinking goes, the immune system may be more likely to overreact and turn on our own cells instead. This hypothesis seems to bear out in the Shuar communities of Ecuador where Cepon-Robins studied disgust sensitivity: helminth infections are common while allergies and autoimmune conditions are rare.

One other potential benefit of a helminth infection is a reduction in the kind of chronic, low-grade inflammation that studies have increasingly linked to obesity, cardiovascular disease, and metabolic syndromes in Western societies. By measuring the levels of a biomarker for inflammation in stool samples from Shuar children, Cepon-Robins found that whipworm infections were associated with lower levels of intestinal inflammation.

Consider the remarkably complicated balancing act potentially mediated by small animals wriggling through the intestinal tract. Without them, you might be taller and more energetic and you might develop faster. But you also might be more obese and prone to allergies and autoimmune conditions and cardiovascular disease. Consider, too, that for the bulk of human history, parasitic helminths have been largely unavoidable. It's not yet clear whether most of the dimming is due to the worms hijacking the immune system's regulatory mechanisms or to the immune system imposing some self-restraint after recognizing the worms. "But either way, it's win-win a little bit for both parties because the worms can live comfortably in your gut and you don't really exhaust all your resources trying to fight them," Cepon-Robins said. "It's important to keep in mind that when you're talking about nature, the norm is for things to be infected."

That historical perspective has been a central feature of Mitchell's research on understanding both the pathogenic and commensal organisms that humans have carried around for millennia. If researchers can work out the microbiome of past populations "before we started messing it all up with antibiotics and everything else," as he puts it, he and other researchers argue that fecal transplants might be made more effective by including the missing bacterial strains needed to fully restore a patient's healthy microbiome. In essence, it's a rewilding of our inner ecosystem. But the only way to really know what a "complete" or "healthy" microbiome is, he said, might be to piece together the microbiome of people who lived hundreds or thousands of years ago.

In 2020, Mitchell and colleagues published an extensive analysis of the sediment from two medieval cesspits: one in Jerusalem and another in Riga, Latvia. As might be expected, the samples were teeming with parasites: the researchers had previously identified six separate species in the Jerusalem latrine, and four in its Riga counterpart. But the latrine sediments also allowed the researchers to reconstruct the gut microbiomes of two preindustrial populations. The researchers isolated DNA from all of the microbes found in the latrine soil and then used a process of elimination based on comparisons with known soil microorganisms to determine which ones were likely to have had a fecal provenance.

Most previous studies had reconstructed microbiomes from individuals, such as from the nearly thousand-year-old paleofeces and colon of a mummified woman in what is now Cuzco, Peru. But Mitchell said his group's analysis, the first one done in cesspits, provided a proof of principle that scientists can do the same reconstruction for the more representative microbiome of a community (or at least of those who frequented the connected latrines). "And if people can do that, going back into prehistory, then maybe we can actually work out what the ideal microbiome is," he said.

The fifteenth-century cesspit in Jerusalem was likely used by multiple households while its fourteenth-century Riga counterpart may have been designed for public use. Although much of the microbial

DNA undoubtedly degraded over time, the researchers found that the residents' microbiomes were a hybrid, showing similarities to modern industrialized *and* hunter-gatherer populations. The cesspits' abundant *Bifidobacterium* representatives, for instance, are characteristic of the former and depleted or absent in the latter. Conversely, the *Treponema succinifaciens* spirochete found in the Jerusalem and Riga deposits is enriched in hunter-gatherer populations, as we've seen, but has been lost in more industrialized populations. Overall, the medieval microbiomes clustered together in a group that differed from all modern sources. Of course, what may be ideal or healthy or complete in one time and place may not be so in another; Mitchell's work has already matched distinct bacterial and parasitic diseases to the different lifestyles of our ancestors in the Stone, Bronze, and Iron Ages. In the same way, the microbiome of someone who hunted and gathered food in Mexico more than a millennium ago or farmed in Riga during the Middle Ages is unlikely to be perfectly adapted to my middle-aged Seattle gut. A historical microbiome from a Duwamish or Puyallup inhabitant of the Pacific Northwest might be more useful—if it could be pieced together. But with my vastly different (and considerably less diverse) diet, many of those bacterial enzymes might find little to do with my relatively pedestrian food choices. Future supplements may be more about adding back a handful of endangered or extinct species that can help fill out a weedy gut than trying for a full historical reset.

Treating people with "helpful" groups of parasites could be even more fraught. "People are quite happy with the concept of healthy bacteria, but they're not so happy with the idea of giving worms to people," Mitchell said. "I think if they're big enough to see, then there's that squeamish factor. Whereas if it's a yogurt with a few bacteria in, then people cope with that a bit better, don't they?" They do indeed. Even Mitchell might hesitate if given the chance to jazz up his inner ecosystem. "Am I going to go down the route of giving myself parasites? Nah, maybe not," he said.

Many others have. Just as DIY fecal transplants predated the technique's wider acceptance, patients with autoimmune diseases have

Flush

self-medicated with hookworms and other helminths, though Cepon-Robins cautions against at-home procedures due to the many unknowns and potential for collateral damage. "Worms theoretically will turn off your immune system to some degree," she said. "So if you're having problems with a hyperactive immune system, then taking them—taking a worm—*should* help. But the problem is narrowing down what species is going to be most effective." So far, she said, the few controlled studies to test the idea have yielded very mixed results. Ideally, researchers like her will be able to isolate a specific helminth protein that triggers the dimming of an overactive immune system without provoking the negative side effects associated with an actual infection. If they succeed, future therapies for ulcerative colitis or Crohn's disease might include both bacterial and helminth components.

Our intestinal medicine cabinet may even provide curatives from the most diverse and abundant life-forms on Earth. *The Perfect Predator* details the remarkable story of how epidemiologist and HIV expert Steffanie Strathdee saved psychologist Thomas Patterson, her husband, from a multidrug-resistant superbug he acquired in Egypt in 2015. After the antibiotics failed, her frantic search for a cure for the aggressive *Acinetobacter baumannii* infection led her to phage theory, which posits that the right viral predator can kill every kind of bacteria. Remember those tiny bacteriophage viruses from chapter 1 teaming up on bacteria like a pack of dire wolves attacking a long-horned bison? They're harmless to us but each kind has evolved to infect specific bacterial species and perhaps even individual strains. Unlike other drugs, phages multiply in the body and are rapidly cleared by the immune system after they've hit their targets. And one of the best places to find these viruses is in the sewer, where their potential prey are plentiful.

Phage therapy research is more than a century old and well established in Eastern Europe. In the US, though, it fell out of favor and was relegated to a fringe field of study as antibiotics rose to the fore (sound familiar?). With the relentless advance of antibiotic-resistant superbugs posing an increasing threat, bacteria-killing viruses are getting a second look as a standalone or combination strategy. Strathdee convinced researchers at Texas A&M University and the US Navy's Medical Research Center-Biological Defense Research Directorate in Frederick, Maryland, among the few US labs with phage collections, to join her cause as her comatose husband slipped toward death.

As part of its effort to assemble an effective phage cocktail, the Texas A&M team collected samples from wastewater treatment plants and isolated two promising bacteriophages from the sludge. After testing their pool of candidates, including one from a San Diego–based biotech company called AmpliPhi, the Texas scientists selected four phages with the most promising activity against the superbug strain and sent them to San Diego State University for purification. Patterson's doctors at the University of California at San Diego then delivered the purified viruses to him through three catheters connected to his abdominal cavities.

The Naval lab, which at one point had been on the verge of shutting down its phage program, sent its own cocktail of four promising bacteriophages, all of which had been isolated from raw sewage. Thirty-six hours after the initial infusion, the San Diego medical team began administering the second cocktail intravenously. Patterson started to improve soon thereafter, but then his bacterial infection began to gain resistance to the phages. So the Naval lab tweaked its cocktail to counter the mutating bacteria by adding another sewage-isolated phage. After being treated with the new formulation, Patterson finally cleared his infection; eight and a half months after falling ill, he was well enough to leave the hospital.

No one can definitively say that the bacteria-destroying phages saved his life; other drugs likely helped. But the case energized the push toward clinical trials to formally test the potential of the tiny viruses to slay the world's growing horde of superbugs. Even if phage therapy proves itself, regulatory agencies would have to fundamentally change how they view the viruses for the antibacterial approach to go mainstream. Each phage cocktail would have to be crafted for individual patients based on their source of infection: the epitome of personalized medicine but another regulatory nightmare under existing standards. As with fecal transplants, though, the success story offers a timely reminder of how looking beyond standard or normal considerations can be nothing short of revelatory.

What has value? Our sense of what's normal and what's desirable has long been filtered through a Western perspective. Tom Brady's poop notwithstanding, shit has a way of stripping away the veneer that we've rebranded as the ideal. Throughout history, more frequent and voluminous outputs fueled by high-fiber diets, more diverse microbiomes, and messier arrays of gut infections—parasites and viruses and bacteria living together if not always harmoniously—were the default, the rule instead of the exception, the way things fit together. The relatively short era of antibiotics and processed foods that is fundamentally altering the human microbiome, it turns out, has been a deviation—and one that has come at a considerable cost.

Despite the logistical hurdles and ethical challenges, some of the most promising avenues of medical research lie not in the glorified bowels of celebrities and athletes but in the bulky poop of anonymous donors and rural villagers, parasites and all. The same lowly substance that can warn us of a surge in harmful microbes, it seems, may yet hold the keys to undoing the harm caused by a steady decline in others.

CHAPTER EIGHT

Source

DIANE TANIGUCHI-DENNIS remembers growing up by the ocean in Lahaina, Hawaii, where volcanic soils on Maui turned the water bright red after heavy rains. When she was a teenager, her family sometimes had to boil water or go without when shortages forced moratoriums. "There just simply wasn't enough water to support the growth in the area," she said. Residents regularly talked about the link between storm water runoff and the fate of the island's coral reefs. She remembers the polluting threat of wastewater too: like when her suburban home's cesspool failed and when the community finally opened its first sewage treatment plant in 1975. Taniguchi-Dennis saw how the fate of the water and land and air and people intertwined, and how the arc of development that improved drinking water access and wastewater treatment enhanced those inter-relationships. Her memories of water, both good and bad, helped to shape her future. "All of that, the environment and issues in the environment, became a very big part of who I was as a child and what I aspired to do," she told me. "As a kiddo in the late seventies, I wanted to solve the world's environmental problems."

Trained in civil and environmental engineering, Taniguchi-Dennis is now fully immersed in creating healthy networks and emphasizing the circular use of resources as CEO of Clean Water Services. The second largest water resources management utility in Oregon operates four wastewater treatment plants and serves more than 600,000 people, mostly in Washington County just west of Portland. In her leadership role, she has used her environmental ethic and philosophy to push for a better alignment between science and technology and the power of Mother Nature. It's only fitting, then, that a savvy system designed to safeguard the region's water resources is creating renewable power through the transformation of another natural resource: poop.

In re-thinking our throwaway culture, we'll need to be more thoughtful about making things that last, and more creative about reusing available materials. For years, we concerned ourselves only with how to dispose of what we no longer wanted. Now, Taniguchi-Dennis said, "we need an upcycling mentality." Global warming has only added to the urgency. Amid the growing alarm over climate change, renewable wind, solar, and geothermal power are looking more attractive than ever. But other renewable alternatives to fossil fuels and prime examples of upcycling exist much closer to home. In the human gastrointestinal system, we produce exportable energy in the form of biomass and gas—or more accurately, our microbial residents do—as a by-product of digesting food. (We occasionally swallow air, too, but burping gets rid of most of it.)

A classic 1949 study, "The Quantity and Composition of Human Colonic Flatus," found that the toots of twenty Danish volunteers on a cabbage-free diet consisted mainly of gases like nitrogen, hydrogen, carbon dioxide, and oxygen. Some people also produce trace amounts of hydrogen sulfide, which smells like rotten eggs and tends to clear a room. And in the oxygen-free depths of the colon, some 30 to 60 percent of us host the archaea microbes we know as methanogens. When other microbes ferment carbohydrates, these specialists use building blocks like hydrogen and carbon dioxide to form colorless, odorless methane: the primary component of natural gas. For some biogas producers, biomethane accounts for nearly one-third of every blast. Multiple YouTube pranksters have confirmed that they harbor abundant methanogens by posting videos of their pyroflatulence. (Yes, it's the technical term for lighting a fart on fire. Yes, flammable hydrogen gas probably contributes to the brief flares. And yes, firefighters warn, the party stunt is a dangerously bad idea.)

Not all of us are natural gas manufacturers, but all of us make the raw materials that when combined in, say, a giant digesting vat that mimics the oxygen-free conditions of the human gut and contains methanogens, can safely yield an impressive amount. Some historical accounts suggest that the Assyrians were the first to understand our inner energy

Flush

potential as far back as the tenth century BCE when they used flammable biogas to heat bathwater. Other sources suggest that the anaerobic digestion of solid waste may have provided energy in ancient China as well. The first well-documented biomass digesters appeared in India and New Zealand in the mid-nineteenth century. In 1890s Exeter, England, a sewer sludge digester powered streetlamps. Eighty years later, biogas production gained worldwide momentum when high oil prices during the 1970s energy crisis spurred a renewed search for viable alternatives.

Our inner biomass and wind power may not be as flashy as other sources. In a 2015 report, though, the United Nations University Institute for Water, Environment and Health estimated that if the world captured and repurposed its collective output, the theoretical value of human-derived biogas alone could be worth up to US$9.5 billion annually. Methane is one of the most potent greenhouse gases contributing to global warming, but plant-derived biomethane that has been captured and purified from organic waste like leftover food and poop (sometimes called renewable natural gas) is widely considered capable of yielding carbon-neutral energy or better. When burned, it still releases carbon into the atmosphere, but that amount is roughly canceled out by the atmospheric carbon soaked up by plants before their death and decay. Methane-rich biogas has the added benefit of replacing fossil fuels as a renewable source of electricity or heat, or as a cooking, industrial, or transportation fuel.

The leftover fecal sludge, for its part, can be liquefied into biofuels or dried and converted into surprisingly good cooking and heating briquettes called biochar, thereby slowing deforestation by sparing trees otherwise felled for charcoal or firewood. The UN report, in fact, suggested that the total solids remaining after biogas production could yield more than one hundred million pounds of biochar. A charcoal-like briquette can be made by heating organic matter in the absence or near-absence of oxygen and is essentially the same carbon-rich material you'd see in an activated-carbon Brita water filter. Poop's lifesaving potential, it seems, may include rescuing us from the fatal trajectory of climate change. A concerted push to repurpose the world's poop could simultaneously alleviate some of the world's worst sanitation crises, including an estimated 1.7 billion people who lack access to a toilet at all.

Those heroics may seem a bit overblown until you consider that the same waste-to-power technology being perfected for poop can be applied to our ample supply of food and yard scraps, agricultural leftovers, manure, and other overlooked fuel sources. A 2019 report by the World Biogas Association estimated that the planet was capturing just 2 percent of this considerable potential. By generating renewable energy plus avoiding emissions from deforestation, crop burning, landfills, and livestock waste, the report calculated that ramping up anaerobic digestion to yield biogas could reduce global greenhouse gas emissions by an impressive 10 to 13 percent.

One part of Clean Water Services' upcycling-minded portfolio has focused on how to accelerate the production of biogas by repurposing a natural enemy of treatment plants: the FOGs that can glom together into sewer-clogging fatbergs (yes, the same fatty clumps that can attract particles of the SARS-CoV-2 virus). These notorious FOGs—often a mix of butter, mayonnaise, cooking oil, bacon drippings, and other food waste caught in restaurant grease traps—are regularly sent to landfills to avert fatberg blockages. But gas-producing microbes, it turns out, thrive on the energy-packed waste.

I went to see the FOG repurposing process for myself on an unexpectedly dramatic September weekend in 2020 when ash and dust flung across Portland's southeastern horizon gave the city a *Dune*-like quality. Fires had ignited across the West Coast—fueled by a heat wave, fierce red-flag winds, and bone-dry conditions—and the dark bruise spreading across the sky was deeply jarring. To the west, the evening sun had taken on an eerie orange-red glow as smoke obscured trees in the distance. A pedestrian noticed me taking a picture near the cottage where I was staying. "Beautiful, isn't it?" she asked. I nodded and thought how smoke and dust and pollution tend to make the prettiest sunsets. The next night, I brought a six-pack of Apocalypse IPA beer to a friend's backyard and we laughed nervously about yet another calamity in an unsettling year.

The backdrop, though, was a perfect example of why we should care about where we get our energy. Tens of thousands of customers in northern California were without power due to preemptive blackouts intended to reduce the load on utility lines in especially fire-prone areas. Oregon

Flush

had swung from drought to floods over the past few years, and the wild-fires were scorching parts of the Tualatin River watershed and forcing evacuations in Washington County (larger fires to the east and south would eventually burn more than one million acres in the state). The increasing extremes linked to climate change, Taniguchi-Dennis said, were highlighting the need to design more resilient systems that could handle the huge variability. She recalled how she had laid out the reality in a talk to a group of utility executives. "I said, 'We have to fundamentally realize that our biomes are shifting because humans are now part of the biome.'" In other words, we've helped create the conditions that are contributing to the increasing instability all around us.

Converting our unwanted matter into greener sources of energy and fuel that can reduce and even reverse the instability will require considerable buy-in. But modern poop-to-power strategies have built upon centuries of historical precedents and are picking up speed. By 2015, China alone had tens of millions of home-size digesters producing biogas and more than 110,00 larger systems. Germany, the European leader and world's second-biggest installer of biogas plants, had set up nearly 11,000 by 2017. Wastewater treatment plants, which seem to have become regular tourist destinations for me (hey, some people love roller coasters and others gravitate toward sludge pools), offer particularly good examples of this transformation. At these plants, a disposal process once narrowly viewed as a way to minimize our damage as consumers is being reimagined as a means to expand our usefulness as producers. Some advocates, in fact, have begun calling them "resource recovery" facilities. We are the source of these resources: gases that can become heat or electricity or biomethane, solids that can become heating briquettes or compost or fertilizer, and liquids that can become biofuels or clean water.

In the patchwork of temperate forests, farms, and bedroom communities of western Oregon, Clean Water Services is responsible for safeguarding the eighty-three-mile-long Tualatin River. The name translates as "lazy river" in the language of the Indigenous Atfalati tribe of the Kalapuya, who lived here and who carved petroglyphs in the sandstone farther upstream. The river flows from a spring in the Coast Range east

into suburban Portland, provides drinking water for about 200,000 Ore-
gonians near its headwaters, and receives the effluent from Washington
County's four wastewater treatment plants before it empties into the
larger Willamette River. Combined, the plants discharge about sixty-six
million gallons of treated water every day, on average. Most goes back into
the Tualatin; along with water from an upstream dam, it makes up about
two-thirds of the river's leisurely flow, which is critical during drier summer months. The replaced water keeps the river moving. It is life-giving.

I wanted to see it for myself, so I walked through the Tualatin River
National Wildlife Refuge, one of the few located in a major metropolitan area. On a pleasant late summer day that could have been mistaken
for any other except for the wind and distant smoke, only a handful of
other visitors had ventured out and the refuge seemed unusually quiet;
even the birds seemed to be holding their breath. I walked about a mile
through pockets of woods and a sunny open meadow to an observation
deck overlooking a very dry-looking expanse of restored wetlands. In
the winter and spring, they are part of nature's water-cleansing system,
which can screen and pool and filter out and break down many of the
contaminants before they reach the river. But after a parched summer,
the expanse had reverted to a grassland. I lingered at other spots overlooking the slow-moving Tualatin. One sign read, "The River Is the Life-
blood of the Refuge."

Most newer wastewater treatment plants have taken their cue from
wetlands, rivers, and other parts of nature for some of their water-
cleansing steps. After large screens that mimic the effects of cattails,
rushes, and reeds remove bigger pieces of debris and gravity helps remove
grit and gravel, settling tanks act like natural pools and let some of the
heavier organic matter sink to the bottom. Filters commonly incorporate fine and coarse sand as effective ways to reduce algae, chemicals,
and harmful microbes. Bubblers add oxygen to the water when needed,
re-creating the turbulence and aeration that occurs when water flows
over rocks and boulders. And at the center of advanced plants, beneficial
microbes are the true stars: chomping away at the organic waste and
nutrients to break them down. Some tanks are pumped full of oxygen

to fuel the growth of aerobic bacteria; others are depleted of oxygen to favor the growth of anaerobes that cannot survive in its presence.

The Durham Water Resource Recovery Facility in Tigard is about five miles from the refuge, and a biogas generation system there yields a mixture of 60 percent methane and 40 percent carbon dioxide. To increase its heat and power generation, the plant has made particularly good use of the FOG waste that is normally landfilled. Instead, the plant solicits the waste from restaurant grease traps and pipes it right into two digesters to help feed the microbes. "Put it directly into the digester, and it can double or triple the amount of gas that you're producing," principal engineer Peter Schauer told me. This co-digestion process, as it's known, is like supplementing the microbes' regular meal with calorie-dense candy bars. "Or crack," quipped senior engineer Pat Orr.

As I walked with a small contingent from the plant toward the receiving building that accepts the FOG deliveries, Orr said it would be more of an olfactory experience.

"Oh really?" I asked.

"Well, you'll see," Orr said, grinning. "Or smell."

Inside, heating and grinding and recirculating elements kept a large tank of the stuff warm and well mixed so the liquid wouldn't solidify and block the feed lines taking it to the digesters. A screen helped catch errant silverware from the restaurants' grease traps. It smelled, well, like rancid grease. Not overpowering, but the kind of odor that tends to cling to your clothes. One of the biggest initial challenges, Orr said, was maintaining a happy medium between overly "hot" loads of incoming FOG and a scarcity on weekends. The microbes would get acclimated to the fattier food, he said, "and then you kind of take away all their sweets." The plant began giving incentives to drivers to even out their deliveries, though those deliveries plummeted when many restaurants closed during the COVID-19 pandemic.

When the system is humming along, both of its engines fed by full loads of the biogas, Orr said the plant provides about two-thirds of its own electricity needs: most from the engines and a smaller fraction from an on-site array of solar panels. Heat recovered from the engines, in turn, supplies about 90 percent of the plant's hot water needs, including the microbial digesters, workers' showers, and other heating systems. Biogas that isn't stored and can't be used by the engines has to be flared; at some other treatment plants, the excess gas is regularly burned off.

Adding FOG to the mix at the Durham plant has increased its production variability, but a nearby dome-shaped building, which Orr called a storage bubble, acts like an accelerator pedal to keep the system in sync. As gas production increases and the volume inflates a liner inside the bubble, the engines go full tilt in powering an electricity generator. As the gas volume decreases and the liner deflates, the engines back off. "It's matching our use to our production, which is a very elegant way to do it," Orr said. He pointed to a nearby flare stack. "You don't see anything coming out of that flare. We use every cubic foot of gas we make." No excess gas at all? I asked. Not even a mouse fart, he replied, laughing.

● ● ●

Similar upcycling innovations have the potential to fundamentally overhaul disposal practices that have changed little in decades. In the United States, we still bury about 22 percent of our poop; doing so means carting it off by truck or train to vast landfills. In 2018, an uproar over the nauseating stench of a stranded "poop train" carrying ten million pounds of treated waste from New York to Alabama shone an uncomfortable spotlight on the disposal of urban waste in rural landfills—often hundreds of miles away. Coastal communities barred from dumping their sewage in the sea are finding that inland jurisdictions don't want it either.

At landfills that still accept it, highly concentrated organic and inorganic compounds can leach away from the sewage and other decomposing matter with the help of percolating rainwater and contaminate surrounding land and water if not properly captured and treated. Within the decaying mass itself, the same anaerobic digestion that creates biogas in the human gut yields landfill gas that wends its way to the surface. In 2019, solid waste landfills accounted for 15 percent of all human-associated methane gas emissions in the US, equivalent to the greenhouse gases released by more than twenty million cars driven for a full year. Unlike the Durham facility that uses all of the biogas it produces, landfill operators often burn off the escaping methane in a process similar to the flaring of excess natural gas at thousands of oil production sites around the world. I like to think of it as a massive bout of pyroflatulence that needlessly sends hundreds of millions of tons of greenhouse gases into the atmosphere. Despite the 2020 economic slowdown caused by the COVID-19 pandemic, global carbon dioxide and methane emissions still jumped; the spike in methane was the largest since researchers began keeping records in 1983.

More advanced landfills are recapturing and using the gas instead of letting it escape; a massive one in Washington State's Klickitat County that handles more than twelve million pounds of garbage every day is producing enough renewable natural gas to meet the daily needs of furnaces, kitchen stoves, and water heaters in 19,000 homes. Nearly everyone seems to agree that converting trash into gas is a good thing. But only about 25 to 30 percent of the Roosevelt Regional Landfill's trash is organic matter capable of producing methane, meaning that efforts to improve composting and recycling could eventually squeeze off the gas supply.

We dispose of another 16 percent of our poop by burning it. Drying sludge in order to fully incinerate it often requires more energy than it produces. The resulting ash, maybe a tenth of the original volume, is free of pathogens but still needs to be disposed of or reused and the incineration still releases carbon dioxide into the atmosphere. Some engineers have demonstrated the potential to instead turn poop into the charcoal-like briquettes cited in the UN report through a high-temperature, oxygen-free process called pyrolysis.

In Kenya, the government banned logging and timber harvesting in all public and community forests in 2018 to curb deforestation blamed largely on the production of charcoal briquettes for cooking and heating. But with limited enforcement capacity and few alternatives, charcoal production continued. As a greener option, sanitation-focused companies Sanergy and Sanivation, both based in Kenya, are using poop to make fuel briquettes. Sanergy uses the biomass, collected from urban toilets, as a feedstock for a vast farm of larvae from the black soldier fly. A larval by-product, called frass, can be charred and pressed into briquettes. Sanivation, for its part, uses large parabolic mirrors as solar energy concentrators to dry and transform human feces and rose waste from flower farms into another form of fuel briquettes. The sun-drying process lowers costs though it also limits how much poop can be converted at once; each ton of resulting briquettes, the company has estimated, can replace twenty-two trees that would have been felled for firewood or charcoal.

Beyond their tangible products, the efforts may be creating some spill-over benefits. Sanergy employee Sheila Kibuthu told me that the movement has helped people become more conscious of their waste and consider how they might recycle more of it or generate less of it. A combination of strategies may be needed to maximize the potential, but this much is clear: we can no longer do business as usual after doing our business.

● ● ●

In the US, the world's top oil producer by 2017 and one of its top export-ers, annual greenhouse gas emissions are second only to China. But a timely import from Norway has suggested another strategy for scaling up our own biogas production. In Washington, DC, the Blue Plains facil-ity operated by DC Water is billed as the largest advanced wastewater treatment plant in the world and serves about 2.3 million residents. The sewer system collects wastewater through 1,800 miles of pipes that snake across Washington, DC and the surrounding region and directs it to the plant in the southwestern part of the district. A big part of the system, the Potomac Interceptor, conveys wastewater from two coun-ties each in Maryland and Virginia. There are bigger districts, but Blue Plains has adopted the most stringent treatment process, said Chris Peot, the plant's director of resource recovery, when I met him on a cold and cloudy winter morning. The 153-acre facility was treating roughly 300 million gallons of wastewater every day, on average, but had the capacity to treat up to one billion gallons during peak flows.

A self-described "geeky engineer" who wants to save the world, Peot hap-pily dove into the details of the agency's water treatment process. Unlike most US utilities, DC Water allowed its technical team to look abroad for biodigesters that would clean and reduce the wastewater's organic mat-ter while creating biogas in the process. Through a newly refurbished treatment plant in Dublin, Ireland, the team found what seemed like a perfect fit: a system manufactured by a company in Asker, Norway, called Cambi. These Cambi units specialize in thermal hydrolysis, which uses high heat and then a sudden difference in pressure to break open and kill bacterial cells in wastewater. In 2015, Blue Plains installed four of these units in what was the biggest such installation in the world.

Wastewater treatment plants tend to have their own unique design flourishes, and the eight stainless steel tanks in each Cambi unit, connected by a series of catwalks and extensive piping, make an impressive focal point. The first one, the pulper tank, thoroughly mixes and preheats the incoming slurry of roughly 85 percent liquids and 15 percent solids to a bit below the boiling point of water. Pumps send that mix into the six reactor tanks. At any one time, two are filling, two are heating, and two are emptying. When they're full and sealed, these reactors stew the waste at 338 degrees Fahrenheit for at least twenty minutes at a pressure of ninety pounds per square inch (or about sixfold higher than the normal atmospheric pressure at sea level).

The reactors, Peot said, act like oversized, computer-controlled pressure cookers, fed with steam heat from a boiler. The thoroughly cooked slurry then empties into a shorter but fatter flash tank. With the opening of a valve, the intense pressure suddenly drops to normal atmospheric conditions, and the dramatic pressure differential causes the microbial cells to burst open. As a bonus to annihilating any pathogens that had survived the heat, the cells' released nutrients become immediately available to other hungry microbes. If I ate a bag of dried pinto beans, Peot said, my stomach wouldn't digest them very well. But if I first put the beans in a pressure cooker, the heat and pressure-aided breakdown would allow me to easily mash them up with a fork.

The pressure release also lets off a lot of steam, which can be fed back to the pulper to help preheat the next batch of incoming sludge. The outgoing, sterilized microbial mash, once cooled, becomes an ideal food source for other bacteria and methane-producing archaea microbes housed in four giant concrete digesters. The eighty-foot-tall digesters look like fat, flat-topped silos. Together, they can hold about 15.2 million gallons of sludge, or enough to fill twenty-three Olympic-sized swimming pools. Peot said the Blue Plains plant initially received its archaea strains from a sister plant in Alexandria, Virginia, in a process akin to seeding a batch of sourdough starter from a proven source. Over a three-week period, the bacteria and archaea within the oxygen-free tanks feast on the organic matter released by the burst microbial cells, efficiently producing biogas as a by-product of their metabolism.

In the plant's nearby gas building, the captured biogas is purged of moisture and some chemical contaminants, and then compressed. When burned in a power plant, the gas turns three turbines the size of jet engines, though in this case the turbines turn the methane into electricity and heat. Within its first few years of operation, the process was generating roughly one-fourth of the plant's power needs, with the potential to increase that production to about one-third of its consumption.

After the archaeal feeding frenzy, liquid resembling chocolate milk flows out of the digesters and into a dewatering building where belt filter presses squeeze out much of the water. Think of them as a series of giant rolling pins pressing down on a looping conveyor belt. The well-pressed organic matter that comes out the other end, a moist and pathogen-free by-product of the treatment process that the industry has dubbed biosolids, makes an ideal soil amendment. "There is no such thing as waste, only wasted resources," Peot told me, repeating a common refrain in the industry. The plant no longer has to add lime to reduce pathogens and stabilize the organic matter before delivering it to farms or reclamation projects. By reducing the biomass and eliminating disease-causing microbes, the system has saved the plant millions in chemical and hauling costs. We walked past the bays where trucks load up the biosolids and Pamela Mooring, a communications manager for DC Water, recalled how the dewatering building used to make her gag. "It used to be a horribly smelly building," she said. It smelled more like a farm on the day I visited, with a faint ammonia smell.

Flush with success from its gas production, Blue Plains has since added a new nitrogen removal process called anaerobic ammonium oxidation, or anammox for short. Researchers at Delft University of Technology in the Netherlands developed the process using bacteria that had been discovered in wastewater sludge in the 1990s. These microbes have the unusual ability to combine the problematic ammonium with nitrate or nitrite compounds to form nitrogen gas in the complete absence of oxygen. Peot said the nitrogen-removal shortcut, if a pilot test bears it out, could yield even more energy savings at Blue Plains.

For many of us, the most apparent impact of a treatment process that removes nitrogen and phosphorus and pathogens and fine particles may be what we don't see. Algal blooms. Murky water. Dead fish. But along with the roughly 3,500 gallons returned to the Potomac River every second, some changes have begun attracting notice. "The stuff leaving the plant here is way cleaner than the river," Peot said. Anglers in the know position their boats near where the discharge enters the Potomac, especially during fishing tournaments. The clear water allows the bass to see the bait more easily, leading to bigger hauls. It's a not-so-secret tip, Peot said, chuckling. Eagles sometimes fly by, perhaps looking for a catch of their own, and workers are seeing more red foxes and woodchucks on the treatment plant's property.

● ● ●

The act of transformation, of changing bad to good or vice versa, has featured prominently in folklore and mythology from around the world. A frog or beast becomes a prince. A shape-shifting kupua of Hawaiian legends—a demigod like Māui—manifests as a human who is "preternaturally strong and beautiful or ugly and terrible." A kindly old Japanese woman transforms into a monstrous yamauba: a mountain witch who helps travelers and then tries to eat them. So it is in Norway with the story of an "ugly, withered old man with red eyes" who once lived beneath a hill on the outskirts of Oslo. The troll called the King of Ekeberg, as recounted by author Peter Christen Asbjørnsen, conspired with his much more attractive wife and their subterranean servants to steal virtuous and beautiful children from nearby homes and replace them with their own mean and hideous changelings. The troll king and queen, according to an English translation by Simon Roy Hughes, were particularly keen to rid themselves of their natural offspring, "incorrigible screamers with big heads, red eyes, and insatiable greed."

When I ventured into the subterranean Bekkelaget wastewater

treatment plant tucked within the same hill, known locally as Ekebergåsen, I saw a reverse twist on the nineteenth-century folktale. The
caverns now gather up Oslo's unwanted human waste and replace it
with virtuous, if not traditionally beautiful, gas. Every time someone
flushes a toilet in Oslo, some of the captured waste is converted into
biogas—at the time, purified gas from Bekkelaget was powering all of
the city's garbage trucks and about 15 percent of its gleaming red and
green transit buses.

The poop-to-power transformation may help answer the question
of how to make the benefits of biogas more apparent to the general
public. And as engineers expand the methods for turning waste into
energy, Scandinavia—land of trolls and IKEA and ABBA and cozy *hygge*
hospitality—has become a showplace for the sheer potential. At the
tail end of a weeklong trip through central Norway on our way to the
2016 Eurovision Song Contest in Stockholm (squeeeee!), Geoff and I had
driven east to Oslo. We had rolled down the windows of our rental car to
savor the unseasonably warm weather (sixty-eight degrees in May!) and
smelled the familiar tang of farm country in the springtime as shirtless
farmers sprayed manure on their fields. Oslo itself was carpeted in tulips
and sunbathers. We visited the sleek white opera house that seems to
be rising out of the Oslofjord, the long inlet that connects the capital
city to the North Sea, and meandered through Frogner Park to admire
its marvelously eccentric collection of nude statues by sculptor Gustav
Vigeland. And then for an encore, as the temperature began climbing
toward a July-like seventy degrees, I spent much of the next morning in
a glorified troll cave housing the Bekkelaget treatment plant.

Kathrine Kjos Five, a project manager at BEVAS, the company that
runs Bekkelaget for the city of Oslo, met me with a cheerful smile at the
main door of the plant. Not many tourists stop by, she told me in her
nearly impeccable English. Local schools sometimes sent tour groups
and the kindergartners seemed to enjoy their visit, Kjos Five noted,
though the twelve- to thirteen-year-olds tended to make their disgust
fairly clear. She directed me to a small second-floor conference room
for a short pre-tour presentation; fluorescent yellow and black safety
jackets hung on pegs by the door while a blue tin of Danish shortbread

Flush

cookies sat invitingly on the conference table. On the wall across from me, five tinted photos overlaid with colored spheres depicted panoramas and close-ups of an underground cavern housing one of the plant's large settling tanks. Not your typical subject for landscape photography, perhaps, but it was a strangely compelling statement that wastewater treatment can indeed be elevated to an art form.

It's been a tough climb even here. Until the early 1900s, most homes in the city then called Kristiania had privies in the backyard or rear stairwell. Parts of Norway, though now equipped with sewers or septic systems, still lack comprehensive wastewater treatment facilities. Instead, some smaller communities filter out only the larger debris before using deep fjords to flush out and dilute the rest. The Oslofjord's geography, though, has made it especially sensitive to untreated waste. From the North Sea, water sweeps into the larger western basin of the two-lobed inner fjord through a funnel that's less than sixty-five feet deep at its shallowest point, called the Drøbak Sill.

On a map, the inner waterway resembles a pair of wings at rest, with Oslo fanning out from the base of the shorter eastern wing. Before the city's first treatment plant began operating in 1963, "the Oslofjord was just a horrible place," Kjos Five said. Soaring levels of nitrogen and phosphorus fed a booming population of algae, a process called eutrophication, which depleted the dissolved oxygen and choked off other marine life, creating massive dead zones. That same process was happening all over the world: in Lake Washington not far from my house. In Lake Erie and the Gulf of Mexico and the Potomac River and Chesapeake Bay. In Copenhagen's Lake Fure and the Baltic Sea. And on and on.

In the 1960s and early 1970s, scientists finally linked the devastating algal blooms to the explosion of nutrients dumped into freshwater and coastal environments from agriculture and aquaculture (including manure, synthetic fertilizers, and fish food), from industrial flows, and from untreated sewage disposal. After nitrogen and phosphorous levels in the inner Oslofjord peaked around 1970, increasingly strict government requirements for the treated effluent helped to gradually reverse the pollution.

Additional environmental regulations have aided the push for more sustainable solutions. Norway banned the landfilling of biodegradable waste in 2009—eliminating another way to dispose of its poop—and invested in waste-to-energy plants that burn the country's trash to boil water that can provide heat for homes, electric power, and steam for industrial applications. But incineration carries its own drawbacks. Ironically, Norway faced a garbage shortage when neighboring Sweden, with its own fuel-hungry incinerators, began competing for its trash and forced some of the Norwegian plants to import rubbish from the UK and elsewhere. The Fortum Oslo Varme incinerator in suburban Oslo, one of the biggest sources of greenhouse gas emissions in the metropolitan region, demonstrated that it could use chemical solvents to capture some of its carbon emissions from the flue gas as part of a successful pilot project. A full-scale project will capture up to 90 percent of its carbon dioxide, along with about half of the emissions from a cement factory, and stash them within a sandstone formation beneath an oil and gas field under the North Sea. The ambitious Longship Project won't be cheap, though, with expected costs approaching US$3 billion.

The considerable cost of environmentally conscious waste disposal, in other words, can provide big incentives for repurposing poop. Norway, of course, has derived plenty of wealth over the past fifty years from its oil and gas exports, a paradox not lost on critics. In a 2021 interview with *Vox*, Bård Lahn, an oil policy expert at Norway's Center for International Climate Research in Oslo, put it this way: "Ever since climate change has gained importance, there has been broad political agreement in Norway that the country should address the problem. But at the same time, we are still one of the largest fossil-fuel exporters in the world—we're exacerbating the problem we want to solve." As a consumer, Norway actually obtains much of its own electricity from hydropower.

The multiple caverns excavated beneath Ekebergåsen offer an apt metaphor for the country's complex position as both an enabler and leading fighter of climate change: separate chambers near an oil terminal in the harbor store gasoline, diesel, and kerosene for the five biggest oil companies in the country plus the Norwegian government. Just to the south, Bekkelaget's caverns are showcasing advanced technology aimed at reducing the effects of climate change caused by burning those very same fossil fuels.

Flush

For our tour of the caverns, Kjos Five and I donned orange helmets and descended a set of stairs to a long tunnel carved from the rock that led us into the bowels of the plant. To our left, a grayish-tan, fifty-five-inch-diameter pipe was sending cleaned water hurtling back toward the Oslofjord. At the far end of the tunnel, through another door, she showed me the intake, where pipes bring raw sewage into the plant and screens filter out larger debris, then smaller grit. "What's the strangest thing you've ever found caught in the filters?" I asked her, and she had a ready reply. A colleague of hers once found a pair of dentures and the plant manager placed a notice in the newspaper, just for fun. Then a man came to claim them. More commonly, the filters collect tampons, condoms, Q-tips, and washcloths that accidentally slip down utility sink drains. Sometimes, they find drowned rats after heavy rains and once, a small snake (called a "worm" in Norway).

Kjos Five and I walked a few steps to a long, narrow pool—one of three primary settling tanks—where I spied a white life preserver with red bands helpfully affixed to the railing. "We've never had to use it, but it calms people down," she said, laughing. I had a sudden flash of Augustus Gloop falling into the chocolate river in *Willy Wonka & the Chocolate Factory*. A second grouping of three pools accessible by a metal grated walkway also came equipped with a life preserver. The wastewater mixes in these tanks and sludge sinks to the bottom while the greasy FOGs float to the top and can be periodically skimmed off to reduce the risk of clogs.

For its microbe-aided waste digestion, Bekkelaget uses activated sludge tanks called bio-pools. Aeration in the upper portion would normally give the wastewater the look and consistency of liquid dark chocolate (again, Willy Wonka!). But a temporary power outage earlier in the day had sent the solids floating to the top and given the crust an entirely different appearance. "It looks like chocolate cake," Kjos Five said. Consistency issues notwithstanding, the tanks here have borrowed another important page from nature by exploiting a central part of the global nitrogen cycle. Like carbon and phosphorous, nitrogen continuously cycles from one form to another. In this case, nitrogen cycles from the gas that dominates our atmosphere to an organic form that plants and animals can use to build structures like chlorophyll, DNA, and amino acids. Then to a form that predominates in decaying matter, and then to successive conversions that eventually transform it back into atmospheric gas.

A two-step part of that cycle, called nitrification-denitrification, helps treatment plants get rid of excess nitrogen from incoming urine, feces, fertilizer, food processing waste, industrial solvents, and other sources, and prevent it from triggering an overgrowth of algae upon its release. "If we don't do it here, it's going to happen in the fjord," Kjos Five said. Remarkably, experts have figured out how to accomplish the entire removal process within the same tanks using bacterial specialists known as nitrifiers and denitrifiers. Shortcuts, like the anammox process used by Washington, DC's, Blue Plains plant, have further improved the efficiency of nitrogen removal.

As for Bekkelaget's remaining sludge, pipes from the primary settling and activated sludge tanks carry it to thickeners, where belt filters drain off some of the water and centrifuges act like the spin cycle on a washing machine to wring out some more. Then the thickened sludge heads to two large digestion tanks, where anaerobic bacteria and methanogen archaea digest it over two weeks and produce biogas. A gas upgrading system removes carbon dioxide, pungent hydrogen sulfide, and other pollutants. Metal bottles kept within tanks the size of rail cars store the nearly pure biomethane. Each tank reads "Biogas" in green lettering, with a red flower dotting the i. This is the fuel that powers some of Oslo's municipal vehicles.

The gas-yielding digestion process has another key benefit: it kills off the vast majority of pathogens in the sludge. As it reaches the finish line, the cleaner, thicker, and nutrient-packed matter slides down a chute into waiting trucks. Farmers call to ask for it, Kjos Five said; like the ones I saw in the fields of central Norway, they have a better understanding of manure's fertilizing benefits and are used to the idea of reusing waste. It just doesn't seem so gross to them.

The upgraded Bekkelaget plant, which opened in 2000, treats about twenty-six million gallons of sewage every day from the city's eastern half and several smaller kommunes, or municipalities. A larger plant called VEAS handles wastewater from the western half and adjacent kommunes. By mandate from the Norwegian government and a

European Union directive, both are required to remove 90 percent of the phosphorus, 70 percent of the nitrogen, and 70 percent of the degradable organic matter before sending the water on to the Oslofjord. To keep up with the Oslo region's booming population, the city had already begun expanding the Bekkelaget plant by blasting further into the forested Ekebergåsen hillside. Dull rumblings regularly punctuated Kjos Five's workdays as controlled explosions dug into the granite gneiss rock under a plan to nearly double the plant's treatment capacity.

The long-term investment, though, is paying off. By 2012, the success of the Bekkelaget and VEAS plants had allowed the city to begin opening actual swimming areas along the inner fjord, including the immensely popular Sørenga Sjøbad swimming dock near the opera house. Beyond the clean water and biogas, the wastewater treatment plants were creating thousands of tons of dry sludge fertilizer every year and generating excess energy that could be repurposed to reduce their needs. The Bekkelaget plant, in fact, was net-positive in its energy production, meaning that it was producing more than it was consuming every year. That energy surplus has been a key asset in Oslo's pledge that by 2030, it will slash its 2009 greenhouse gas emission levels by 95 percent.

From Oslo, we took a train to Stockholm for the Eurovision contest. If we had been traveling instead from Västervik on Sweden's east coast six years earlier, we could have taken a fifty-four-passenger train nicknamed the Biogas Train Amanda to Linköping just south of Stockholm. Its fuel source: biogas made at a sewage treatment plant and at a separate anaerobic digester fermenting cow fat, blood, organs, and guts from a local abattoir. Had we made the same Oslo-to-Stockholm trip six years later, we could have hopped aboard one of the world's first international biogas-powered buses, which is fueled with super-cooled liquid biomethane. Then we could have transferred to a bus in Stockholm, which in 2018 became the world's first capital city to remove fossil fuels from its *entire* land-based public transportation system. All trains and trams are powered by renewable electricity, while the buses are powered by biodiesel, biogas, or ethanol. Just as Norway has become a world leader in improving biogas production and promoting electric vehicles, neighboring Sweden is arguably unmatched when it comes to applying biogas and other biofuels toward overhauling its own transportation sector. The country has pledged that all road vehicles will be fossil fuel–independent by 2030, with a decent percentage of the necessary biogas supplied by anaerobic digesters at wastewater treatment plants. To help meet the surge in demand, researchers have proposed that co-digesting kitchen scraps and yard waste along with sewage sludge could yield a fourfold boost in Sweden's biogas production.

Demand is rising elsewhere too: if we were to travel to the UK today, we'd be able to ride one of the growing number of biofueled buses in cities like Bristol, Reading, and Nottingham. The environmental calculations are still rough estimates but suggest that the biobuses there may slash greenhouse gas emissions by up to 84 percent compared to their diesel counterparts. The new versions can reduce emissions of dangerous air particulates by even more. That means less air pollution in cities.

If Denmark ever hosts Eurovision again, visitors there might contribute to an even more remarkable transportation option. Researchers at the Technical University of Denmark have developed an electrochemical method to convert biogas from wastewater and other sources into jet fuel. Denmark, some calculations suggest, could make enough fuel from its biogas supply to fill the tanks of every plane in Danish airports, at a premium of only 25 percent above current costs. Imagine the promotional possibilities if they called their early departures *Lortemorgenfly*. Crap Morning Flight.

A different technology called hydrothermal liquefaction is showing considerable promise in using wastewater and leftover food to create an alternative to petroleum-based crude oil. Justin Billing, a chemical engineer at Pacific Northwest National Laboratory, told me that he and other scientists were focusing primarily on refining this biocrude, as it's known, into fuels for the transportation options least likely to be electrified in the near future, such as container ships and planes. Billing and colleagues initially experimented with farmed microalgae as a source for the biofuel, until they discovered that primary sludge from a wastewater treatment plant worked nearly as well. With a feedstock that many treatment plants pay to be rid of, the sewage-to-oil economics suddenly became competitive with fossil fuels.

Flush

The hydrothermal liquefaction process is similar to what happens in the pressure cooker tanks at DC Water's Blue Plains plant. Only in this case, extreme heat and pressure within the hydrothermal reactor mimic the geological process of compressing ocean-sediment plants and animals into petroleum, all at the warp speed of about fifteen minutes. In a second step called condensation, the dissolved molecules reconnect into chains of carbon that form the backbone of biocrude oil. "It's still almost like pure magic in the laboratory to know that we had buckets of human waste and then, after one day's reaction, it's a biocrude product," Billing told me. If scaled up in production, the lab has estimated that the biocrude could be refined into a gasoline equivalent at a cost of little more than $3 per gallon.

A biodiesel equivalent for trains and buses, for its part, could yield a 60 to 70 percent reduction in greenhouse gas emissions compared to standard diesel, Billing said. Part of that calculus includes heating the hydrothermal reactor, and a separate technology developed at Pacific Northwest National Laboratory, called an electrocatalytic oxidation fuel recovery system, can efficiently remove contaminants and generate hydrogen to fuel its own operation. That process could potentially make biocrude refinement carbon-neutral. In Canada, the Metro Vancouver regional district is using the hydrothermal liquefaction technology to create biocrude at a facility connected to its Annacis Island wastewater treatment plant. A local refinery will then convert that oil into biofuels, laying the groundwork for any type of biomass to be fed into a reactor and converted into alternative transportation fuels.

Even the sky is no longer the limit. Space scientists have experimented with converting human poop and its derivatives into interplanetary essentials like rocket fuel. The ability of biocrude to be refined into rocket fuel means there's no reason why astropoo couldn't help future spaceships, er, blast off. The lab of agricultural and biological engineer Pratap Pullammanappallil has already devised a system that could convert astronaut feces into as much as seventy-seven gallons of biomethane per day. The methane gas, produced through an anaerobic digestion process not unlike what wastewater treatment plants use, could help propel a spaceship back to Earth while the water by-product could be reused or split into oxygen and hydrogen. How extraordinary that a lowly object of disgust might be radically remade into an engine of wonder.

• • •

In the finale of the original Norwegian folktale, the stealing of fair and virtuous children apparently ended when the King of Ekeberg could no longer stand all of the "shooting and roaring" that accompanied the prelude to the Swedish-Norwegian War of 1814. The drumming and cannon fire and thundering carriages shook his house and rattled the silverware on his walls. And so he moved his entire household, along with his herd of cattle, to his brother's place in Kongsberg to the west. The locals near Oslo still reported "no shortage of thick-headed youngsters" but at least they could no longer blame the subterraneans of Ekeberg. Today, one can imagine that the blasting deep within the hill itself, fueling a friendlier global rivalry of waste-to-energy initiatives, would almost certainly keep the ornery, red-eyed troll from ever venturing back.

Resource recovery plants, I discovered, have become engines of transformation. When I expressed interest in their work, the engineers at the Durham plant in Oregon lit up and delighted in pointing out multiple features beyond the biogas generation, even amid a fiery natural disaster. Near the edge of the property, I saw how a screen of stones and sand housed other microbes that acted like natural deodorizers to keep the plant's olfactory experiences from wafting over to an adjacent high school. An industrial scrubber would need a "ton of chemicals" to neutralize the foul air, Orr said. "And it still doesn't get out some of the nastiest stuff," Schauer added. From my vantage, one of the four on-site biofilters looked more like a big box of rocks. Beneath the porous rubble field, though, several feet of wet sand provided an ideal environment for microbes to thrive on a diet of hydrogen sulfide and other gases forced up through the rocks. With the blustery wind blowing the filtered air right at me, I couldn't smell anything. "The trick is to do nothing. Let the bugs do *all* the work," Orr said.

It occurred to me later that all of the biogas and biofuel and biofilter applications may have partially solved a question posed by Benjamin Franklin in 1780. In a letter to the Royal Academy of Brussels that year, Franklin poked fun at the academy's intellectual contests for including a math-based question that he dismissed as having no practical value. So the venerable statesman and writer made his own humble suggestion for a different prize question: "To discover some Drug wholesome & not disagreable, to be mix'd with our common Food, or Sauces, that shall render the natural Discharges of Wind from our Bodies, not only inoffensive, but agreable as Perfumes."

It's true that the output of biogas digesters may not rise to the level of a perfume and that the deodorizing comes on the back end. But I'd like to think that Franklin would have been mighty impressed by the ability of a giant box of rocks and sand to make foul air more agreeable. And I'm pretty sure that the inventor of the lightning rod, bifocals, and Franklin stove would find that turning big turbines, fueling buses and rockets, and regulating engines with a giant gas-filled bubble are eminently practical applications for the "natural discharges of wind from our bodies."

Taniguchi-Dennis said Clean Water Services was exploring whether more of its biogas could be converted to energy via engine generators or to biomethane as a natural gas substitute for industrial or transportation applications. The gas would have to be scrubbed of carbon dioxide and other impurities to make it "pipeline-ready," she said, "but what a beautiful thing we could do: not just creating electricity but actually creating a biogas that's natural gas." It may not be the standard definition of beauty, perhaps, but replacing unwanted and unloved human waste with a dazzling array of useful and renewable products would be a remarkable makeover indeed.

CHAPTER NINE

Catalyst

About 17,000 years ago, a massive sheet of ice some 3,000 feet thick buried our front yard. It buried the land that now slopes from a ridgeline a few blocks above our street down to the western shores of Lake Washington. It buried Seattle and all of Puget Sound, and a sixty-mile-wide expanse that reached from the Olympic Mountains to the Cascade Range.

Then the Puget lobe of the Cordilleran ice sheet began to melt and retreat—the most recent of at least seven icy incursions into western Washington over the past 2.5 million years. As the glacial lobe advanced and receded, the ice carved out the region's largest lake below our house and the distinctive series of ridges and valleys that torture Seattle drivers whenever it snows. The glacial till deposits here have given us fairly young soils of gravelly sandy loam atop *very* gravelly sandy loam. With the ice's retreat, the bears and wolves and wildcats returned. So did a succession of plants and trees that colonized the slope and formed a forest, filled with vine maples and western red cedars, evergreen huckleberries and western sword ferns.

This is the traditional land of the Duwamish, or dxʷdəwʔabš in the Lushootseed language spoken by Coast Salish Indigenous peoples. Archaeological records of their villages date back more than 10,000 years, to a time when the Ice Age described in ancient stories like "North Wind, South Wind" loosened its grip and gave way to a temperate coastal climate graced by mild, rainy winters and cool, dry summers. The Duwamish harvested Pacific herring and salmonberry. Coho salmon and camas bulbs. A list of traditional food sources compiled from archaeological surveys and the recollections of elders includes nearly 300 plants and animals.

When I read through the list in early 2021, I realized to my shock that our yard alone harbors at least fifteen of these edible plants and trees. *bubx̌ǝd*. Horsetails that we had dismissed as prehistoric-looking weeds. *k̓ayuk̓ayu* and *t̓aqa*. Berries of the native kinnikinnick and salal plants that we had reintroduced but valued only as ornamental groundcovers. *čǝbidac*. Douglas firs whose tender tips can be steeped to brew a citrus-flavored tea.

Geoff and I hadn't considered any of those native plants in March 2020 when a landscaper and his crew installed two retaining walls of yellow-gray blocks and leveled off much of our long-neglected front yard. We had saved a huge pile of gravelly sandy loam from a backyard construction project. The fill, topped with a four-inch layer of dark brown compost, was just enough to even out the front yard. Our new walls ended at a set of stairs and a curving walkway that replaced their crumbling concrete predecessors, and we joked that this was our West Coast version of a Brooklyn stoop.

The walls also created space for two eastern-facing terraced gardens. We decided to fill the lower level with a butterfly- and bee-friendly mix of herbs and ornamentals, while we would try our hand at growing vegetables in the upper level. With a neighbor, we set out to transform a third plot in a dilapidated city easement behind our house that basked in the afternoon sun. COVID-19 was sweeping through the Seattle region and a gardening project seemed like a good way to calm our jitters, focus on something positive as the city locked down, and maybe even grow some of our own food.

The first thing I noticed were all the robins. The worms and insects in our front yard's new layer of yard waste compost appeared to be keeping the daily visitors well fed. In April, we began planting sugar snap peas and red shallots, French breakfast radishes, and Rosie red romaine— the first of dozens of vegetable and ornamental varieties in the three gardens. The upside to planting so many living things from seeds or starts—perhaps less than advisable for first-time urban gardeners—is that something is bound to grow. We noticed that two crows had begun hanging out on the telephone wires in front of our house, watching our progress. Geoff named them Spot and Chicken.

Around the same time, I discovered what I was sure would be the perfect mulch to give our young plants a healthy boost. Since 1976, a local bark and landscaping supplier called Sawdust Supply had been buying organic biosolids from King County's three wastewater treatment plants. Biosolids, by-products of the wastewater treatment process we just learned about, are made up mostly of microbe-digested poop and other organic debris mixed with bacteria and sand. The county's plants churn out about 130,000 wet tons of this nutrient-rich matter every year.

King County treats its solid waste in giant vats heated to the temperature of the human body. Over twenty to thirty days, anaerobic microbes feast on the organic matter and kill off most pathogens in a step that mimics the human gut's digestion. "It's completely a biological process, where the microorganisms from our own bodies, really, break down that material," Ashley Mihle, the county's compost project manager, told me. If one digester "gets sick" due to an imbalance in its microbes, engineers can reseed it with the microbes from a healthy, balanced counterpart to reset the population—basically, a fecal microbial transplant. Afterward, a centrifuge wrings out the water, leaving behind what Mihle calls a "cakey, Play-Doh material."

Through that material, the county has reclaimed 100 percent of its solid waste and turned it into a soil amendment product. Like the wastewater treatment plants I visited in Norway; Washington, DC; and Oregon, King County also converts some of its wastewater into biogas— in this case to produce heat and electricity, biomethane, and liquid biofuels. Diverting thousands of truckloads of organic output from landfills or incinerators and reintroducing them to the region's soils helped the county department that includes the wastewater treatment division become carbon-neutral in 2016.

After quietly creating its soil product for nearly forty years, the county began branding the organic matter as Loop in 2012. Its slogan: "Turn your dirt around." Loop is designated by the US Environmental Protection Agency (EPA) as a Class B biosolids product, which means that disease-causing pathogens have been greatly reduced but not necessarily eliminated and that it can't be applied to public access areas like city lawns or gardens. After spreading it on their soils, farmers and other recipients are required to keep people and livestock away for at least thirty days to allow the vast majority of remaining pathogens to die off from natural exposure to sunlight, heat, and competing microbes. That process usually takes several weeks. Loop is very dark and spongy and smells a bit of sulfur and ammonia. "I actually think it smells quite nice. Perhaps not everyone would agree, but it's sort of an earthy odor," Mihle said. Often, Loop also contains sparkly particles of struvite, a fancy name for the magnesium ammonium phosphate mineral that settles out of wastewater during the treatment process. The effect is like seeing glitter in a brownie mix. Every year, King County produces about 4,000 truckloads of glittery Loop. Nearly 80 percent goes to dry-land wheat fields and other agricultural sites in eastern Washington. Another 20 percent or so goes to help amend the soil in forests.

I was after the remaining bit of Loop that went to Sawdust Supply, which composted it in a one to three ratio with sawdust waste from local sawmills in a process called aerobic digestion. Unlike the anaerobic digestion of organic matter in which archaea microbes release methane-containing biogas as a by-product, aerobic digestion by bacteria and fungi during composting gives off heat, water, and a bit of carbon dioxide. As temperatures spike for weeks at a time in the steaming piles, the slow pasteurization-like process effectively eliminates the pathogens, though too much heat can kill off beneficial microbes and disrupt the composting process. Sawdust Supply found the sweet spot for its popular mix and earned a Class A biosolids designation, marking it safe even for home gardening and landscaping.

After curing its blend for a year, the company distributed a nutrient-rich soil conditioner it called GroCo to regional stores that in turn sold it to a devoted fan base. Theoretically, I could have been contributing to the compost for years. To get some for our own gardens, though, I would have to hurry. The company's owner had recently died, and his family was shuttering the beloved 108-year-old institution, with its "Bark now or forever hoe your weeds" slogan. Only a few sixty-pound bags remained of what I imagined to be a slightly pungent but potent plant food, so I hopped into my Subaru and set out in search of my own recycled poop.

• • •

Had I lived in early seventeenth-century Japan, farmers might have come knocking in search of my output. More precisely, they sent hired agents door to door in the city of Edo, now Tokyo, to seek out a commodity so valued that it yielded a substantial price. The farmers called this precious material *shimogoe*, or literally, "fertilizer from the bottom of a person." Given the opportunity for some extra income, landlords of rental units even laid claim to their tenants' regular deposits. Economist Kayo Tajima writes that shimogoe became one of the main sources of fertilizer for agricultural villages dotting the city's urban periphery.

In her book, *The Reality Bubble*, science journalist Ziya Tong describes how this human "night soil" became highly profitable for some entrepreneurs. "Landlords could increase the rent they charged if the number of tenants dropped in their building, because with fewer defecators to pad an owner's income, running the property became less profitable. As a business, managed through private agents and not the government, shimogoe prices were set by the landlords, leading to conflict with farmers, who were often gouged with high prices." Farmers, for their part, understood that not all crap was created equal. "There was also good shit and bad shit. Rich shit surely stank as much, but it was more highly prized. As the rich ate more diverse diets, this resulted, according to the farmers, in better nutrients in their feces," Tong writes. By the 1800s, the price of primo poop had soared so high that stealing it could land the thief in jail.

The city's urban dwellers, Tajima writes, were the main consumers of fresh farm produce grown in the surrounding villages as well as the main producers of its fertilizer, thereby creating a loop of interdependence that brought other unexpected benefits. "Through bringing human excrement out of the city, this practice is considered to have greatly helped in preventing pandemic diseases," she writes. "While growing foods with human night soil posed risks of oral infection or waterborne diseases, the custom of cooking foods thoroughly and drinking boiled water (or tea) kept this risk minimal."

Similar practices abounded throughout the world: Imperial China's Jiangnan region also boasted a thriving trade in night soil. And in South America, the Amazon is stippled with *Terra Preta de Índio* or Amazonian dark earths. Researchers believe the unusually fertile swathes of land were created between 2,500 and 500 years ago by Indigenous residents who mixed their feces and other organic waste like chicken manure and fish carcasses with plant-derived biochar to nourish woodland gardens. Some scientists think the Amazonian deposits, which contain far more carbon than the surrounding soils, were created intentionally. Others suggest that the organic mix could have been a side effect of adding biochar to waste deposits as a deodorizer.

Either way, biochar expert Hans-Peter Schmidt writes that mixing decaying organic material with the highly porous and expansive surface area of natural charcoal activates its unique ability to soak up water and dissolved nutrients like a sponge while providing innumerable niches for soil microbes. Biochar, it turns out, can be a remarkably stable fertility aid. It also excels at binding positively charged ions such as ammonia and ammonium, making them more available to plants and soil microbes and less apt to leach from the soil.

The mass extinction of Pleistocene megafauna that Chris Doughty and colleagues have studied hit South America the hardest; beyond their decimated capacity to distribute nutrients, land giants were largely unavailable to humans as livestock or beasts of burden. Instead, Schmidt writes, Indigenous communities relied on "native wild fruits, small animals and fish or horticulture in the woodland gardens to cover their food demand." And in the absence of livestock manure, "the digestive tracts of the population at large" provided much of the necessary organic fertilizer. This early form of geoengineering created rich and lasting layers of humus—a stable organic material that gives soil its dark color and feeds its microbial residents—and dramatically improved the soil fertility while steadily increasing yields. The sophisticated if labor-intensive system that combined nutrient cycling with cultivating diverse crops in mixed groupings, Schmidt and other researchers now believe, helped feed surprisingly large populations once deemed impossible. In essence, the Amazon supported a network of self-fertilizing garden cities.

Centuries later, we're facing a daunting combination of agricultural crises. Topsoil loss. Drought and floods. Clouds of locusts devouring crops. Recent projections suggest that the world is headed toward severe food shortages worsened by global warming. We'll need a concerted effort to feed everyone on Earth, not to mention a major overhaul of our food security systems. Genetic modifications and agricultural efficiencies may help boost crop yields, but our throwaway culture is squandering more than a billion tons of food every year and rejecting the proven and plentiful fertilizer of human waste that has been used to grow crops for centuries. To correct our course, we will need to convince each other (and ourselves) that there's no duality between us and nature, our bodies and the natural resources we exploit. We can build a circular economy in which we become our food's food by judiciously and safely using our waste to nurture life. If that seems radical, it's only because we've forgotten or ignored our own history.

In continent upon continent, our ancestors understood a universal truth: good shit makes things grow. In much of Europe and North America, though, the concept of using human feces as fertilizer has had a more conflicted past marked by cycles of acceptance, revulsion, and cultural amnesia, as Dominique Laporte's *History of Shit* reminds us. In nineteenth-century London and New York, disposing of night soil required the services of well-compensated crews who worked at night to keep the cities' unwanted waste out of sight. Some of the cartloads eventually made it to farms, just like the deliveries of "gong farmers" or "rakers" had helped to fortify fields supporting Europe's medieval towns. Often, though, the waste was simply dumped into the nearest body of water.

Misguided ideas about public health hastened the redirection of human poop from farms into streams, rivers, and bays. In *The Ghost Map*, Steven Johnson describes how the prevailing miasma theory that had blamed cholera on malodorous air precipitated a disastrous public health campaign in London that hastened the fouling of the Thames. One particularly influential 1848 law, ironically named the Nuisances

Removal and Contagious Diseases Prevention Act, ordered the contents of thousands of cesspools to be discharged into the river in the name of better sanitation. "If all smell was disease, if London's health crisis was entirely attributable to contaminated air, then any effort to rid the houses and streets of miasmatic vapors was worth the cost, even if it meant turning the Thames into a river of sewage," Johnson writes. After a succession of devastating cholera epidemics, hot weather in the summer of 1858 famously filled the city with an overwhelming stench from the crap and industrial waste dumped in the river. The Great Stink, as it was called, finally prodded Parliament to invest in a comprehensive sewer system.

Just as shit has elicited vehement expressions of revulsion throughout history, so has it inspired fervent evangelism for reusing waste as fertilizer, attracting its own supporters and detractors. In Victorian England, journalist and reform advocate Henry Mayhew wrote enthusiastically about the potential economic gain and virtuous circle of life. In France, a parallel movement of hygienists and physiocrats—the latter of whom maintained that agriculture was the sole source of national wealth—likewise drew bright-eyed optimists like philosopher Pierre Leroux, who argued that the state-sponsored reuse of human waste could help end poverty. During a trip to London, Leroux even delighted in making his own soil, which he apparently used to grow green beans. His list of ingredients:

River sand from the Thames, pounded into a fine dust
Pounded charcoal
Coal ashes from the hearth
Pounded brick
Urine
Excrement

Laporte's book skewers Leroux's grandiose descriptions and arguments but includes a quote from him that suggests the good philosopher may have at least understood the essence of a circular economy:

Flush

"By nature's law every man is at once a producer and a consumer, and if he consumes, he produces."

I wasn't prepared to go quite that far with my own gardening. But I was intensely curious about the moist chocolate-brown clumps from Sawdust Supply that I might have played a small part in producing. I brought my nose close to a bag and smelled a faint tang of wood with some rich undertones. I was so surprised that I took several deep breaths. I'm not sure what I had been expecting, but the composting had done its job—no sulfur, no ammonia—and the four bags of GroCo I luckily snagged from a garden store were soon overwhelmed by the sharp scent of bark from the mulch I bought afterward.

I gave one bag of GroCo to friends down the street, with instructions to observe the effects on their pole beans and tomatoes and raspberries. And then I parceled out the remainder on our own gardens over the next few weeks, partly as a compost and partly as a mulch on top of the soil to control weeds and retain moisture. By mid-May, we had installed two trellises for the peas and filled the upper front garden with twenty-eight rows of new plants. Lettuces and arugula and kale. Yellow onions and garlic. Curious passersby quizzed us on what we were doing and seemed both bemused and delighted. A vegetable garden in full view of the sidewalk was still a novelty in the neighborhood.

In the garden out back, we planted sun-craving vegetables like jalapeño peppers and tomato varieties named Bloody Butcher and Black Prince, Amana Orange and Pink Berkeley Tie-Dye. We had tried to grow tomatoes before, and sympathetic women from the neighborhood just smiled and shook their heads. No one grows good tomatoes in Seattle, they had told us. And here we were, hands in the moist soil, gardening through a pandemic, giddily tempting fate. On the last day of May, we celebrated our first garden salad: a few green and purple leaves of lettuce, romaine, and wild garden mix, garnished with a few pink chive flowers. In the first days of June, I pulled out the first ruby red radishes, little gems that nearly glowed with color.

● ● ●

For the gardens and farms of Kenya, Nairobi-based Sanergy has adopted a more indirect strategy of reusing residents' poop to feed their plants. The same company that creates fuel briquettes from poop-fed fly larvae to address deforestation also specializes in bringing sanitation services into urban areas that lack a conventional sewer system by safely collecting and transporting the sewage. In 2011, it began installing Fresh Life Toilets in informal settlements around the capital city. Inside the bright blue structures, a waterless squat toilet drains into separate collection chambers for urine and feces, which can then be hauled away. By the end of 2021, Sanergy had installed more than 3,600 of the toilets as part of a franchise model in which the company sells individual units to Kenyan entrepreneurs. Another arm of the company, a waste management service, uses compact trucks to navigate the narrow streets, empty the toilets, and transfer the waste to a central processing plant. For two similar services, Sanergy empties pit latrines and in-home squat toilets. This container-based sanitation, as it's called, can serve otherwise hard-to-reach communities.

Sanergy representative Sheila Kibuthu told me how the company's processing plant has made Nairobi's relatively low-cost container-based sanitation more economically attractive by creating multiple agricultural products from the collected waste. For one of them, Sanergy mixes food and farm waste with the feces it extracts from the toilets and latrines as the feedstock for its huge farm of black soldier flies. The flies, which normally hang around livestock and decaying matter, eat little if anything during their fleeting adulthood of a few weeks. Their ravenous larvae, on the other hand, are unique in their ability to convert the biomass they eat into fat and protein. Tons of it. The wriggling, feasting larvae store so much protein that it can account for an astounding 40 percent or more of their dry body weight. "We have a massive, massive colony of flies that we rear," Kibuthu said. "In fact, the fecal waste that we get is barely enough."

The poop-loving larvae, in turn, form the nexus for products like an animal feed supplement. Sanergy harvests the larvae, pasteurizes them, and sells them to millers to be ground into protein-packed feed for farmed fish, poultry, and pigs as well as pets like dogs and birds.

The protein supplement can replace fish meal, most of which has been unsustainably harvested from a small fish called omena in Lake Victoria, one of Africa's Great Lakes. Seasonal shortages in the fish meal have compelled millers to import more expensive protein sources like soybeans. A homegrown source created from repurposed waste could help solve a major agricultural problem and boost food production to boot. Compared to traditional feed, some researchers have found that the larvae-based protein can boost egg production in chickens and weight gain in farmed tilapia and catfish.

As they chow down, the black soldier fly larvae make the by-product known as frass, which converts the complex biomass into a simpler form. "So it's kind of pre-decomposed," Kibuthu said. That frass, which Sanergy has used to make fuel briquettes, can also be mixed with calcium and magnesium-rich lime plus food and farm waste to form yet another kind of fertilizer. The organic mix, Kibuthu said, can address the root cause of declining soil fertility and crop yields by feeding the soil (and especially soil microbes) in ways that plant-focused synthetic fertilizers cannot. For improving Kenya's yields of important but threatened crops like maize, then, the flies may offer an unexpected ally. A study by the International Centre of Insect Physiology and Ecology in Nairobi found that of twenty-eight insect species, the black soldier fly larvae offered the best solution for sustainable development aimed at alleviating poverty and food insecurity.

A growing number of American cities have taken a more direct approach to turning biosolids into soil amendments. A few decades ago, though, human manure was almost a clandestine, underground commodity to be exchanged in private. "It was hidden. It wasn't talked about," Mihle said. Treated biosolids might show up in an unmarked truck in the field of a farmer that had been convinced to take it, often for free. "There's just been this transition towards being really proud of this valuable resource and this commodity, and setting a price for it, and putting it in a truck that says what it is," she said. Well, almost. A flatbed labeled "Recycled Poop!" might still be a bit off-putting, but the Loop brand has polished some of the rough edges with a more visually appealing and gently humorous approach.

Just to the south, Tacoma has its own highly popular biosolids-based product, TAGRO (short for Tacoma Grow). A high-temperature drying process kills off all of the pathogens, meaning that the Class A–designated soil amendment can be used as is for home gardening. Milwaukee has sold its biosolids-based pellet fertilizer, Milorganite, since 1926. At DC Water, Chris Peot told me that the utility's own brand, Bloom, was named as an homage to Loop, though the pathogen-free product has a wider range of applications, like TAGRO. Bloom's tagline is "Good soil, better Earth," and when I sniffed a bag in his office, it did smell like rich soil, tangy but not unpleasant. One bonus of being in Washington, DC: in collaboration with the US Bureau of Engraving and Printing, the utility has experimented with mixing Bloom with shredded passports and currency, like old twenty-dollar bills. The shredded paper, a good source of carbon for soil microbes, works well because the ink is vegetable-based, Peot said. "All we're doing here is accelerating nature." And perhaps proving that there really is money in composting.

Our output, of course, contains much more than that. Because we take in a wide range of minerals like zinc, nickel, molybdenum, and selenium from our diets, drugs, and the environment, we also poop them out. In small amounts, these minerals are beneficial plant nutrients. For communities with combined sewers that transport sewage, rainwater runoff, and industrial waste through the same pipes, however, some of the wastewater minerals come from more concentrated soil deposits or chemical sources. Arsenic appears in geologic formations but also in pesticides. Mercury can flow from batteries and dental fillings, cadmium from pigments and solar cells. Lead can leach from older pipes, copper from newer ones. The same is true for pharmaceuticals and chemicals from personal care products. Depending on the source of the wastewater awaiting treatment, its constituents can say quite a bit about the local population, infrastructure, and pollution. If some chemical or mineral concentrations are too high, they can also present a problem for municipalities hoping to reclaim their solid waste.

When I first talked with Sally Brown, one of the best-known researchers on the safety of biosolids and composting, she was whipping up a batch of granola in her Seattle kitchen. Oats, cashews, sunflower and pumpkin seeds, coconut, brown sugar, maple syrup, butter, and peanut oil. Cooking has been a central part of her life for decades, and she jokes that one of her crowning achievements was making French toast for actor Warren Beatty while a chef at New York's Soho Charcuterie in 1981.

Brown hoped to strengthen the Big Apple's ties to agriculture by delivering locally grown produce from truck farms on the urban periphery. Unlike sixteenth-century Edo, there was no return stream of composted food scraps to fertilize the farms, let alone biosolids. There was a historic precedent, though. In the mid-1800s, a New York City company had collected barrels of waste and made dried shit bricks that it sold as fertilizer to help rejuvenate fallow farms in the Northeast. But by the late 1800s, after the city built and expanded its sewer system, it was dumping most of its sewer sludge into the Atlantic Ocean (and would continue to do so in one form or another until the US Congress finally put a stop to it in 1992). When she flipped through an obscure magazine called *Bio-Cycle*, though, Brown realized that plenty of people were talking about greener possibilities. She found her calling, enrolled in graduate school at the University of Maryland in 1990, and helped devise some of the first biosolid safety and quality regulations for the US Department of Agriculture.

Brown has studied how contaminant metals like lead and cadmium interact with soils that contain compost made from human waste. "Cadmium was a big concern because people can get sick and die if they eat foodstuffs with too much cadmium in them, especially if they have a poor diet," she said. Her research, though, suggested that biosolids made metals like cadmium and lead *less likely* to be taken up by plants. "So biosolids were actually protecting you from metals." But how? Studies suggest that the mineral component of biosolids can physically bind up other metals. During the wastewater treatment process, for example, minerals like iron, aluminum, and manganese form clays that carry a high surface area of nooks and crannies and act a bit like Velcro to grab onto other metals. Biosolids also contain plenty of zinc, which is both an essential plant nutrient and a close cousin of toxic cadmium on the periodic table. When given the choice, plants overwhelmingly prefer zinc.

"So if you have sufficient zinc, you reduce the plant uptake of cadmium," Brown said. "Rather than focusing on biosolids as a point for metals to enter the food chain, I realized, Oh, my, you can use this to *protect* the food chain in contaminated sites."

The same general principle applies to lead. Biosolids add stable molecules that bind to the metal, essentially "locking" it in the soil and making it unavailable to plants as food. Instead of digging up and hauling away tons of lead-contaminated soils, Brown and other researchers found, the technique could help decontaminate the ground. She compared the mechanism to padlocking your freezer to block access to a tub of ice cream. "You don't get fat from the ice cream in your freezer if you don't eat it," she said.

Beyond her research on the safety considerations, Brown's work has focused on how biosolids like Loop might in fact help improve the nutritional content of some foods. One planned research collaboration between her lab and King County will examine the link between soil health and the yield and nutrient content of kale, Swiss chard, carrots, and broccoli grown in plots amended with GroCo, TAGRO, and food scraps composted by inmates at Washington's Monroe Correctional Facility. Many native plants grown by Indigenous communities, Mihle said, carried far higher micronutrient concentrations than the food we eat today. Soil health seems to account for at least some of the difference, and the joint research project is investigating whether returning our own organic matter to the earth might help reclaim some of that tilth, or its suitability for growing crops.

If so, the work could boost efforts to decolonize food cultivation, something that Brooklyn dietician Maya Feller described to me as a way of reaffirming indigenous foods that by their very nature are whole and minimally processed. "They're always the ones that are the heirloom varieties, they're always the ones that are the most nutrient dense, and they fall naturally into the slow-food category and they're so good for us," she said. Collard greens, for example. Even avocados—now ubiquitous in hipster cafés and health-conscious cookbooks—were once disparaged as fatty and unwholesome before mainstream America "discovered" the virtues of a fruit that Latino, Black, and Indigenous communities had long valued as part of their history and culture.

Instead of the full suite of macro- and micronutrients that such plants require, modern inorganic fertilizers tend to focus on a few essential elements—most often nitrogen, phosphorus, and potassium. With their high concentrations, inorganic fertilizers give plants a quick boost that can become a burn if unwary farmers or gardeners use too much. The optimal ratios depend upon local soil conditions and plant preferences: for a lawn in western Washington, experts might recommend three parts nitrogen, one part phosphorus, and two parts potassium. For vegetables and annuals, a balanced ratio of one-to-one-to-one is better. And for the region's trees and shrubs, nitrogen alone is usually sufficient.

• • •

For ancient farmers who first extracted nitrogen from nature, bird guano must have seemed like a miracle. Archaeologist Francisca Santana-Sagredo and colleagues found evidence that communities in northern Chile's famously arid Atacama Desert, sometimes used by scientists as a stand-in for Mars, grew an impressive range of crops with guano fertilizer more than 1,000 years ago. With the precious "white gold" hauled in from coastal colonies of pelicans, boobies, and cormorants that dined on nitrogen-packed fish, the extreme desert environment bloomed. Amaranth and quinoa. Chili peppers and popcorn. Maize, squash, and beans.

A form of natural fertilizer subsequently discovered within the Atacama Desert itself provided a rich alternative to the guano: deposits of the whitish mineral sodium nitrate, also known as caliche or saltpeter. But then a new source of nitrogen appeared, literally from thin air. A synthetic method called the Haber-Bosch process was widely hailed as a technological triumph when German chemists Fritz Haber and Carl Bosch developed it in the early 1900s. Under high pressure and heat and in the presence of a catalyst, Haber discovered, humdrum nitrogen gas from Earth's atmosphere will combine with hydrogen to form ammonia.

Replacing nature's nitrogen cycle with a synthetic shortcut launched a new era of ammonia-based fertilizers such as ammonium nitrate that quickly replaced closely guarded caches of nitrogen-rich mineral and guano deposits. As environmental journalist Elizabeth Kolbert writes, "Haber had, it was said, figured out how to turn air into bread." The chemist earned a Nobel Prize for his ingenuity, though his reputation was forever stained by a separate project in which he developed poison chlorine gas and oversaw its deployment by German troops on the front lines in Belgium during World War I.

Even as the Haber-Bosch process has transformed global agriculture, several stains have tainted its reputation as well. Most nitrogen fertilizer factories now source the hydrogen ingredient in the process from natural gas, though some chemical plants still derive hydrogen from gasified coal or petroleum coke. In the air, ammonia-based fertilizer production has been blamed for emitting more greenhouse gases than any other single chemical-producing reaction. On the soil surface, inorganic fertilizers can form a crust that impedes the absorption of water into lower layers. And within the soil, the negatively charged nitrate ion from ammonium nitrate-based fertilizers tends to stick poorly to clay or humus particles, meaning that it's more water-soluble and can move easily through the soil.

These latter two attributes help explain why rain or irrigation can leach the added nutrients and wash excess fertilizer into nearby storm drains, streams, and other bodies of water. Too much nitrogen can acidify the soil and contaminate groundwater. A burst of phosphorus can trigger explosive algae blooms in which the plants block incoming sunlight and use up the available oxygen, snuffing out other aquatic flora and fauna. The resulting dead zones have wreaked environmental havoc from Lake Washington to the Gulf of Mexico to the Oslofjord.

Manufacturers often source the phosphorus and potassium for inorganic fertilizers from natural mineral deposits, though as Chris Doughty and others have warned, accessible phosphorus stores are beginning to empty out, worsened by increasing depletion from eroded soils. Doughty and other researchers have raised the specter of "peak phosphorus," or a looming era of major shortages around the world that could dramatically worsen food insecurity. In a follow-up to his study describing how megafauna and other long-distance travelers acted as the world's major phosphorus distributors during the Pleistocene, Doughty joined two colleagues in proposing a new strategy to help mitigate the element's loss. Establishing a global trading and recycling scheme, they suggested, could allow countries to partially achieve their recycling goals by revitalizing the "natural phosphorus pump": conserving wildlife habitat. "By restoring wild populations of whales, seabirds, anadromous fish, herbivores, scavengers, and filter feeders, we can enhance the retention of phosphorus across ecosystems," they wrote.

Another strategy is to recover more phosphorus from the things we throw away. Here again, wastewater treatment plants have taken the lead. In 2008, Oregon's Clean Water Services opened the first commercial nutrient recovery facility in North America by partnering with a spinoff from the University of British Columbia called Ostara. (The goddess Ostara first appeared in the 1835 book *Deutsche Mythologie* by German folklorist Jacob Grimm and is associated with fertility, renewal, and the spring start of the agricultural cycle.) The Oregon utility is essentially extracting phosphorus from wastewater and converting it into fertilizer pellets that Ostara sells back to farms and nurseries. "We're removing something valuable that would normally just get sent to the ocean or be wasted," said Brett Laney, an operations analyst who manages the project. Laney walked me through the phosphorus extraction process at the Durham Water Resource Recovery Facility the same day I learned about the site's biogas production. In a way I hadn't expected, the two kinds of resource extraction are closely linked.

Most wastewater treatment plants are plagued not only by the FOG fatbergs that can clog pipes, but also by glittery struvite. The mineral can settle out of wastewater and form a concrete-like buildup called scale on the inside of those same pipes. For part of his show-and-tell, Laney picked up what looked like a large shard of white-gray pottery. It was a half-inch-thick chunk of struvite that had to be blasted from a pipe with a plant-based compound called citric acid, which lowers the pH and encourages the mineral to dissolve. Failing that, workers have had to descale some pipes with chisels.

Next to the shard, Laney showed me a Tupperware container with a bit of sand in it. I took a small bit in my fingers to get a closer look. No, glitter mixed with sand. Under a microscope, Laney said, struvite is a triangular-shaped crystal, and the grit can settle into the corners of the biodigesters that convert biosolids into biogas. Struvite slows them down and adds unwanted bulk to the biosolids. Minimizing it in those digesters, where it's wasted and dulls their efficiency, has allowed the plant to maximize it in its fertilizer.

"So you want less glittery poop," I said.

"Exactly."

Ely O'Connor, a communications specialist shepherding me through the tour, couldn't resist. "It's fool's gold!" she said.

Removing the struvite before it causes headaches and transforming it into a valuable fertilizer has mirrored the repurposing of FOGS as a bacterial food to increase the biodigestion efficiency. To handle the first part of the bane-to-boon process for struvite, the Durham facility has harnessed the unique ability of another group of microbe specialists called polyphosphate-accumulating organisms. When grown in aerobic conditions, these bacteria can be coaxed to take up a large amount of phosphorus and magnesium as they feed and stash it away within their cells. But when switched to anaerobic growing conditions, Laney said, the bacteria start releasing the minerals back into the water. Relieved of some of their cargo, the microbes continue on to the biogas-producing digestion tanks while the captured phosphorus and magnesium head to the on-site fertilizer factory.

This isn't your typical fertilizer, though. The ingredients flow into a large funnel-shaped vat through pipes near the bottom. The stream of mineral-rich water and added magnesium comes in one side. In the other flows a stream of ammonium and phosphorus-laden water drained from centrifuges that have spun-dry the digested sludge, along with a caustic compound to increase the pH. The pipes deliver the building blocks needed to form more struvite, and the higher pH encourages it to settle out of solution. The same mineral blamed for the hated scale buildup is intentionally formed here and accretes onto a mass of tiny seeds. Known as prills, these struvite seeds attract layer upon layer of the mineral being made all around them. Once they reach the right size, they can be harvested, dried, and bagged into a product that resembles a sea of pearls. Clean Water Services, in turn, has mixed some into its own line of pellet fertilizers for flowers, vegetables, and lawns.

Laney let me inspect two plastic jars of the finished pearls—the smallest of which were less than the diameter of a pinhead. The plant makes them in four sizes; smaller is better for maximizing the fertilizing efficiency. Each pearl contains 28 percent phosphate, 10 percent magnesium, and 5 percent nitrogen. That may seem like *a lot* of phosphate—it is—but the fertilizer has another twist that gives it an unusual advantage. The pearls are highly water insoluble, meaning that the minerals aren't easily flushed out with rainfall. This physical property prevents the leaching and runoff into streams and rivers that often plague inorganic fertilizers. Instead, the pearls slowly release their nutrients over time. Remember how a high pH can cause struvite to settle out of solution? A low pH can begin dissolving the mineral back into water. Plant roots, when they require more nutrients like phosphorus, release a compound that lowers the soil pH around them. That compound is citric acid, the same one that treatment plants use as a descaling agent. The fertilizer pearls, in other words, respond to the plant's natural feed-me signals and release more minerals in response to the root-directed drop in soil pH.

As more complete plant foods, organic fertilizers like manure and guano typically supply a fuller selection of nutrients but in smaller concentrations and over longer periods. That's because the soil bacteria and fungi need time to further decompose the organic matter and convert the nutrients into plant-available inorganic forms. These fertilizers are sometimes called "slow release" versions; they also tend to cling more tightly to soil particles, meaning that they won't wash away as easily.

Though Loop acts much like an organic fertilizer, it's technically billed as a soil amendment. While its mineral nutrients feed the plants, its organic material feeds the microbes and conditions the soil. As carbon-rich matter, Mihle told me, Loop can effectively reduce emissions by sequestering more carbon in the soil. Aided by its nutrients, larger-growing plants can suck up more of that carbon from the earth and pull in more carbon dioxide from the atmosphere, storing it in the plant mass and further adding to the carbon sink. And as a fertilizer alternative, Loop can replace the fossil-fuel intensive process used to make synthetic ammonia.

When mixed with sawdust to form GroCo, Loop becomes a less concentrated but more versatile fertilizing and soil-amending compost. Microbes can process the nitrogen and other nutrients while feeding beneficial insects and worms that burrow through the dirt and keep it aerated. The organic humus formed by the decaying compost physically and chemically alters the soil as well: by binding particles into larger clusters, it loosens the soil structure and creates more channels and pockets and pores that retain water, air, and nutrients. "And so when it rains, water goes into the soil instead of across it," Mihle said. That greater water-holding capacity helps reduce runoff.

With their mix of organic matter and minerals, products like Loop and Bloom and TAGRO have helped to regenerate damaged land with depleted topsoil. They have controlled erosion. When applied to forested land they have increased the growth rate of timber, leading to a faster harvest. King County has been conducting agricultural research in eastern Washington for more than twenty-five years in collaboration with three universities and two generations of a farming family. Mihle recalled walking across a series of experimental plots after a spring snowstorm and marveling at the difference. Plots with synthetic fertilizer tend to be harder and crunchier than untreated soil, she said, while those with Loop feel spongier. "It's so weird how easy it is to see and feel," she said. With the new snow, the difference in the three experimental plots was even more apparent. A camera showed that the snow melted fastest on the plot with biosolids, perhaps because the soil had been warmed by increased microbial activity, she said. The results prompted the county and its research collaborators to conduct a new line of research on how Loop might be altering the soil's microbial communities.

By the end of June 2020, our own garden was thriving. We harvested sweet sugar snap peas that we ate from the vine and starchier purple pod shelling peas that stained our teeth and tongues. So many peas—I joked that the vines were like a clown car of vegetables. I learned that the radish greens were nearly as good as the surprisingly tasty roots, and by snipping just a few outer leaves at a time from the lettuce and romaine plants, new leaves would keep appearing from the middle as the plants grew taller.

Several Eastern cottontail rabbits discovered the spinach and parsley but left the lettuces alone, and I brought multiple bags to neighbors and a local food bank. The kale began to take off and our tomato vines reached nearly four feet in height. Our peppers, celery, and fennel looked encouraging, and our neighbors said their plants were flourishing as well. Neighborhood kids stopped by to nibble chive flowers and chocolate peppermint, and braver ones tried the pea flowers.

All around us, we began to see how our neighborhood's flora and fauna were beginning to change. The parsley mowed down by the rabbits came back with a vengeance, an elegant example of compensatory growth that I hadn't anticipated. Our arugula and endive had begun to flower and seemed like bitter mistakes until I saw honeybees and bumblebees swarming their light purple, yellow, and white blossoms. The unruly bunch of greens we had nearly uprooted became as much of a bee magnet as the lavender and echinacea we had planted specifically for them.

We began eating salads nearly every day and discovered that Black Prince tomatoes were the best we'd ever tasted. We talked about where our food was coming from and what we liked and what we might do differently. Neighbors donated dahlia bulbs and sword ferns and gathered to talk or have a drink and unwind on our garden wall and stairs. In the midst of a pandemic, our yard had become a living room, a kitchen, and a gratifying link to the neighborhood around us.

● ● ●

Farmers and gardeners who have seen the advantages of Loop and TAGRO firsthand tend to be its most ardent admirers. Mihle and her colleagues call them the "poop champions," at least internally. "I think there's been a huge shift in public perception. It's one of the things I'm most optimistic about," she said. Likewise, Brown said the people she talks to seem to increasingly understand that human waste can be a resource. Some, of course, still don't want to hear or talk about it. "And you know what? There are three times as many people that are like, 'Ewww, really? OK, I'll try it!'" she said.

Not everyone is a fan. A 2019 article in *The Guardian*, for instance, blasted the use of "toxic sewage sludge" as a public health–jeopardizing and money-making scheme in which the waste management industry had resorted to "repackaging the sludge as fertilizer and injecting it into the nation's food chain." It's a depressing thought: we're so hopelessly contaminated that what comes out of us must now be burned or buried instead of reused. But what such accounts leave out, beyond the research suggesting that properly treated biosolids can help *detoxify* contaminated land, is the reality that the chemicals of concern, ones we've created, are nearly everywhere. In our bodies and in the soil, yes, but also in the water, and the air, and the toothpaste and weed killers and preservatives and nonstick pans and odor-reducing socks we use every day.

Some recent reports suggest that "forever chemicals," the giant family of poly- and perfluoroalkyl substances (PFASs) used in products like firefighting foams and pans and dental floss, have infiltrated aquifers, wells, and biosolids used as fertilizers. The high levels in some biosolids, potentially linked to industrial sources that contaminated the waste through combined sewer systems, have led to more scrutiny and efforts to restrict the use of biosolids as fertilizers. The "forever chemicals," though, have also appeared in some compost made with food packaging, prompting calls to ban their original sources, like the coating on some compostable paper products.

Minimizing the long-term risks will require thoughtful conversations about whether and how we can shift away from using toxic substances—and test more of them to understand their consequences. We may need to think about diverting industrial waste for separate treatment as a matter of long-term safety instead of comingling it with our poop as a matter of short-term efficiency. Some of our soils and biosolids are contaminated precisely because of their proximity to industrial sites that have produced dangerous compounds. Turning off these chemical spigots and cleaning up the primary sources of pollution, no easy task, will help minimize their continued spread. Due to the potential risks, biosolids are tested more often than other sources of compost and fertilizer for the presence of metals and chemicals, meaning that we're more likely to hear about them because they've become de facto environmental indicators. "I mean, we have decades and decades of data," Mihle told me. Some of the experiments, calculations, and risk assessments have been used to set quality and safety thresholds for other amendments, like food waste compost and cow manure.

For biosolids used as fertilizer or compost, the US EPA has established safety limits for nine metals while the European Union regulates six (where they've been imposed, European standards are more stringent except for a curiously high threshold for lead). Like other biosolids, Loop is tested every month to make sure its metal concentrations pass muster; so far, they have easily done so. King County tests Loop for nine additional metals as well, and for a select range of fecal-associated bacterial pathogens, parasites, and viruses.

A separate risk analysis commissioned by the county summarizes the number of years that a farmer, gardener, hiker, or child in the region would have to work with or play in biosolids or compost made with them to equal routine exposures to eleven pharmaceuticals and other chemicals in personal care products like hand soap. A typical farmer working with biosolids, for example, might ingest the equivalent of one tablet of the antibiotic azithromycin after 23,309 years; a gardener would do so after 965,819 years. The same analysis found no evidence that wheat fertilized with biosolids had taken up any of the eleven compounds. With growing concern about the ubiquity of microplastics in the environment, King County and other treatment plant operators have begun testing for them as well.

Research by Brown and other scientists has helped assuage concerns over soil contamination in communities like Tacoma. For nearly a century, a smelter in the city separated copper from mineral deposits and contaminated the surrounding soils with lead and arsenic. In acidic environments, to which the smelting waste contributed, lead releases more easily from soil particles and becomes more mobile and available to plants. Biosolids raise the soil pH and add the metal-binding molecules that can help keep it locked away. The added phosphorus, for its part, can selectively outcompete toxic arsenic for preferential uptake by plants. Research has suggested that like Loop, Tacoma's TAGRO is low enough in heavy metals that it dilutes their concentration in urban soils. As part of his doctoral thesis, a Kansas State University agronomy student who examined arsenic- and lead-contaminated soils at two gardens in Tacoma and Seattle concluded that treating them with biosolids could help *increase* gardening safety.

Wastewater treatment agencies have learned to pick their battles. Some people, they've found, will never be swayed by facts. Other initial skeptics, though, have been won over by seeing the results for themselves. On a warm, early October afternoon in 2020, a mix of fog and smoke from distant wildfires had just lifted when I met Kristen McIvor at a community garden in Tacoma. McIvor sported a cheery yellow face mask festooned with bees, which matched the dozen or so groupings of sunflowers scattered throughout the one-acre garden. Other dots of color still decorated the plot as well: dark purple grapes and dahlias, pink tickseed, deep red tomatoes, vibrant green onion leaves, and other vegetables that I couldn't immediately identify. The residents who tend this garden, adjacent to the city's Swan Creek Park and Salishan neighborhood of mixed-income homes, are among the most diverse in the county. They hail primarily from Southeast Asia, Eastern Europe, and northern Africa and speak seven languages among them, McIvor said. It seemed more than a little fitting that the soil that bound them all together here was something that they had all personally contributed to.

For gardening and farming, TAGRO has been developed into several blends. One of the most popular is a potting soil mix that includes biosolids, sawdust, and aged bark. As a graduate student in Brown's lab in the aughts, McIvor was interested in urban food production and food sharing, and focused her research on using the soil product to create a community gardening program in Tacoma. Once she began working with the organic matter, she had a revelation: "Oh my gosh, this is the best thing ever when it comes to gardening in areas that have poor soil, or anywhere really. It's like gardening for dummies." The potting soil worked so well that McIvor and her colleagues built instant gardens atop gravel parking lots. There's even a TAGRO blend for green roofs, and another used by some growers in the state's marijuana industry, a veritable pot-to-pot medium.

As her academic work progressed, McIvor worked at the TAGRO facility and started organizing in support of policy changes to improve community health. In 2010, the government-supported Pierce Conservation District launched a new program called Harvest Pierce County, with McIvor as its director. A decade later, she had overseen the opening of more than eighty community gardens and a food forest. Each is managed by the gardeners themselves, who come up with their own rules and charter. The organization has been careful to offer options for its free soil deliveries: either TAGRO or a more traditional compost made from yard waste. Only a handful have opted for the latter, McIvor said; many hesitant gardeners were quickly won over by demonstrations showing how quickly a neglected urban lot could become a bountiful vegetable garden.

Since 2012, Harvest Pierce County's Share the Harvest Project has distributed more than 140,000 pounds of garden produce to food banks. Some of the more ambitious sites, including a gardening group that has drawn heavily from the US military's Joint Base Lewis-McChord and a former Intel campus, have experimented with greenhouses, solar-heated hot water, smartphone apps, and other methods to get a jump on the growing season and maximize their donations. "Most people when they talk to me about what they like about their community garden, it's not about tomatoes or having more vegetables," McIvor said. Some are indeed seeking more food security for their own families, but most want to supplement their diet while doing something fun and interesting and getting to know their neighbors. Healthy soil, in other words, can build communities both in the ground and above it. Some friction is understandable when diverse groups work together, but gardening has provided valuable lessons in democratic decision making, however messy, for groups that are unaccustomed to it. And being outside and physically active, connected to both food and to nature, she said, can be incredibly healing.

For some gardeners, beloved plots have become essential parts of their lives. When a community garden at Tacoma Community College was threatened by a construction project in 2015, a columnist at the *Tacoma News-Tribune* filmed a video segment that captured the distress of Roza Nichiporuk, a gardener who tended her plot daily. When asked about her reaction to the news, an interpreter and fellow gardener replied on Nichiporuk's behalf: "She thought she get, have heart attack. So upset. She didn't sleep three nights."

"What will you do if you can't have a garden?"

"She says, 'I will die.'"

The journalist interviewing them, incredulous, laughed. "No, no, that's not true!"

The two women smiled. But the interpreter conveyed Nichiporuk's profound attachment: "She is here, two, three times a day. I see her morning, evening, and sometimes day. She works hard. She loves. It's her life." McIvor and Harvest Pierce County were able to intervene and develop a governance structure to ensure that the garden remained intact.

The program has excelled at forming partnerships, including with the Tacoma-Pierce County Health Department's Eastside Family Support Center, whose clients are mainly Latino. McIvor and I stopped at a small adjacent community garden. Most of the crops there had been harvested, but I spotted some corn, tomatillos, and squash. She pointed out an unused pile of TAGRO, and we both stuck our hands in it. Like slightly warm, loose dirt. A faint ammonia smell; most of it had dispersed with time.

McIvor had something else to show me. From the success of the small community garden, the center's director and a core group of gardeners began brainstorming uses for a long and narrow vacant lot down the hill, a fenced-off acre and a half on the other side of a bare-bones park and past a mass of invasive Himalayan blackberries and spiderwebs. On what was once Puyallup tribal land that was now owned by Tacoma Public Schools, gardeners from the Puyallup tribe were partnering with a group of Purépecha people from Mexico. The Indigenous groups wanted a space for people to learn about precolonial foods and traditional medicines and plants. A place to teach those Purépecha and Lushootseed words to their children, including the ones at the Puyallup tribe's Grandview Early Learning Center on the other side of the lot. From the ground up, they were replanting their words and wisdom and values and culture. "I feel like this garden is the culmination of everything we've learned over the past ten years," McIvor said.

Six gardeners from each group had jointly enrolled in a traditional medicine course and had just finished their first class together when the pandemic struck, slowing the project's momentum. Even so, we saw two different kinds of squash grown by one of the Purépecha gardeners. *Purhu*, they're called. And we saw patches of vivid orange where he was growing marigolds—*apátsicua*—in anticipation of Day of the Dead. During the Mexican holiday, families remember and venerate their dearly departed, with the marigold's bright petals and fragrance luring the souls back to visit their relatives. The gardening site itself had been blessed before the gardeners broke ground, though both groups agreed that another cleansing ritual was necessary to clear away the negative energy that had accumulated in the lot over the decades.

Other projects were pressing onward as well. Despite the budget crunch triggered by the COVID-19 pandemic, Mihle said King County was hoping to launch a small-scale composting pilot at one of its three treatment plants, pending its ability to meet odor control and other air-quality standards. If it took off, Mihle said she could envision future partnerships with smaller treatment facilities that could bring their own biosolids to a regional facility for conversion into garden-safe compost instead of shipping it to faraway landfills. Locally produced compost, Mihle mused, could grow equity and social justice as well by helping to fill in urban food deserts and expand green spaces.

Repurposed waste clearly won't solve everything, but I marveled at how it had been a catalyst that brought so many people together. In our own yard, Geoff and I bickered over inconsequential things like how to trim the ferns and whether to plant more peppers and *Oh my God, why are you putting that there?* We were humbled by our gardening failures and cried at the mounting toll of a mentally exhausting year. On Election Day in 2020, not knowing what else to do, I began planting the first of more than one hundred tulips.

But good shit makes things grow. And in the spring, the tulips put on a stunning show for the neighborhood: pink and lavender and deep purple and double pink-and-white that played off the wallflowers and azaleas and Japanese andromedas with their delicate spring leaves. Nearly everything, it seemed, was growing like a weed, including the actual weeds that were overtaking some of the vegetable beds.

Maybe we would scale back just a bit on the food crops, which was just as well given our new project in a sloping section of the side yard beneath a red cedar and two Douglas firs. The soil had badly eroded over time, leaving a barren patch surrounded by a few native plants that we had tried to nurture and by more invasive plants that we were continually beating back. After leveling off parts of the slope and bringing in better soil and compost, we set to work picking out more native plants to fill in the space. Evergreen huckleberry and Cascade Oregon grape. Red-flowering current. Nodding onion and Oregon iris and Western spirea and false Solomon's seal.

This time, I knew the perfect compost to give our new plants a healthy boost. To get some TAGRO, though, I would have to hurry. The distribution center was closing soon for the workers' lunch break, and I wanted to load up on bags of the slightly pungent but potent plant food ahead of a weekend of gardening. And so I hopped into my Subaru and set out in search of someone else's recycled poop.

CHAPTER TEN

Bounty

I CELEBRATED A MAKESHIFT
Oktoberfest at a neighbor's picnic table on a sunny September
evening. They contributed a plate of
green olives, Manchego cheese, and
liver paté, and I contributed a cold four-
pack, judging scorecards, and pretzels as
a palate-cleanser. The light blue beer cans I had
brought for the occasion, each prominently labeled "Pure Water Brew,"
held sixteen ounces of handcrafted ale made for a unique brewing competition. "Quality not History," the cans read. And "Dam good" next to
the illustration of a beaver. The beers were indeed good, with an American pale ale edging out a light hybrid as the winner of our decidedly
amateur scoring.

The evening taste test might have been unremarkable if not for the
fact that we were participating in an urgent water conservation effort.
Additional text on each can explained that the beer had been brewed
with "100% *pure* RECYCLED H20." Some fine print running up the side
offered additional context: "(1) All water has been consumed before and
will be consumed again. (2) Made with the purest water on the planet.
(3) Clean water is closer than you think."

In this case, the clean water had come from sewage effluent treated
by Oregon's Clean Water Services utility and then run through a separate three-step purification process. The demonstration project's recycled water is so clean that brewers add minerals back to replicate the
water of famous brewing cities: a Burton-on-Trent profile for a dark
Burton ale, say, or the water of Munich for a brown dunkel lager or the
profile of the Czech Republic's Plzeň for a pale Pilsner. "Brewers are
water quality experts, and they really couldn't believe that they could

have this blank slate," Mark Jockers, the utility's government and public affairs manager, told me. Initially, water regulators couldn't quite believe it either. When it launched in 2014, the brewing project failed to secure permission from Oregon's Department of Environmental Quality to use purified wastewater as its starting material. Could the project instead purify water from the Tualatin River immediately downstream of the treatment plant's effluent discharge site? "Now the regulator said, 'Oh, sure! That's fine,'" Jockers recalled. "We call it the waters of amnesia: once you take effluent—cleaned wastewater—and you put it in a body of water, it magically becomes water again. Whereas if you didn't put it in the body of water, then, oh my God, it's like nuclear waste, right? People are so anxious about this stuff."

So are animated reindeer, apparently. In one of our household's favorite Disney movies, *Frozen II*, Sven the reindeer is drinking from a stream when Olaf the snowman shares one of his many bits of trivia, though this time he pairs a scientific myth with a truth. First the myth: "Water has memory." It does *not*, in fact. Water is agnostic as to what was previously in it or what it was previously in. That's good because the second part of Olaf's wisdom-dispensing is widely considered true: "The water that makes up you and me has passed through at least four humans and/or animals before us." Sven then promptly spits out the water.

Estimating how many living things a water molecule has passed through is a bit tricky, but research suggests that yes, the water within us has been in multiple other living things in the past. But because it has no memory, the slate can be wiped clean with every reuse, leaving nothing more than two hydrogen atoms to every one oxygen atom in H_2O. Scientists believe that all water on Earth was originally seeded from cosmic gas clouds, space dust, or meteorites that delivered the necessary hydrogen and oxygen billions of years ago. Since then, the planet's water cycle has been a continuous closed loop of use and reuse.

Charles Fishman, author of *The Big Thirst*, explained the take-home message in an interview on National Public Radio:

> So all the water on Earth—the water in your Evian bottle, the water in your glass of water, the water you use to boil a pot of spaghetti—all

that water is 4.3 or 4.4 billion years old. No water's being created on Earth. No water's being destroyed on Earth. And what that means is the whole debate about reusing wastewater is kind of silly, because all the water we've got right now has been used over and over again. Every drink of water you take, every pot of coffee you make is dinosaur pee, because it's all been through the kidneys of a *Tyrannosaurus rex* or an *Apatosaurus* many, many times, because all the water we have is all the water we have ever had.

On a shorter timescale, downstream cities routinely reuse river water that has passed through the kidneys of their upstream neighbors before being treated and released back into the river.

Those of us who are lucky enough to drink clean tap water aren't normally forced to consider where it's been and contend with our own disgust. We don't have to, just as we don't normally consider how many of the bacterial and yeast strains we depend upon to ferment foods and beverages have also passed through the intestinal tracts of our ancestors or other animals. Like the other things we expel, they too can be reused. Scientists and engineers, in fact, are borrowing a page from nature to demonstrate how we can safely reclaim drinking water and the starter cultures for fermented and probiotic foods. Purifying and repurposing the by-products of what we drink and eat could help ease potable water shortages worsened by global warming and pollution and address nutritional deficiencies worsened by poor sanitation and food insecurity. The main limitation to doing both isn't technical, but as Sven the reindeer suggested, psychological.

Consider the curious case of a new kind of fuet, a Mediterranean-style sausage made from cured pork and fat. Professional tasters agreed that a specialty version handcrafted by researchers at the Institute of Agrifood Research and Technology in Girona, Spain, was especially tasty. Yet the fuet didn't attract a single company interested in commercializing it. The problem, it seems, derives from the provenance of the bacteria used to ferment the meat and give it its zesty tang: baby poop. The study, "Characterization of Lactic Acid Bacteria Isolated from Infant Faeces as Potential Probiotic Starter Cultures for Fermented Sausages," certainly attracted notice, if not eager sausage-makers.

Why use such an unusual source for the fermentation process? Beyond its profusion of biological markers, baby poop is unusually rich in the probiotic *Lactobacillus* and *Bifidobacterium* species more widely associated with yogurt. Researchers at the institute isolated and grew the bacteria from forty-three healthy infants' soiled diapers and tried a variety of curing combinations to see which species might work the best. The microbes then feasted on carbohydrates added to the pork, producing the preservative lactic acid as a by-product. This fermentation process, the researchers reasoned, could turn the sausage into another probiotic food. To work, though, the microbes would need to survive the acid trip down the human intestinal tract. So the food scientists focused on species that are already present in our intestines and relatively easy to get (and baby poop is, as we know, readily abundant). If the end product wasn't exactly a hit, the process helped confirm the intrinsic value of our inner fermenters.

Based on the same rationale, researchers at Wake Forest University in North Carolina developed a probiotic "cocktail" composed of ten *Lactobacillus* and *Enterococcus* strains that they had likewise collected from infants' dirty diapers. When fed to mice, the baby-bottom mixture raised the rodents' production of short-chain fatty acids, which are critical for a healthy gut. The science of probiotics is still in its, well, infancy, and many health claims have withered under closer inspection. But some of the emerging results suggest that bacteria derived from healthy infants could be used as probiotics to aid adults—potentially opening the door to an entirely new line of specialty food.

When we hear about how water and microbial cultures can be recycled from our poop, though, we run up against the same cognitive dissonance that prevents us from reimagining how a substance long associated with disease could be a medicine or a soil amendment. As with food, the line of questions about what form of reclaimed water we might accept becomes a kind of Dr. Seussian exploration of disgust, Jockers said. "Would you drink it in a beer? Would you drink it in a vodka tonic? Would you drink it as water? Would you drink it in baby formula?" Even a scientist might hesitate on the final question, he said, despite knowing intellectually that it's the same water. Multiple municipalities have struggled with the third question, as vocal "toilet to tap" opponents of water reuse projects—including Miller Brewing Company, ironically—harnessed the power of disgust and stigma to drown out explanations of the benefits and assurances of safety. In Australia, opponents called it "sipping sewage."

The blowback caught advocates off guard and scuttled multiple reclamation projects in the 1990s and 2000s. Plans failed in California's San Gabriel Valley, where a Miller brewery drew its water downstream of a proposed groundwater replenishment system. "They didn't want the potential stigma of someone using it against them that their beer was being made out of toilet water," Mike Markus, general manager of the Orange County Water District, told me. Despite a revision that would recharge the groundwater *downstream* of the brewery, the plans still failed. In Los Angeles, the East Valley Water Recycling Project became a political football during the 2001 mayoral race. A pipeline and plant had already been built to send treated water to percolate through recharge basins in the San Fernando Valley as potable water; after only a few days of operation, though, the political pile-on brought the $55 million project to an abrupt halt. In San Diego, some opponents falsely claimed that questionable water would be sent primarily to disadvantaged communities, helping to a kill a recycling project there too.

Clean Water Services initially focused on the first question: Would you drink it in a beer? Unlike Miller, the utility bet that people would say yes. They did, though Jockers noted that a carefully considered public relations strategy was critical to its success. Although the novelty aspect of reusing water for beer garnered plenty of initial publicity, the utility claimed naming rights. Pure Water Brew became an umbrella term for the beer after the utility nixed some more colorful suggestions like Loose Stool Pale Ale, I Pee IPA, and Brown Trout Ale.

The popularity of adult beverages, as I learned, may be a secret weapon for breaking the mental blockade. Using the recycled water for beer and liquor means that children aren't drinking it, and the brewing and distillation processes both kill pathogens. In Stockholm, brewers made PU:REST Pilsner as an attention-grabbing novelty. In Berlin, they brewed Reuse Beer. And in San Diego, a brewery came out with Full Circle Pale Ale. By early 2015, Oregon's Clean Water Services had secured a state permit that allowed it to distribute its reclaimed and highly purified effluent to members of a home-brewing association called the

Oregon Brew Crew. The program was a hit and by 2020, the utility and multiple partners had planned to take another big step by distributing reclaimed water to twelve commercial brewers to create their own small-batch beers as part of Portland's annual Oregon Brewers Festival. COVID-19 scuttled the event, so the Oregon Brew Crew held another homebrewers' competition instead, and the four-pack that my neighbors and I taste-tested was a leftover from the contest.

Jenn McPoland, who crafted the winning American pale ale (and our unofficial Oktoberfest champion), met her husband at an Oregon Brew Crew meeting in 2006. They had since outfitted their garage in Portland with a ten-gallon brewing system, a walk-in cooler, and eleven taps. "Neighbors really like us," she told me, laughing. McPoland learned about the Pure Water Brew reuse project through the homebrewing club and has entered the contest nearly every year since its inception. "Ultimately, it was all dinosaur piss at one time, right?" she said. That factoid has evidently struck a chord.

Maybe it was due to the circles they travel in, but McPoland said neither she nor her husband had received any pushback on the brews they created from the reclaimed water. "For the most part, when we've talked about it, people get excited about what we're doing," she said. "Everybody likes beer." In a city nicknamed "Beervana," it's hard to argue her point, and it helps that she and her husband have wowed friends and judges with concoctions like a rhubarb Berliner Weisse and a Mexican chocolate porter. For her winning 2020 entry, she started with the water profile of Munich and then experimented with adding more salts to "kick up" the hops. She called it Salty Bitch Pale.

McPoland told me she thought it was a brilliant idea to have homebrewers make beer with the recycled water because of the chance to strike up a conversation that wouldn't have happened otherwise. "More than anything else, people are curious," she said. Feeding that curiosity—and perhaps a hankering for a pale ale with a nice finish—creates a unique opportunity to educate a fellow beer enthusiast about the bigger picture. To get us past the yuck factor, "humans need an experience," agreed Diane Taniguchi-Dennis, CEO of Clean Water Services. "You can talk to them, and you can try to educate them, but the action of drinking the water or having a safe conversation over a beer has some cachet. In fact, when we go to countries where we're worried about the water, what do we do? We drink the beer, right?" We've been doing so for thousands of years: just as fermentation preserves food, it can preserve water as beer or wine.

Beer made with what began as human waste, in fact, may have at least one historical precedent from the seventh century CE in what is now Ecuador. In tombs unearthed by the Quito International Airport, archaeologists found offerings for the afterlife that included earthenware jars once filled with chicha made from fermented corn. Remarkably, biologist Javier Carvajal Barriga resurrected viable yeast by scraping the insides of the jars and coaxing the cells back to life. None of the revived 1,350-year-old cells were common brewer's yeast. Instead, he discovered that they were most closely related to *Candida* yeast strains found in human saliva and feces and more commonly associated with infections. Human-derived yeast, in other words, may have helped jump-start the ancient fermentation process for chicha before it was killed off by the rising alcohol concentration. In a further twist, the yeast from Ecuador is identical to a strain found in Taiwan. Human-associated yeasts from archaeological sites around the world could be yet another marker of human migration, in this case providing more evidence for a potential trans-Pacific route between Polynesia and South America.

It's one thing to learn about our coevolution with fermenting microbes or the cyclical rebirth of pure water, though, and another to experience the food or water in person. Taniguchi-Dennis described how reuse demonstrations at water industry meetings have featured artisanal vodka and an ice sculpture of a leaping salmon—both made with reclaimed water. If Clean Water Services wins over regulators by demonstrating its purification equipment's ability to consistently meet safety standards, Jockers said, the demonstration project could expand to other specialty beverages like coffee and kombucha.

The project has expanded its geographic reach as well, given that the entire purification system is portable. At the utility's facility in Forest Grove, Oregon, operations analyst AJ Johns gave me a tour of his "baby," a small bus-size unit called the Pure Water Wagon. Since its debut in

2019, the "cleanest water on wheels" has toured multiple cities; at each stop, a door and large awning windows on one side of the trailer open up to reveal its three-step purification system. On the back of the trailer, a beaver seems to sum up the message for the Svens of the world: "Water: too precious to use just once."

Johns, wearing a blue Pure Water Brew T-shirt and a green John Deere baseball cap, talked me through the process above the thrum of pumps and motors. From the highly treated wastewater effluent we learned about back in chapter 8, a small portion is diverted from its normal route to the Tualatin River and instead piped through a sand filter to remove some of the remaining particles. Then it's channeled into the Pure Water Wagon and sent to a cylindrical Plexiglas holding tank. A pump pushes the brownish-gray water, reminiscent of what you might find in a pond, through a narrower side cylinder packed with hollow fibers, each resembling a thick strand of hair. Tiny pores in each fiber trap microbes and larger molecules while allowing water molecules to be sucked up like soda through a straw. "What it's doing is called size exclusion: it's basically just blocking anything that's bigger than the size of the pores," Johns said. "So it will remove particulates, bacteria, protozoa, and viruses." The unit uses the same technology as water filters for backpackers, "just kind of industrial scale."

The filtered water, now a dull yellow, flows to the next purification step in a small cluster of white columns filled with reverse osmosis membranes. During osmosis, water molecules pass through a partially permeable membrane to equalize the concentrations of solutions on either side. Science classes often demonstrate the concept with food coloring: while the dye is initially concentrated on one side of the membrane, osmosis eventually balances the color on both sides. Reverse osmosis instead adds pressure to deliberately *create* an imbalance: high-pressure pumps force water molecules through to the other side of the membrane while protozoa, bacteria, viruses, larger salts, and organic compounds like pharmaceuticals remain trapped on the near side in what becomes an increasingly concentrated solution. Reverse osmosis also effectively removes the PFAS family of "forever chemicals" that have become ubiquitous in groundwater, well water, and our own bodily fluids (other potable water reuse projects have used ozone and activated carbon to remove larger PFAS compounds and other contaminants).

The same general concept can work for desalinating seawater, though its high salt content means that the reverse osmosis process normally requires intense pressure and thus considerable energy to keep the brine on one side of the membrane and salt-free water on the other. Beyond the challenges of preventing marine life from being sucked into the intake pipes and properly disposing of the concentrated brine, existing desalination plants use up to threefold more energy than water reclamation plants to recover the same amount of water.

For its final treatment step in the Pure Water Wagon, the nearly clear water courses through a horizontal tube filled with two synergistic purifiers. The first, ultraviolet light, penetrates bacterial membranes and viral capsids and destroys their inner DNA to prevent any pathogens from replicating. The light rips apart chemical bonds and effectively dismantles small organic molecules like 1,4-dioxane, a likely carcinogen found in industrial solvents, and NDMA, a probable carcinogen found in burnt foods like grilled meat and roasted coffee and once used in liquid rocket fuel. And in its coup de grâce, UV light bombards hydrogen peroxide, the second purifier, to create highly reactive molecules called hydroxyl radicals that attack and degrade other organic compounds. "By the time it comes out, we have removed any sort of organics, heavy metals that might have been there, and pathogens," Johns said.

The purified water, crystal clear, was gushing through a hose and over the edge of a white pail at a rate of about five gallons per minute, or up to 7,200 gallons every day if you don't count the periodic downtime for cleaning and maintenance. "As we like to say, that's the purest water on the planet, right there," Jockers said. It's so pure, in fact, that it's not an ideal source of drinking water on its own. Under the principles of osmosis, drinking water that has been stripped of everything is more dilute than the slightly salty water in your cells. To regain a balance between the extracellular and intracellular concentrations, water flows into your cells and dilutes out the electrolytes; like distilled water, in other words, it can effectively pull some minerals from your body. Does the recycled water need to be so pure? The process might be overkill, Jockers conceded, but getting regulatory agencies and the public comfortable with water reclamation has required utilities to address both safety and psychology.

In 2016, microbiologist Ian Pepper and colleagues at the University of Arizona teamed up with multiple public and private groups to win the Water Innovation Challenge sponsored by the nonprofit Arizona Community Foundation. As part of their winning entry, the team built their own portable water treatment system and toured it throughout the state on the back of a semi. Similar to Oregon's Pure Water Brew project, the researchers gave the recycled water to local microbreweries and hosted a beer-tasting contest at a national conference. With a certification from the Arizona Department of Environmental Quality attesting that the water was free from pathogens and other contaminants, the team also gave out samples of bottled water. Beer proved to be much more popular. "You can treat water and make it purified, but then you ask someone, do they want to drink the water? No, they don't," Pepper told me. "But if you use it to make beer and you ask someone, 'Do you want some free beer?' the answer is yes."

Eventual acceptance may hinge on three main factors: how the water is described, how much people know about its treatment, and how much they trust the water's purveyor. Australian researchers, for instance, found that people who were given information about the production process for desalinated or recycled water were significantly more likely to be willing to use it. Early advice on choosing the right words came from a water industry–commissioned study by consultants Linda Macpherson and Paul Slovic. Beyond outright disgust, Slovic and other researchers had previously shown that stigma can flow from the subtler feelings of affect, or a "faint whisper of emotion." Whereas stigmatizing words like wastewater and sewage tended to deter water reuse acceptance, Macpherson and Slovic found, terms like "pure" tended to enhance it. In her presentation of their results, Macpherson chastised the industry for using educational materials that were "equivalent to the fine print of an insurance policy." Interactive tours and fun learning experiences could help sway the undecided, she added, but a minority who are really bothered by water reuse tend not to change their minds even with more information. Just like those dead set against the reuse of biosolids for compost and fertilizers, it seems, some water reclamation opponents will *never* be convinced.

Paul Rozin, the disgust expert who conducted the "doggie doo" experiment in toddlers, has examined the attitudes of US adults toward water reuse as well. After a brief description, his study asked about 2,700 adults waiting in train stations or other public spaces in five major US cities, "Would you be willing to drink certified safe recycled water?" Thirteen percent of the respondents said they wouldn't, 49 percent said they would, and 38 percent were uncertain (though presumably persuadable). The survey then took the participants through its own Seussian progression of questions assessing their willingness to try fourteen different kinds of water, ranging from raw sewage to bottled spring water. The near universal rejection of the former and acceptance of the latter was unsurprising, given what we've learned about our behavioral immune system's focus on avoiding perceived danger. Among the 13 percent who rejected certified safe recycled water, though, even "sewage water that has been boiled enough to destroy all microbes, and then is evaporated, and then condensed and collected as pure water" failed to sway the vast majority.

A psychological concept called "the magical law of contagion" holds that whenever two things come into contact, they're forever linked in someone's mind. In other words, Rozin and his colleagues explain, "anything that touches something disgusting becomes disgusting." A contaminant viewed as a "material contagion" can be neutralized by washing or purifying whatever it touched, but a "spiritual contagion" can leave an indelible stain no matter how hard you scrub. For a minority of people, Rozin and his colleagues conclude, wastewater contamination may behave like a spiritual contagion: no amount of persuasion or purification will ever convince them to try it.

At a certain point, even die-hard opponents may be forced to come around. Of all water on Earth, less than 3 percent is in the form of freshwater—and much of that is locked within glaciers and polar ice caps. In regions like sub-Saharan Africa, the Middle East, Australia, and the US Southwest, recent drought cycles worsened by global warming have heightened concerns over the future of the dwindling freshwater supply. Within the next century, a study in *Nature Climate Change* estimated, an eye-opening 44 percent of the world's aquifers will be endangered by climate change–wrought declines in rainfall. The necessity of importing from other regions has already pitted city against city and fueled a sense of urgency, especially around the issue of equitable distribution from over-tapped rivers like the Colorado and the Rio Grande in the United States.

In that context, reusing wastewater has become a matter of economic growth and local autonomy. Cities may not be able to control the amount of rain that falls or snow that melts into rivers and reservoirs, but they can always generate wastewater. If pejoratives and pushback have killed past water reclamation projects, sheer necessity may be helping to revive them. "Well, the thing that overcomes the yuck factor more than anything, you know what that is? Shortage of water," Pepper said.

• • •

Water is essential for life. Fermented and probiotic foods may be helpful, but how necessary are they? One of the big questions in the field of nutritional sciences, fermentation expert Robert Hutkins told me, is whether probiotic bacteria are important constituents of the microbiome—or should be—beyond early childhood or merely temporary residents after a meal of yogurt or kimchi or kombucha. "As you get older, you start to lose some of those *Bifidobacteria*, and they're a relatively minor component of your microbiota," he said.

One argument suggests that we simply outgrow the need for the microbes; another suggests that the bacterial depletion is evidence of a Western diet and lifestyle. Adults who regularly eat fermented foods that are rich in lactic acid bacteria shed them in their poop as well, suggesting that the microbes grow and multiply in the gut. But after a gap of three or four days, evidence of their existence almost completely disappears. "So they'll wash out pretty quick. They're not persistent. They don't establish residency," Hutkins said. Even so, the "near-commensal" bacteria, which he compares to long-term visitors, can confer benefits through their metabolites, proteins, and influence on other microbes. In my case, the lactic acid bacteria *Streptococcus thermophilus* from my daily yogurt did indeed show up in my poop, though *Bifidobacterium lactis* and other species from my daily probiotic supplement did not.

Hutkins and colleagues have traced the lineages of multiple microorganisms used in traditional fermented foods and concluded that many likely derived from the guts of our ancestors or those of other animals. Perhaps an animal pooped near a cabbage plant and some of its microbes adapted to that new niche and helped convert the cabbage into sauerkraut. *Streptococcus thermophilus*, for instance, is a close relative of strep throat–causing *Streptococcus pyogenes* and pneumonia-causing *Streptococcus pneumoniae* but seems to have lost its own virulence genes as it slowly adapted to living in milk.

Scientific papers touting the benefits of certain gut microbes in warding off physical or mental ailments date back more than a century. But early proponents were saddled with the baggage of autointoxication as a vague catch-all term for intestinal poisoning that initially focused on constipation as the main culprit and spawned the raft of pseudo-scientific cure-alls we saw back in chapter 5. After Pasteur's germ theory of disease provided a better explanation for that poisoning, though, researchers began suggesting other ways in which the contents of the gut might influence the rest of the body. In the early 1900s, French pediatrician Henry Tissier reported using "good bacteria" as a successful therapeutic, which he collected and purified from the poop of healthy breastfeeding infants and then gave by the spoonful to infants with diarrhea. Tissier linked their recovery to the restoration of normal gut flora. One of the "good" microbes he isolated from breastfeeding babies is now called *Bifidobacterium bifidum*—one of the most common probiotics in the human body (though less so in adults).

Russian microbiologist Élie Metchnikoff was another avid investigator of our intestinal microbes. Metchnikoff was a cowinner of the 1908 Nobel Prize in Physiology or Medicine for discovering the immune defense and maintenance system known as phagocytosis in which specialist immune cells devour bacteria and debris. He also hypothesized that age-related decline could be caused by poisoning from certain gut microbes, a more modern spin on the concept of autointoxication. In this case, though, he posited that the infection or poisoning provoked the housekeeping phagocyte cells to speed the deterioration of tissue. "As in his theory of immunity, his beloved phagocytes were yet again granted a star status—but surprisingly, they were no longer heroes but villains," science journalist Luba Vikhanski writes in her biography of Metchnikoff. "Aging appeared to him as a Darwinian struggle of sorts between stronger and weaker cells."

Metchnikoff became a germaphobe, urging his audiences to avoid raw food and instructing them in how to keep it germ-free. But after he was introduced to Bulgarian sour milk and its supposed link to centenarians who drank it regularly, a new idea took hold. He proposed a preventive diet of the fermented milk, made by the lactic acid bacteria *Lactobacillus bulgaricus,* as a way to "fight microbe with microbe" and thwart the harmful gut bacteria he deemed responsible for aging. Metchnikoff reportedly drank the sour milk daily and his diet caught on with the public, at least for a while, but his larger ideas about the healthful benefits of some intestinal microbes took decades to gain traction and accumulate convincing evidence.

Although fermented foods have risen in popularity over the past decade, Hutkins cautioned that few randomized controlled trials have been conducted on ones beyond yogurts and cultured dairy products. While other foods can contain probiotic microbes, the designation is limited to those with living microorganisms that have clearly demonstrated clinical benefits, he said. Most studies have focused on the potential health benefits of yogurt mainly because researchers can more easily control and randomize which bacterial cultures are added or subtracted from the mix. Results so far have suggested some preventive benefits for heart disease, blood pressure, type 2 diabetes, and certain cancers, which Hutkins said he had found rather convincing. Even so, the research hadn't fully distinguished causation from correlation, meaning that the fermenting microbes might be actively driving the health effects or simply along for a ride that's being driven by something else.

Flush

When consumed regularly, though, yogurt has demonstrated another, potentially more profound ability: decreasing the incidence and severity of diarrheal diseases in children, as Tissier's work first suggested. Around the world, diarrhea is the second-leading cause of death in children under the age of five, commonly through severe dehydration and fluid loss or intense bacterial infections. In the Netherlands, researchers began investigating the antidiarrheal potential of a bacterial strain called *Lactobacillus rhamnosus* yoba 2012, a generic version of the well-studied probiotic bacterium *Lactobacillus rhamnosus* GG that was first isolated in 1985 from the feces of a healthy adult. Over beers, appropriately enough, nonprofit cofounders Remco Kort and his high school buddy, Wilbert Sybesma, hatched the idea of producing a probiotic drink for the developing world and settled on *L. rhamnosus* as the best candidate. Working with other scientists, Kort and Sybesma then decided to mix their generic version of the bacterial strain with a second microbe, *Streptococcus thermophilus* C106, isolated from an artisanal Irish cheese and able to efficiently ferment milk into a probiotic yogurt drink. The two-strain starter culture allowed the *L. rhamnosus* strain to grow more efficiently in milk and create a probiotic fermented drink the collaborators called Yoba. The starter culture can also create fermented drinks from soy, wheat, sorghum, millet, maize, and even baobab fruit, which are then sold by dairy co-ops or individual distributors.

The nonprofit, Yoba for Life, has touted Yoba as a cheap and effective way to help ward off childhood respiratory tract infections (like the common cold), balance the gut microbiome, and reduce the intensity and duration of rotavirus-associated diarrhea. Not every analysis has reached the same conclusions about its health benefits, and *L. rhamnosus* has failed some trials; on its own, for example, it has performed no better than a placebo in treating existing C. diff infections. But separate studies have supported its effectiveness against rotavirus diarrhea, irritable bowel syndrome, urinary tract infections, and atopic dermatitis, among other conditions. Since its founding in 2009, Yoba for Life has helped distribute the starter culture to more than ten countries in Africa and Asia. Hutkins said he's not completely unbiased since he has coauthored studies with Yoba for Life–affiliated researchers. But lactic acid bacteria have demonstrated their ability to reduce the foothold of other pathogens such as *Salmonella enteritidis, Listeria monocytogenes,* and *E. coli,* and Hutkins said using a locally made fermented food to improve a population's health strikes him as a great approach.

No single probiotic can do it all. But a ramp-up in research has begun to build a firmer base of evidence, like the finding that other probiotic formulations might do what Yoba cannot and help reduce the risk of diarrhea associated with antibiotics or C. diff infections. A systematic review of more than thirty randomized trials found that giving study participants probiotics along with antibiotics reduced their risk of C. diff–associated diarrhea by an average of 60 percent. A variety of formulations, such as the yeast *Saccharomyces boulardii* or the bacterial strains *Lactobacillus acidophilus* plus *Lactobacillus casei,* yielded a large protective effect. In my case, the research suggests that even though my poop contained only partial evidence of the nonpermanent bacterial strains I was consuming through yogurt and capsules (the latter of which contained both the *L. acidophilus* and *L. casei* strains), they still may have helped me ward off the effects of the C. diff bacteria that had colonized my gut.

Canada's Bio-K+ uses three patented bacterial strains in its own probiotic products, including its original fermented milk drink. French microbiologist François-Marie Luquet originally isolated those lactic acid bacteria from human gut cultures in 1960. "We give them a place to live; they provide us with health benefits," a company blog explains. Perhaps to assuage the squeamish, the blog makes clear that although the bacteria were initially isolated from humans, "we do not collect them from humans, nor do they contain any human by-products." Like other modern producers, Bio-K+ grows its bacterial strains in artificial conditions that approximate their former environment. So what can they do when re-ingested? Lab experiments conducted by company scientists and researchers at Quebec's Institut National de la Recherche Scientifique suggested that the three-strain Bio-K+ concoction was particularly adept at inhibiting C. diff through separate lactic acid–dependent and independent mechanisms. Those results may help explain some of the probiotic's apparent effectiveness in preventing both antibiotic-associated diarrhea and C. diff–associated diarrhea in a randomized, double-blind, placebo-controlled trial among hospitalized patients in Shanghai, China.

Beyond the better-known lactic acid bacteria, researchers are exploring the potential of other "next-generation" probiotics such as *Akkermansia muciniphila*, which my gut apparently lacked but which digests the mucin in intestinal mucus and can strengthen the gut's lining. Another candidate, *Faecalibacterium prausnitzii*, accounted for more than 13 percent of my intestinal community but is depleted in Crohn's disease patients. *Bacteroides fragilis* and *Parabacteroides distasonis*, each of which contributed a bit less than 2 percent to my gut flora, may offer a more complicated Jekyll-and-Hyde tradeoff. While some of their strains have been tagged as opportunistic pathogens, others are emerging as potential probiotics that could be added to food.

Recall the eye-opening study led by Stanford's Erica and Justin Sonnenburg that suggested a diet rich in fermented foods can remodel the gut microbiome to yield considerably higher microbial diversity. On its own, as we've seen, more diversity isn't necessarily a slam-dunk for better health. But the Sonnenburgs' study found that the fermented foods in this case had a kind of calming effect on the body, with less activation of multiple immune cell types and a decrease in multiple markers of inflammation. "Fermented foods may be valuable in countering the decreased microbiome diversity and increased inflammation pervasive in industrialized society," they concluded. One remarkable suggestion of the study is that our own guts, if properly conditioned through a diet of microbially modified food, could produce the agents of our own salvation.

Hutkins, in fact, has mined the fecal samples collected by his lab as potential sources of other probiotic bacteria. One approach, he said, is to feed volunteers prebiotic foods that can nourish their endemic communities of probiotic bacteria "and then mine what comes out the other end." Reintroducing a health-promoting microbe along with its preferred food source, a combination called a synbiotic, could yield the "perfect match" for maximizing its effect on the body and lead to personalized probiotics. "Because don't forget, a lot of people that consume a probiotic or prebiotic are what we call nonresponders: the prebiotic doesn't change anything in their microbiota," Hutkins said. "And it could be that they don't have the right microbes, and so that's why we're doing it this way: well, if you don't have the right microbes to respond, we'll give it to you."

So would you eat it in the form of a fuet sausage, or a cup of yogurt, or fermented milk? Probiotics benefit from the general acceptance of a name that means "for life," and time has blurred the history of many repurposed bacterial fermenters. In that regard, a product label that merely lists the included species may be a much easier sell than reused water.

• • •

Some public relations efforts have tried to elide recycled water's origins. Others have taken the opposite tack and pointed out that all water is basically "toilet to tap": the dinosaur pee line of persuasion. Rozin and his coauthors put it more bluntly when they summarized a geologist colleague's calculations that "it is impossible to drink a glass of water in Europe without ingesting at least a few water molecules that passed through the body of Adolf Hitler!" As they later noted, convincing people that all water is contaminated "could potentially lead to some personal crises, but the desire to survive would presumably dominate."

Normalizing water reclamation as a continuation of the planet's circular economy may be a somewhat gentler way to sway holdouts. A shift from "novelty to everyday familiarity," Rozin and colleagues write, may decrease the perceived threat. Highly treated water that is injected back into aquifers or allowed to percolate through reservoirs is arguably cleaner than what it touches; reintroducing it to a natural water source doesn't offer further physical purification. But Rozin and his colleagues hit upon a potential psychological advantage by proposing that "the reintroduction of treated wastewater into natural systems could potentially serve a 'spiritually purifying' role of removing consumers' historical association of the clean water with its origins as wastewater." Emphasizing the perceived purity of "nature," in other words, may wash away at least some of the stain.

It would be hard to find a more aggressive outreach campaign than the one accompanying the 2003 introduction of NEWater by Singapore's Public Utilities Board. The city-state's government viewed the water reclamation effort as a national security issue since Singapore's population density and land scarcity have limited its ability to catch and store its own rainwater. To make up the shortfall, a series of water agreements linked to its independence from Malaysia in 1965 have allowed it to import water from the Malaysian state of Johor through 2061. Since 2000, though, the agreements have been a source of friction between the two neighbors due to price disputes and Johor's fluctuating water reserves. NEWater has become an increasingly vital part of Singapore's "Four National Taps" water security strategy and by 2018, it was meeting about 40 percent of the total demand. Treated wastewater effluent undergoes the same basic purification steps as the Pure Water Wagon, though on a vastly larger scale.

After its purification, NEWater earmarked as drinking water heads to reservoirs, where it intermingles with captured storm water. The blended water then heads to Singapore's waterworks, where it is treated again before being piped to taps across the city. The accompanying public relations blitz, widely hailed as a success story, included a comprehensive outreach campaign, the construction of a popular NEWater Visitor Centre, and the rebranding of sewage as "used water" and sewage treatment plants as "water reclamation plants." Unlike in the US and Australia, though, Singapore's government didn't have to contend with serious political opposition. As the US National Research Council noted in a 2012 book on the potential for wastewater reuse, "it is difficult to ascertain if the absence of domestic opposition to the NEWater program is because of the successful visitor center, positive press coverage, cultural differences, national policies that limit civic discourse, or all of these reasons."

Could something similar be done in the US? In fact, it already has. A major water reclamation project in California has suggested how extensive outreach and educational efforts combined with indirect reuse can have a normalizing and spiritually purifying effect that dampens the opposition. Since 2008, the Orange County Water District's Groundwater Replenishment System in Fountain Valley, California, has been the world's reigning champion for indirect reuse, even larger than Singapore's program. From the secondary effluent treated by an adjacent wastewater treatment plant, the water district has been purifying and pumping reclaimed water through fourteen miles of pipes into recharge basins. Over a month or so, the water percolates down into a large underground aquifer, where it mixes with groundwater. Closer to the coast, the district is injecting reclaimed water directly into the ground to create a ridge of freshwater that acts as a barrier against the influx of contaminating seawater. In 2018, the Orange County facility set a world record by recycling more than 100 million gallons of wastewater in a single day. At its current peak capacity, the Groundwater Replenishment System can put enough water back into the ground to meet the needs of about 850,000 people.

The treatment system follows the same general microfiltration, reverse osmosis, and ultraviolet-advanced oxidation process as the Pure Water Wagon. Separate buildings house the successive purification steps, all connected by big blue pipes. A new addition was under construction when I toured the Orange County plant in October 2021. Once completed in 2023, the water district estimated that it would top out at about 130 million gallons of reclaimed water every day, or enough for about one million people. The vast scale also helps improve the efficiency. Of the wastewater that enters the treatment plant, the next-door water replenishment plant can reclaim about 80 percent. "So everything's a resource if you have the wherewithal and the time and the money and the effort and the technology, but we don't feel like we're doing anything that's super revolutionary," Mehul Patel, the water district's executive director of operations, told me. Rethinking how we can convert a perceived waste into a resource is never easy, he said, "but it's definitely doable."

Water district general manager Mike Markus said the Orange County utility is essentially a groundwater wholesaler also tasked with protecting the flow of the Santa Ana River. That means negotiating with the county's nineteen retail water agencies—collectively serving about 2.5 million people—about how much water they can draw from the underground basin through municipal wells, from the river, and from other imported sources. Drought has complicated the calculus; when rain does fall, Markus said, it's more often in the form of intense storms that hamper the county's ability to effectively capture and store the storm water. Unlike other regional rivers that are dry for much of the year, the Santa Ana always has water in it, if only because upstream cities discharge their treated wastewater into it. Downstream Orange County communities then recapture and treat the river water as part of their own water supplies. More than 75 percent of the retail agencies' water, though, comes from groundwater. The recycled water that replenishes the groundwater basin, Markus said, offers a high degree of reliability for the supply: the county will always have wastewater. "It's a source that could be considered drought-proof."

The infrastructure built to capture that resource, though undeniably expensive, solved another problem faced by the Orange County Sanitation District, the water district's neighbor. Modeling had suggested that a second outflow pipe would be needed to safely discharge water into the ocean during peak storm events that were swelling the volume of its incoming wastewater. But building it would be expensive and logistically challenging in the densely populated county. Instead, the district took the money it would have spent and put it toward an expanded water reclamation plant that could handle its treated wastewater. That collaborative agreement became the genesis for the groundwater replenishment system and helped pay for half of the project's capital costs.

There were other hurdles to consider. When the water and sanitation districts began sketching out their joint project in 1997, one of the first steps was to hire an outreach consultant. Meeting by meeting, the project secured the support of local governments, environmental organizations, the medical community, and all nineteen water retail agencies that would use the water. Project planners met with members of the county's sizable Vietnamese and Latino communities. A speaker's bureau recruited engineers and other staff members, who collectively made more than 1,200 presentations over a decade. Unlike its neighbors, Markus said Orange County's Groundwater Replenishment System ended up with no organized opposition.

To ensure that the reclaimed water is safe, the facility tests for the presence of more than four hundred compounds (116 of which are required by the regional water quality control board). In its entire history, Markus told me, the purified water it released had never exceeded the permitted limit for any of them. Municipal wells, on the other hand, have routinely struggled with chemical contamination—including some that draw from a portion of the same underground aquifer in Orange County. Excessive levels of PFAS compounds alone had forced the closure of nearly one-third of the county's two hundred wells. "People just don't understand the infrastructure that it really takes to get water to their houses," Markus said. In a poll commissioned by the water district, though, respondents rated it as doing a better job and being more believable on water issues than the county board of supervisors. Maybe people tend to take water for granted, he said, but the trust could help build support for reclamation efforts as a way to ensure long-term reliability of the local water supply.

Since its debut, the Orange County facility has hosted regular tour groups so people can see the science and scale of the purification steps for themselves. Sandy Scott-Roberts, program manager for the groundwater replenishment system, guided me on an in-depth tour of the existing campus and partially completed addition. Between the expansive reverse osmosis building and an installation of UV purifiers—a collection of steel tanks filled with the type of lamps you'd find in tanning beds—Scott-Roberts led me to a triple stainless-steel sink with three running taps. The middle tap was filling a sink with pale yellow water that had been through the microfiltration process. That water, sometimes called "purple-pipe" or tertiary treated water, is free of most pathogens and large contaminants but hasn't yet been through the reverse osmosis step. To its right, a sink full of coffee-colored water provided an unmistakable contrast. That concentrate, she said, was full of the contaminants removed by the reverse osmosis membranes. And on the left, a third tap was filling a sink with the perfectly clear water that had made it through the membranes and then past the ultraviolet purifiers. Scott-Roberts pulled two cups from a dispenser for an impromptu taste test.

Flush

"We probably have about 95 percent of the people that will do it," Markus had told me earlier. The preceding tour undoubtedly helps. He has also been known to publicly shame holdouts into taking at least a sip. Scott-Roberts didn't have to prod me. Maybe it was the visual contrast, but the clear water looked positively inviting compared to its tinted counterparts. "Cheers!" we said, laughing. We tapped our glasses in a toast and drank. "And then the question is, what does it taste like?" Scott-Roberts asked me. "Tastes like water," I replied, having been primed by the blue lettering on the cup that read, "Tastes like water... because it is water!" But it was true. No hint of salt or other minerals, no taste at all. A blank slate.

The water was nearly equivalent to distilled water, Scott-Roberts explained, with maybe one-tenth the concentration of dissolved solids that most tap water contains—so little that the water could corrode other mineral-containing surfaces. "If you leave distilled water on concrete, it will start to take out the calcium in that concrete," she said. Likewise, drinking a lot of it would start to remove calcium and other minerals from my cells. To protect its equipment, the plant removes some acidifying carbon dioxide and adds back some lime to raise the pH and level of dissolved solids, not unlike what McPoland and other brewers do to match city water profiles.

The Orange County plant has benefited enormously from existing and added infrastructure: the adjacent wastewater treatment plant that delivers the treated effluent, the multiple recharge basins that accept the reclamation plant's output via other pipelines. And the region is blessed with a huge groundwater basin that can accommodate the percolating water. Its success in surmounting both technical *and* psychological barriers, though, has made waves throughout the region; it became the precedent that other utilities could point to when drawing up plans or dusting off others that had been shelved through prior opposition. New water reclamation systems were being planned or built in Santa Clara, Monterey, Los Angeles, and San Diego, among other cities. The San Diego system would pipe the reclaimed wastewater to surface reservoirs. Los Angeles dreamed of a system that would dwarf Orange County's.

For parts of the world that lack expensive infrastructure, water reclamation and sanitation solutions may need to be largely self-contained. For direct recycling of human waste, one of the most promising strategies is a unit called the Janicki Omni Processor, which is based on seventeenth-century steam engine technology repurposed for the twenty-first century. Steam engines have existed in some form since Thomas Savery's steam pump of 1698, initially designed to remove water from mines. Subsequent improvements increased the engine's efficiency and broadened its applications: drawing water from a well, then replacing horses to power sawmills. Then powering cotton, wool, and flour mills. Then driving locomotives and farm equipment. Commissioned by the Bill & Melinda Gates Foundation, a Washington State company called Janicki Bioenergy (since rebranded as Sedron Technologies) further expanded the list of applications to include a combination waste incinerator, steam power plant, and water filtration system. Its prototype Omni Processor, built in eighteen months, converts biosolids into sterilized ash, renewable electricity, and clean water in a largely self-sustaining loop. "We didn't really invent anything," mechanical engineer Justin Brown told me. "We took very well proven, existing technologies and packaged it all together."

Brown, who had worked on the Omni Processor since 2014, said the initial plan was to conduct a pilot test on the prototype at the company factory. But it worked so well that the Gates Foundation decided the pilot test would be more meaningful if it was conducted in the field. So Sedron Technologies shipped the unit to Dakar, Senegal, in 2015, where it began processing about one-third of the city's solid waste in a test case of whether the strategy could provide a workable sanitation solution. Wet sludge trucked to the processor from septic tanks or pit latrines is first sun-dried to reduce the wastewater down to about 50 percent solids. A small wastewater treatment plant handles the black water that leaches from those solids as they dry. To convert the sludge into a useful fuel, a conveyer belt carries it into the Omni Processor's drying tube, which boils the biomass to remove most of the remaining water. The evaporated steam released from the biomass courses through a filtering system to create vapor that is condensed back into pure distilled water.

A purification unit can then send the water through further treatment steps until it exceeds international safety standards for potable water.

The dried biomass burns in a combustion chamber that converts the fuel into sterilized ash. In an adjacent boiler, the intense heat turns evaporated water from the biosolids into high-pressure and high-temperature steam that runs a steam engine. The engine sends exhaust steam back to heat the initial sludge dryer and turns a generator that produces enough electricity to power the entire processor. "The whole vision for it is to be grid independent," Brown said. Initially, the installation requires some form of start-up electricity, such as an on-site generator. "But then the minute you reach steady state, there is sufficient power being produced by the plant to power itself," he said. "And in most cases, there is excess available depending on how wet the material is."

Shortly before I talked with Brown in 2021, the last shipping container had left for the coastal village of Tivaouane Peulh, just northeast of Dakar. When it arrived, workers would begin installing the second Omni Processor near the beach; once operational, the full-scale model would occupy roughly the same footprint as a regulation-size basketball court. Estimates suggested that the compact power plant would be capable of processing more than 66,000 pounds of fecal sludge and generating about 1,850 gallons of clean water every day. The modeling, though, was tricky given the difficulty of finding representative sludge in Washington. Brown said developing countries tend to have far more inorganic content in their wastewater due to contamination from sand, dirt, and other particles when it's collected from latrines or septic systems and transported to the processing site. The higher inorganic content makes the biosolids less energetic, "because there's a ton of stuff in there that just can't burn." Because fiber absorbs more water and increases the overall bulk of biosolids, its relative abundance in poop can impact combustion efficiency as well, though Brown and fellow engineers didn't look at differences based on diet.

Other independent modeling suggested the full-scale unit's estimated capacity to handle waste from 250,000 people might be too low; Brown said he suspected its true limit might be 500,000 or more. In fact, the new unit could have a capacity gap that its operator in Senegal,

Delvic Sanitation Initiatives, would need to fill with other biomass such as food and agricultural waste. Brown hoped the better-than-expected ability would provide an incentive for more collection of solid waste, thereby improving the region's sanitation. As for the leftover ash, initial studies and market analyses by Delvic Sanitation have suggested that mixing it with compost could enhance the soil amendment properties. Alternatively, the ash could be used for construction as an additive to mortar and concrete mixes.

The power of disgust threw up obstacles here too. In a 2015 promotional video touting the Omni Processor's potential, Bill Gates held a canning jar in his right hand as Peter Janicki, the company's chief executive officer, filled it with water from a spigot on the machine. Just five minutes earlier, the reclaimed water had been fecal sludge. Gates took a sip as spectators applauded. He announced, "It's water," and laughed.

"The processor wouldn't just keep human waste out of the drinking water; it would turn waste into a commodity with real value in the marketplace," he wrote in a subsequent blog. "It's the ultimate example of that old expression: one man's trash is another man's treasure." Gates may have convinced himself from studying the machine's engineering that the reclaimed water would be both safe and highly valued, but convincing the public proved to be a much harder sell. Instead of a triumphant unveiling of the technology, echoes of the "toilet-to-tap" pejorative appeared in multiple media accounts of the promotional video. "Bill Gates Drinks Water That Was Human Feces Minutes Earlier," read one headline. "Bill Gates Drinks Water Made from Human Poop," read another. On *The Tonight Show Starring Jimmy Fallon*, Gates playfully tricked Fallon into drinking the "poop water," after which Fallon fell over and pretended to gag. "It tasted really good!" he said eventually, though he suggested that Gates needed to change the name.

Brown even encountered some resistance from family friends when he told them what he was working on. "Some of them were like, 'Well that's gross. I wouldn't want to drink water that was previously poop,'" he recalled. As Olaf the snowman reminded us, they almost certainly already had as part of the planet's continuous recycling loop. "At the end of the day, all we're doing is kind of speeding up the process," Brown said. From a chemical and safety perspective, tests confirmed that the Omni Processor's reclaimed water was no different from nature's, if not even better. But a five-minute transformation, though a technological feat, may be too rapid to dislodge the irrational but deeply planted conviction that the water is somehow still contaminated. As with fecal transplants, we may require some layers of separation and abstraction to blur its past. "Maybe those additional steps are just fundamentally necessary for the psychology of it," Brown said.

Despite the lofty idea that African communities would embrace reclaimed drinking water, a similar phenomenon played out across the Atlantic. Brown said the Senegalese public enthusiastically received the Omni Processor itself, though he questioned whether some of its abilities were oversold in initial marketing pitches that conjured up images of a "magical machine" that could fully replace a wastewater treatment plant. "I think a lot of the limitations weren't well communicated," he said. And despite the technology's heavy promotion as an efficient way to create clean drinking water from human waste, the pilot project's reuse plan encountered some stiff resistance. As in the US, it seems, the yuck factor was too high of a hurdle for many Dakar residents; misinformation that accused Bill Gates of using Senegalese people as "guinea pigs" added to the pushback. "So that full-scale unit will not be making drinking water, even though it's completely, technically, capable of doing it," Brown said. Initially, at least, the clean water would be used for commercial and industrial purposes instead.

● ● ●

If researchers are to convince the public of both the safety and necessity of pursuing a more natural and circular economy, perhaps the hostile environment of space can help us here, too, in understanding what can and should be done within a closed loop. Astronauts aboard the International Space Station are already using a bioreactor system that recycles urine, sweat, and excess moisture into drinking water—not quite a real-life incarnation of the form-fitting still suits of *Dune* that capture and recycle precious bodily fluids on the harsh desert planet Arrakis. But Frank Herbert's science fiction is looking increasingly prescient.

For the closed system of a tin can hurtling through space or a future colony on Mars, researchers are now turning their focus on how astronaut poop might be similarly recycled into useful products, as we saw in chapter 8.

A biofuel for the spacecraft will be invaluable. But what about fuel for the astronauts themselves? To help us boldly go where no one has gone before, environmental engineer Lisa Steinberg and Christopher House, director of the Astrobiology Research Center at Pennsylvania State University, took a crack at the reuse conundrum with a compact microbial reactor that acts like an artificial stomach to break down astronaut poop and pee into methane gas and acetate salt. In their prototype of a combination waste treatment–food generation system inspired in part by aquarium filters, the methane gas could fuel the growth of a separate batch of edible bacteria to yield a beige-colored "microbial goo" not unlike Marmite or Vegemite spreads made with brewer's yeast. On its own, the protein-rich goo may not win any culinary awards. But it solves the problem of how to cleanly and safely separate waste treatment from food production in space, at least in principle. In the two-step process, methane produced by archaea microbes becomes the food source for a type of bacteria that in turn becomes the food source for humans. The indirect reuse strategy, if it pans out, could help keep astronauts alive on extended missions into deep space while showing engineers back on Earth how to grow protein-rich food and treat organic waste more efficiently.

To test the potential of their anaerobic digestion scheme, the researchers fed their reactor a slurry of synthetic wastewater spiked with simulated astronaut poop (Steinberg told me that they used a concoction of dog food, cellulose, glycerol, and salts). The anaerobic reactor is essentially a miniature version of the methane-producing tanks at wastewater treatment plants. In this case, though, House and Steinberg filled their mini-reactor with the same one-inch plastic balls that are commonly used in aquarium filters. On the surface of the balls, they allowed two bacterial strains to form biofilms. One fermented the waste into fatty acids and acetate while the second transformed the fat into biomethane. After recirculating the simulated waste through the reactor, the researchers calculated that their system removed about 97 percent of the organic matter.

That left the question of how to use the methane by-product to grow food. "I stumbled across some work in the eighties where they were growing this *Methylococcus capsulatus*, this methane-consuming microorganism, on natural gas and using it for animal feed because it's very high in protein," Steinberg told me. The unusual methane-loving bacterial species, first isolated from the Roman thermal baths in Bath, England, seemed like the ideal candidate for the second part of their reactor.

Although the bacteria hadn't been cleared for human consumption, California-based Calysta Energy bought the rights to the technology and has mass-produced a protein meal made primarily from *Methylococcus* grown in methane-filled fermentation vats. The company markets its dried bacterial pellets as a food source for livestock, fish, and pets and a sustainable alternative to more resource-intensive protein sources like fish meal and soy concentrate, just as Kenya's Sanergy has. Calysta asserts that its manufacturing process uses no agricultural land and very little water, though it has partnered with petrochemical company offshoot bp ventures, which is seeking out new markets for its own natural gas—like, say, a microbial feedstock.

Of course, we've seen that methane is a more sustainable fuel if it's made from biogas. Nor is *Methylococcus* the only potential candidate for a foodstuff grown in deep space. Separate testing by House and Steinberg pointed to other contenders like one bacterial species capable of growing in a high pH solution and a second that can withstand temperatures of 158 degrees Fahrenheit. Taking a similar tack, engineer Mark Blenner has experimented with using astronaut poop, pee, and exhaled carbon dioxide as the starting food stocks for genetically engineered yeast. That yeast could then be crafted into food, vitamins, or plastic materials.

Michael Webber, an expert in clean energy technology, has instead conducted experiments on making algae-based proteins from excess natural gas in a microalgae biorefinery. Given the imperative for imaginative solutions in the tight confines of a spacecraft, a two-part reactor that converts waste streams into food makes a lot of sense, Webber said. "One way to think about this is just closing the loop on waste management: turning your waste into valuable resources," he said. "It's only waste if you run out of ideas."

Once again, the lessons and technology behind these space-age research projects could offer some critical solutions to the mounting challenges facing our own world. "Some materials that we start with are easier to imagine recycling than others," NASA environmental scientist John Hogan told me. Sustaining human life means reusing essential water and nutrients, though, and the closed loop of a small spacecraft—where downstream can quickly become upstream—makes these necessary connections exceedingly clear. Figuring out how to extract valuable molecules safely and efficiently from human-generated waste and convert them into forms that are reusable or even edible, Hogan said, could "bear good fruit" both far away and much closer to home.

Despite the challenges, Brown said his overall experience with the Omni Processor has made him "incredibly optimistic" about its ability to improve sanitation and convert waste into value. In the digital age, we're used to start-ups that develop technology virtually overnight and quickly gain immense value from selling a popular product. Infrastructure operates on a far different level, he said. "It takes a lot of time and a lot of money to get an industrial hardware-intensive technology to scale. It's not a consumer product."

Developing countries may be more open to "leapfrog" technologies, like mobile phones instead of landlines and solar panels instead of traditional electrical grids, as long as they align with cultural values. But given the up-front costs, no municipality wants to risk public money on an untested new application, he said. Until it has proven itself over time, that tech may require the backing of federal infrastructure grants, development banks, or nongovernmental organizations like the Gates Foundation. But if done right, an installation like the Omni Processor could generate long-term sales from the electricity, water, and ash outputs. Simply breaking even, Brown said, would be an exciting step forward given the energy-intensive demands of wastewater processing.

Brown said Sedron Technologies is hoping to place other Omni Processors in Senegal and elsewhere in western Africa. The company developed the technology under contract with the Gates Foundation, which is working with other license holders to commercialize it in India and China as well. In the meantime, Sedron has developed another technology called Varcor that may eventually replace the open-air sludge driers. The machine requires an external power source but can use thermal evaporation and water vapor compression to yield clean water, dry solids that can be used as fuel or a soil amendment, and liquid ammonia that can be repurposed as fertilizer.

Beyond thinking about the infrastructure we could build in places that lack it, we may need to re-think how to use the infrastructure we already have. Environmental engineer David Sedlak is increasingly focusing his attention on those industries that still use the sewer as a disposal facility. In a 2020 opinion piece, "Protecting the sewershed," he and colleague Sasha Harris-Lovett argued that looking to wastewater as a potential source of drinking water will require thinking about the sewer in the same way that we think about a watershed and what goes into it. Getting people to accept what comes back out of it, he told me, requires legitimization. It's the same process that has allowed the public to get past their initial suspicion of other unfamiliar new technologies, whether the steam engine or commercial air travel, and California has provided an important demonstration of its power. For legitimization to work, Sedlak said, the public needs to trust the institutions responsible for regulating the new technology, recognize the benefit it provides, and then become familiar with the process.

It occurred to me that efforts to generate health-promoting water and food are learning from each other. The probiotics field has begun backing its effective marketing with the rigorous science needed to support health claims, while water reclamation efforts, long grounded in sound science, have been buoyed by a sharper focus on marketing. As home brewer Jenn McPoland told me, "The more people talk about it, the more people accept it." In the end, the secret to getting past our psychological barriers may be to demystify the natural recycling process that exists within all of us and every other living thing, and to demonstrate how we can safely tap into it, sometimes literally.

CHAPTER ELEVEN

Balm

FROM THE LIMESTONE LEDGE, I stepped down a rusty swimming pool ladder, adjusted my mask and snorkel, and kicked off into a surprisingly clear and vividly blue bay— nothing like its name might suggest. I had imagined Cuba's Bay of Pigs to be murkier, maybe. More mysterious or battle-scarred or mindful of its troubled place in history. Instead, the hazy bay of history books, of the failed 1961 invasion by US-backed Cuban exiles intent on overthrowing the communist government of Fidel Castro, appeared like an artist's palette stretching toward the horizon. Robin's egg blue first, by the limestone. Then a turquoise becoming navy where the seafloor dropped down to meet the bay.

The first corals appeared scarcely ten yards from shore: patch reefs that began as small, forested citadels rising from the white sand, each with its own palette. Brain corals and stoplight parrot fish. Yellow tube sponges and blue tangs. The reefs here, a short drive in a '54 Chevy Bel Air from the beach town of Playa Larga, had lost some larger predators due to overfishing farther out in the bay. But each patch teemed with small and midsized fish. A few flipper kicks out, the patches grew closer together until they rose up to become low mountains before disappearing down into a pool of deep blue. Just beyond the drop-off, I spied a passing school of yellow jack.

Across the highway from the sea, a flooded cenote called Cueva de los Peces appears like a natural swimming pool that belies a narrow but surprisingly deep underwater canyon. The limestone caves and reefs here connect a chain of coastal habitats that extend from mangrove swamps to the deeper ocean. Just to the west, the vast Zapata

Swamp is the biggest and most biodiverse wetland in the entire Caribbean. On two birding trips there, Geoff and I saw rarities like the tiny
bee hummingbird and the endangered Zapata wren. Farther to the east,
a less accessible underwater wonderland called Gardens of the Queen
National Park, or Jardines de la Reina, harbors a bounty of sharks, sea
turtles, and goliath groupers the size of grizzly bears. The archipelago
of reefs, keys, and mangroves is so unspoiled that some visitors have
dubbed it "The Crown Jewel of the Caribbean."

It might be easy to dismiss Gardens of the Queen as a splendid freak
of nature—a largely intact ecosystem protected by its distance from the
mainland. But researchers have documented surprisingly good conditions in the coastal wetland and many near-shore reefs as well, like
those in the Bay of Pigs and others along Cuba's southern shore. Coral
expert Daria Siciliano told me that Cuba's extensive reefs shelter coral
larvae and other inhabitants that may drift with the prevailing ocean
currents to help reseed depleted populations to the north. The reefs' relative health has been a rare bright spot in an otherwise brutal spate of
bad news about sick and dying corals around the world, from the Florida
Keys to Australia's Great Barrier Reef.

What's different? The widely hated US trade embargo, or el bloqueo
as it's known in Cuba, has dampened large-scale development and associated pollution that can harm marine environments, though it has
also blocked access to foreign loans and technology that might aid the
country's environmental research and management capabilities. Cuba
has set aside nearly a quarter of its own shallow-water marine areas
as protected parks and sanctuaries, though critics would contend that
the government's conservation record has been spotty when it comes
to enforcement. But the prevailing story is about how an abrupt change
in what flowed into Cuba's soils profoundly altered what flowed into its
seas.

In places like the Bay of Pigs, scientists have linked environmental
recovery to an economic calamity. The collapse of the Soviet Union
in the early 1990s turned off the spigot for Cuba's regular shipments
of synthetic fertilizers, pesticides, gasoline, farming equipment, and
food. With the US trade embargo cutting off other supply routes, Cuba's

fishing and sugarcane industries collapsed. Abandoned tractors rusted in the fields. Seafood catches, inorganic fertilizer use, and agricultural production all plummeted. In the painful years that followed, euphemistically called "The Special Period" by Cubans, many farms became organic more by necessity than choice when the lack of capital and chemicals led to lower-impact agricultural methods. From 1961 until its peak in 1989, Cuba's inorganic fertilizer use had risen by nearly 900 percent, reaching some of the highest nitrogen application rates in the world. Then the flood became a trickle; by 2000, the country's average use of inorganic fertilizer had fallen back to about one-fifth of its peak—with most of that earmarked for sugarcane production.

Over the past three decades, Cuba has provided a compelling lesson in what's possible when we allow nature to heal. Recall that rainfall can wash fertilizers into drainage channels that feed streams and rivers on their way to the ocean. Nutrient pollution can trigger the explosive growth of phytoplankton and harmful algae. Among other consequences, algal overgrowth can take over reefs and rob the ocean of oxygen when the algae die and decay, creating the dead zones researchers have documented around the world. A report on the status and trends of Caribbean coral reefs suggested that threats from both land and sea had wiped out more than half of the region's corals since 1970. In Cuba, though, the decline in live corals covering its reefs began to reverse course in the mid-90s. With the steep drop in phosphorus and nitrogen-rich runoff, Cuba's corals and coastal wetlands gradually rebounded. Data are limited, but that recovery roughly coincided with an uptick in key indicators of health like parrot fish, which graze on algae and help prevent the overgrowth of corals' main competitors.

Economic catastrophe and widespread hardship are poor trade-offs for environmental rehabilitation, of course. Cuban marine biologist Jorge Angulo-Valdés acknowledged that conservation failures there may send economic ripples throughout the region. "But we still need to make a living; our people still need to have food," he said. Cuba's accidental experiment, though, suggests a deliberate way forward. If we view environmental protection as necessarily linked to our own well-being instead of a competing interest, can we improve both? And might better application of our own by-products help get us there through restorative practices that enhance food security, mitigate natural disasters, and even remediate pollution?

Here's where poop could again be a valuable asset. We've seen how it can sicken us and foul the environment if handled carelessly. But we also know that when administered wisely, it can heal an unbalanced gut ecosystem. Why shouldn't it help do the same for troubled lands and waters? The splendid reefs I marveled at while floating in the Bay of Pigs raised another hopeful question: If all this is possible by giving nature the space and time to heal, what might be possible if we actively help to restore it?

Canada's Sechelt Mine has provided a compelling example of how a scarred or barren space might be nurtured back to health. One of the largest sand and gravel mines in North America, it extracts a highly sought-after sand for cement and sits within the traditional territories of the Sechelt First Nation (shíshálh Nation) on the southern coast of British Columbia. Extracting and processing sand leaves behind a dusting of leftovers, or residuals known as fines. John Lavery, a senior environmental scientist with the Canadian consulting firm SYLVIS Environmental, said the sterile fines wouldn't even grow a dandelion. Lavery has worked with industrial and municipal residuals for more than twenty years. "It's kind of the island of misfit toys," he told me, and SYLVIS specializes in finding new uses for them: ash, wood, and food waste, and the leftovers from pulp and paper processing, mining, and wastewater treatment.

Through a contract with the mine's owner, Lehigh Hanson Materials, SYLVIS developed a kind of two-for-one reclamation plan that would reuse our own residuals—poop—to find a new use for the sterile fines residuals. The strategy was to muffle the sounds of heavy equipment and crushing rocks, protect vistas from the town of Sechelt below the open pit mine, and ultimately create a functional soil system that could nourish fast-growing poplar trees as a valuable cash crop. Sechelt and two other towns in the region supplied biosolids from their wastewater treatment plants. Some of that biomass created earthen berms, or ridges, to muffle the sound and hide the mine from view, and stabilized slopes planted with native grasses and other vegetation. SYLVIS blended more of the biosolids and pulp and paper sludge from a local mill with the fines to create a soil base for two poplar plantations, with more biosolids providing regular infusions of fertilizer. "And then we'd just kind of let the ecosystem do its thing," Lavery said. At its peak, he estimated, the project encompassed roughly one hundred acres.

The soil has since developed a healthy profile of mycorrhizae, or fungi that form symbiotic relationships with plant roots. It has attracted small and large invertebrate creatures, and soil bacteria that help move carbon and nutrients through the soil and into the roots. "We started with sterile aggregate sand and fines and we ended up with a soil in those spaces," Lavery said; a process that might have taken hundreds or even a thousand years to develop from the fines, he figured, had done so in less than twenty. The Sechelt First Nation then assumed ownership of the poplar plantation and harvested the mature trees.

After the project ended, area entrepreneurs began finding long-term uses for other residuals. In 2010, Sechelt First Nation member Aaron Joe opened Salish Soils just beyond the mine's borders. As part of its ambitious recycling portfolio, the company creates three main composts that can be used for different blends: from residential and commercial food waste; from the blood and guts and other leftovers of regional salmon and steelhead farms; and from human biosolids delivered by two of the same wastewater treatment plants that participated in the mine reclamation. "At Salish Soils, we believe in the renewable spirit of nature," the company website says. "We honour the earth by empowering people, while healing lands—collecting, receiving and composting waste resources into high quality organics." The lesson that our by-products can help nature heal is one that Indigenous communities have passed down for centuries.

SYLVIS expanded its reach as well. For two related projects in Alberta, it used biosolids to remediate marginal farmland and a shuttered coal mine. Instead of returning the reclaimed land to more traditional agricultural uses, though, the environmental firm decided to establish short-rotation, high-density willow tree plantations. "Why are we doing that? Well, because we can't think of a system that gives more back to the environment or to society," Lavery said. The strategy uses the biosolids to return mineralizable nutrients to the soil, thereby improving its quality, suitability for growing plants, and ultimate productivity. Then workers plant about 6,000 to 8,000 willow saplings per acre (a standard forest might have 80 to 160 stems per acre). Belowground, the trees sequester carbon with their root systems. Aboveground, they provide homes for birds, rodents, and small mammals, which bring more species to the sites. "So we start to see resident raptors, resident predators," Lavery said. "The willows themselves are also a generic forage for ungulates, so we start to see pronghorns and deer and moose." The end result has been a new ecosystem island within the province's prairie lands, an ecological concept known as island biogeography.

After just three years, the trees can be harvested on a rotating basis to preserve some of the habitat. "When you harvest them, they just grow a whole bunch of new shoots up from their base," Lavery said. "So when you harvest the aboveground material, the belowground root system remains alive and just simply pushes up more stems. It's kind of like mowing a big woody lawn." The renewable woody material can be ground up and mixed with biosolids to create more compost, say, or used as a source of energy and alternative fuels. Our poop, yet again, has become a reliable partner—this time for turning willow biomass into a renewable commodity that simultaneously aids the environment. Lavery is particularly excited about achieving carbon sequestration in Canada's home of oil and gas production amid a gradual phase-out of fossil fuels. "We're busy trying to kick-start a renewable biofuels and biomaterials economy in a place that over the next twenty to thirty years probably needs it most," he said.

• • •

In Cuba, more immediate hardship provided the impetus for dramatic change in the form of a more environmentally friendly kind of agriculture. A movement of scientists, farmers, and activists began to promote and expand an alternative based on agroecology. "It opened that political landscape for them to become leaders in expanding this type of agriculture on the island," Margarita Fernandez, executive director of the Vermont Caribbean Institute, told me. The concept emphasizes methods such as crop rotation and diversity, organic compost, mulch, and biological pest controls.

The extent to which Cuba remains dependent on food imports is a bitter point of contention between supporters and detractors of the country's organic and agroecological experiment; food shortages are common throughout the island, and they worsened during the COVID-19 pandemic. But Raúl Reyes Posada, the Cuban farmer I met in 2016 in the western town of Viñales, was proud of the more sustainable and environmentally friendly approach on his government-certified organic finca. In the farm's kiosk, covered in mementos left by tourists from around the world, he pointed out a few of the selections on offer. Fresh mangoes, bananas, and pineapples. Coffee beans in recycled water bottles. Homemade hot sauce. Hand-rolled cigars. The more intentional agricultural methods had allowed him and other farmers to make do with much less, a necessity amid the US trade embargo and years of economic stagnation. To circumvent the unreliable transportation system, urban farms known as organopónicos began popping up on city peripheries.

It occurred to me that Melissa Meyer's Rose Island Farm on a former horse pasture on the outskirts of Tacoma, Washington, might be considered an organopónico as well. When I met her in late June 2021, she was watering the one-acre plot of raised herb, flower, and vegetable beds in anticipation of a brutal heat wave descending upon the Pacific Northwest. We retreated to an umbrella-shaded picnic table behind her house, sipping water from mason jars and shaking our heads at the oppressive heat. A few days later, the weather station at Seattle-Tacoma International Airport would log the hottest day ever recorded: 108 degrees Fahrenheit.

Meyer named her demonstration farm after the Tsimshian Nation village of Lax kw'alaams (which means "Island of Wild Roses") in northern British Columbia where she grew up. The day I visited, she had been on the land with her family for exactly one year. Already, it had fundamentally changed her. In Canada, she had regularly gone salmon fishing with her family on the Skeena River, which provided communal harvests and essential vitamin D and unsaturated fats. When she moved to the Puget Sound region, she lost that ready access to salmon. "What does a sustainable relationship with my food look like then?" she said.

Meyer was establishing a new partnership, plant by plant. She had brought some of her coastal village's mainstays to her new gardens, like the salmonberry that still grows wild by salmon runs, and the thimbleberry that her husband adores. The land around her new farm likely used to have salmon runs as well, and abundant cedars and other shade trees that offered protection to the berries and other native food and medicinal plants. All of the mature Douglas firs had been cut down, leaving only a line of stumps and a single young fir. Meyer pointed out a native spruce that was shading part of the yard, and a struggling cedar by an invasive laurel hedge at the back of the property; the tree normally grows with alders, she said, whose canopies can protect cedar saplings from being scorched. In the unusually harsh heat, the young cedar would need her help to survive, but a few others that she had seen in the neighborhood were reason for optimism.

The landscape wasn't the same as what she knew, or the same as what it had been. But the remnants suggested that a healthier landscape might yet reappear with a little help. She had planted a maple, the first of many natives that she hoped to reintroduce. Read the land and you can see what used to be here, she said. "You can tell who *wants* to be here." Creating a community for the trees and giving them water and mulch to start would help them hold their own until they could begin to return the favor. When mature, she said, the trees would help to cool down the neighborhood and provide good habitat. "It's too dang hot. Our houses are heating up faster, then we're wasting dollars trying to cool them down. I mean, we have natural ways to do this, and they *want* to be here."

To help bring them back while providing food and medicine for her family, Meyer had adopted several complementary gardening methods. One is called companion planting and takes advantage of the help that certain plants offer each other when grown together. Trees can provide a protective canopy for heat-sensitive plants and herbs. Green beans can fix nitrogen in the soil and aid the growth of cornstalks, while the stalks supply a trellis for the beans. With the summer heat made worse by the city's concrete, the trees and plants needed a community to thrive. "We don't do well in isolation. You know, they're no different than us," Meyer said. "And nothing grows in isolation here; everybody has a community, so you want to mimic that."

Another gardening strategy, a no-till method called Back to Eden, was popularized by a 2011 documentary film but Meyer told me that it has its roots in indigenous agriculture, too. The method uses abundant wood mulch to rebuild the soil, retain water, and suppress weeds. Because much of her land had been degraded, Meyer used Tacoma's biosolids-based soil amendment, TAGRO, to add back nutrients before mulching the planting beds. It's called Back to Eden because it mimics what the forest is doing, Meyer said. "The forest is your number one generator of soil in the world, and it takes about thirty-five years for any land to go back to forest. It'll just naturally want to do that. That healing is already built in." Meyer was essentially replicating the forest's production of soil and promotion of biodiversity while encouraging the native producers to return. In so doing, she was demonstrating how we might use regenerative agriculture to move beyond modern farming methods and help heal the environment.

We slowly walked around her property, and she pointed out some of her cultivated plants and their uses. "I'm growing for medicine, I'm growing for food, I'm growing to feed the pollinators." Lavender. Borage. Feverfew. Meyer was due to start a new job as an herbalist in a few months and many of the medicines she hoped to use in her practice would come from these gardens. She could also see what didn't belong, like the hedge of laurel that had become a breeding ground for white butterflies, likely cabbage whites, whose worms were attacking her vegetables. The longtime horse farm had deep-rooted weeds that she was systematically killing with tarps. To some of the soil, she would then apply a heavy layer of compost, "like the forest." For a full year, the compost would do its work under the tarps, feeding the soil. Letting it rest and regenerate.

Language matters, Meyer said, when we have yet to discover the usefulness of something or have forgotten the value of something else. "When we call it waste, it changes our relationship with it." Bringing back farming to her corner of Tacoma had been difficult but exciting, she said, especially in how it was "waking up that wisdom in people." I suggested that it was a reclaiming of both land and knowledge, and Meyer gently corrected me. Sometimes white people will say they're reclaiming productive land, when the indigenous wisdom was there all along but just erased. *Reasserting* ownership of that knowledge is a better way to describe it, she suggested.

In Mesoamerica, centuries of Indigenous farmers have practiced the Milpa system of sustainable permaculture (sometimes called Maya forest gardens), in which corn, beans, and squash are planted together to help each other thrive. For each plot, eight fallow seasons follow two planting seasons to let the soil and forest regenerate without the need for synthetic pesticides or fertilizers. The system works with perennial shrubs and shade trees, and eventually allows the forest to reestablish itself on former planting sites. According to the MesoAmerican Research Center, "As long as this rotation continues without shortening fallow periods, the system can be sustained indefinitely." That knowledge system spread north, Meyer said. Then came the Three Sisters of North American tribes: again, the mutual support of corn, beans, and squash.

Meyer was using some of her land to assist the partnership between the Puyallup tribe and Purépecha immigrants that had been nurtured by Harvest Pierce County's community gardening program. One of the Purépecha farmers had planted his own version of the Milpa and Three Sisters crops in a corner of her farm. Meyer was growing other indigenous vegetables with deep-seated cultural significance in the region. The makah, or Ozette potato, is low in starch and healthier than other potato varieties. The nodding onion often grows next to Douglas firs and has been an indigenous food staple for centuries. Bringing other Black and Indigenous people to her farm, she said, was about helping them remember the knowledge and reassert their right to it. The wisdom was often never recorded; it didn't have to be because it was passed down orally. "So there is this wading through the information and remembering what's yours," she said.

Meyer saw the same dynamic at work with Salish Soils. "What I love about Salish Soils is he's not doing anything outside of his natural culture, his natural way of being on the land," she said. "You're working with a resource, the salmon, and depositing it back on the earth where it naturally would have gone if we hadn't interrupted so much of this natural cycle. So he's helping it along, and I love it. I love it." Recall that migratory fish like salmon have ferried phosphorus, nitrogen, and other nutrients up rivers and streams for eons, and that a range of fish-eaters beyond bears and eagles have relayed some of those nutrients to forests and meadows. One study found that salmon directly or indirectly support nearly 140 animal species in Oregon and Washington alone, many of them terrestrial. We are now the main determinants of whether salmon will continue to feed the soil.

Meyer felt like the Black and Indigenous farmers who had visited her farm were taking to the idea of working *with* nature instead of against it. Why work so hard to grow food in a way that would require more machinery and more fertilizer, she said? "I just feel like nature's got this really beautiful model. Work with it, work with it. And part of it is waking up that wisdom in people to go, 'Oh, yeah.' To pause, to get slow, to get quiet, to read the landscape, and get back into that rhythm."

In a moment of quiet in my own yard, I was reading *How to Do Nothing* by Jenny Odell when I had a sudden flash of déjà vu as she described her interactions with two neighborhood crows that she had befriended. I looked up, and there were Spot and Chicken, sitting on the electrical wires in front of the house. They starred in their own nature documentaries on the stage that had become our yard, a more curated form of nature but one that refused to stay within the bounds of what we had naively prescribed. In her book, Odell meditates on what we have traditionally seen as valuable or productive and how gardens and other open spaces always seem to be under threat, "since what they 'produce' can't be measured or exploited or even easily identified," despite neighborhood residents all readily understanding their immense worth. In ecological systems and in our attention, we have tended toward aggressive monocultures, she writes, where components that aren't seen as useful or able to be appropriated are the first to go.

"Because it proceeds from a false understanding of life as atomized and optimizable, this view of usefulness fails to recognize the ecosystem as a living whole that in fact needs all of its parts to function." Even an animal carcass or pile of poop has intrinsic value apart from what we deign to give it. Nature does not care whether we're curious or repulsed or dismissive. The organic matter will continue to feed microbes and fungi and flies and beetles and scavengers like crows. Where we allow it, plants and trees will grow in the fertilized patches and the cycle will repeat.

Our own gardens, nourished with biosolids, provided a humbling lesson in attending to the whole and allowing it all to function. We had catalyzed the start of a new ecosystem by allowing life to gain a better footing. And it had, spectacularly, but not in the way we had envisioned. By providing the rich nutrients to transform what had been the barren monoculture of a half-dead lawn and an eroded bank into a hybrid space of flowers, vegetables, and native and nonnative ornamentals, nature took over and begin to knit together its own narrative. A cloud of lovely house finches, with their red-blushed bodies, alighted on our neglected kale that had grown into four-foot-tall medusas of stems and seed pods. We had let it go, assuming that it no longer had value, and then the finches proved us wrong. Our unloved arugula and endive had likewise bolted, and then I noticed the impressive buzzing of bees around their flowers.

The Eastern cottontail population exploded in our neighborhood, devoured our second crop of peas, and seemed to coax the coyotes that might have dens down the hillside to reemerge after an absence of several years. A mole delighted in defying our careful boundaries of garden and lawn, digging holes in both, until I came upon Spot furiously pecking away at the soil in a flowerbed. And then I saw the dirt *vibrating*. He was after the mole. Do crows really hunt moles? I wondered. I had my answer the following day when I saw him on the electrical wire, tearing apart a small creature dangling from his talons: it was a partly eviscerated mole. The rodents evened the score, a bit, when a plump rat raided a bird feeder meant for the finches.

By spending more time in the yard—in the dirt—we were able to see more of the mini-dramas unfolding around us. The ability to observe and identify the life that surrounds us, Odell writes, helps us appreciate how they and we all interconnect. We are active participants with a role to play. If we had a hand in setting up the scenes, though, nature left no doubt about who was directing it all.

• • •

On a larger scale, the science of working with nature to help it regenerate has given agronomists and soil scientists a better read on how our recycled by-products can not only aid but also replenish farmlands and rangelands. Cities throughout Colorado, for instance, own thousands of acres of agricultural plots. Since 1982, the South Platte Renew wastewater treatment plant owned by the Denver suburbs of Littleton and Englewood has collaborated with scientists at Colorado State University to calculate the ideal amount of treated biosolids to spread over their soils. The long-running research project on about 160 acres of test plots has examined how the soils respond to treated biosolids on a physical, chemical, and biological basis.

From the years of data, soil health expert Jim Ippolito and his colleagues have given landowners a recommended sweet spot that balances the nitrogen needs of their plants against the potential detriment of applying too much phosphorus. Wheat fields do best with two to three dry tons of the cake batter–like matter per acre every other year. For rangelands like those owned by the city of Fort Collins, the magic number is five dry tons per acre every ten years. Add too little and you may not prevent plant degradation from cows, sheep, and other grazers, which tend to bunch together far more than the free-ranging mammoths and other megafauna of the past. Add too much and you may increase plant productivity but also decrease species diversity. Excess phosphorus can disrupt critical associations between fungal communities and plant roots and give way to more bacterial-dominated communities in the soil, Ippolito told me.

Fort Collins had taken a conservative approach on its rangelands and was applying one to two tons of biosolids per acre every ten years, he said. Even so, the fertilizer seemed to be giving the land a boost by promoting the growth of more edible plants like western wheatgrass and Indian rice grass instead of less digestible species like prickly pear and ball cactus. In a modern-day version of the Pleistocene grasslands, the approach was essentially supplementing inadequately dispersed herbivore poop with some of our own.

A mid-October storm with "thunder snow," two days after the temperature had climbed to nearly eighty degrees, kept us from visiting those city-owned rangelands. Instead, Ippolito and I talked in his office about his decades of research on the safety and effectiveness of biosolids and his related research on the potential of biochar. Measuring all of the effects on something as complicated as soil is no small feat; the Soil Health Institute in North Carolina, a collaborating partner, lists eighteen top indicators and another twelve secondary indicators.

One test measures how well a microbial enzyme called beta-glucosidase can degrade cellulose. The complex carbohydrate and source of organic carbon provides the structural integrity for plant fibers and cell walls and is composed of long chains of glucose sugar. At several test plots, early results pointed toward improved activity of the cellulose-degrading enzyme, Ippolito said, a precursor of organic carbon building up in the soil. After years of humans taking nutrients out of the soil, giving some of them back was beginning to improve its health.

Other methods for converting unwanted leftovers into environmental remediators may help transform liabilities into assets. Kandis Leslie Abdul-Aziz, a chemical and environmental engineer with an avid interest in sustainability, has experimented with upcycling a carbon-rich waste called corn stover, which refers to the leaves, cobs, and stalks left over after harvesting. Ethanol produced in the United States has relied heavily on corn stover and replaced about 10 percent of the nation's gasoline. Controversy, though, has raged over whether the environmental benefits exceed the costs: one recent study suggested that the sheer volume of corn planted for use as a biofuel has increased both water pollution *and* greenhouse gas emissions. Every year, corn stover accounts for roughly 250 million tons of agricultural waste in the US alone. "About one-third of solid waste that we make in the United States is actually from corn harvesting," Abdul-Aziz said. Burning it can introduce even more greenhouse gases into the atmosphere. Her lab is instead investigating how to convert the stover biomass into activated carbon, which is essentially biochar that has been treated to maximize its ability to filter out contaminants. The charred matter, usually in the form of a coarse black powder, can be made by the high-temperature, oxygen-free process of pyrolysis. Another method, called hydrothermal carbonization, mixes the stover with hot pressurized water to break down the biomass and transform it into the carbon particles.

To activate the carbon, Abdul-Aziz said, it can be mixed with a strong acid, a caustic base, or even steam to etch tiny pores into its surface. Creating all of those nooks and crannies dramatically increases the surface area, transforming each bit of carbon into a mini-sponge that can soak up contaminants. She and her colleagues found that the activated carbon they made from corn stover using the hydrothermal carbonization method was particularly effective at soaking up the compound vanillin. You may know vanillin as an extract from the vanilla bean, but it's also an industrial by-product and thus a useful stand-in for other environmental pollutants. When Abdul-Aziz and her team poured water spiked with vanillin through activated carbon made from stover, their filter removed 98 percent of the pollutant.

Just as activated carbon can be made from carbon-rich corn stover, it can be made from carbon-rich switchgrass or sawdust—or poop. That means our poop, transformed into bits of activated carbon, could be used like industrial-scale Brita filters to remove other contaminants. Environmental engineer Josh Kearns has developed and shared a design for simple furnaces that create biochar from organic material (which could include human or animal feces) and shown how DIY biochar can help filter contaminants from water in rural communities. It's the ultimate loop: using transformed poop to treat poop.

It gets better: "After we make the activated carbon, we can attach things to the surface of it that makes it even more effective," Abdul-Aziz said. Adding nitrogen-containing molecules called amino groups can enable activated carbon to capture carbon dioxide from the air. Adding iron nanoparticles to the surface can magnetize activated carbon so that after it has soaked up contaminants—vanillin, say—engineers can use magnets to pull the bits from water like fishing out a pile of iron filings.

Another potential application of activated carbon as an environmental scrubber dates back thousands of years. In the second century BCE, Roman soldier and historian Marcus Cato—often called Cato the Elder—formalized some accumulated wisdom on farming into his *De agricultura*, the oldest surviving work written entirely in Latin prose. Cato recommended fertilizing crops with pigeon, goat, sheep, and cattle dung. He included an early recipe for compost and instructed readers to make charcoal from firewood as a soil additive. He even described how to make a lime-kiln for producing the pH-raising soil additive quicklime. One of his most intriguing recipes, though, was a veterinary prescription for oxen, "if you have reason to fear sickness." As a preventive, he recommended the following combination:

Three grains of salt
Three laurel leaves
Three leek leaves
Three spikes of leek
Three of garlic
Three grains of incense
Three plants of Sabine herb
Three leaves of rue
Three stalks of bryony
Three white beans
Three live coals
Three pints of wine

The rule of three seemed to carry a special magic for the superstitious Romans, and Cato instructed the person administering the mixture to fast and remain standing throughout its preparation and delivery. Nonetheless, the acidic wine (with a pH that typically varies between three and four) could have etched pores into the pulverized bits of live coal, or charcoal, and created an early version of activated carbon.

Flush

Unlike herbs, vegetables, and wine, activated carbon travels through the digestive tract of both cows and humans relatively unchanged. Because its tiny pores carry a negative electrical charge, they can attract positively charged gases, toxins, and other compounds that stick to the pocked surface through a process called adsorption. That means the carbon can soak up unwanted things in the body, too, and effectively remove them through poop. For that reason, activated charcoal is a mainstay of hospital emergency rooms as a treatment for drug overdoses. The same basic concept works in soil and water: activated carbon can soak up substances like heavy metals, most of which carry a positive electrical charge. Which brings us to cow burps and farts. Globally, our livestock account for more than 14 percent of human-induced greenhouse gas emissions. More than 60 percent of that total comes from beef and dairy production alone, including methane gas produced in the rumen and large intestine that escapes into the atmosphere through cow eructation and flatulence. Countries like Ireland and Denmark have imposed "cow flatulence" taxes in a bid to lower the emissions, while Irish farmers have scoured the coastline in search of seaweed that when fed to cows can dampen some of their blasts.

An alternative solution might come from activated carbon. After feeding a powdered form of it to 180 cows on a commercial dairy farm in Australia's Queensland, researchers calculated that the dietary supplement reduced the cows' methane emissions by an impressive 30 to 40 percent and their carbon dioxide emissions by another 10 percent. Gut microbiome sequencing showed that the cows' intestinal populations of methane-producing archaea significantly dropped as well, while other microbial species increased to take their place. Their daily milk production, in fact, ticked up slightly.

The results confirmed previous lab experiments by separate groups in which researchers added biochar to cattle feed and incubated the mixture in rumen fluid taken from cows. Those *in vitro* studies showed that the dietary supplement could reduce methane emissions by more than 10 percent, while a small field trial in six steers reported a reduction in methane emissions of 10 to 18 percent (red seaweed and other supplements like oregano, some research suggests, can do even better).

The exact mechanism by which biochar reduces methane isn't known—perhaps it adsorbs the gas or favors a shift away from methane producers in the gut. But the results point to another way in which biomass often dismissed as waste could be a potent environmental ally.

• • •

For some forms of pollution, researchers have envisioned an extraction process in wastewater treatment plants that could remove unwanted contaminants from reclaimed water and biosolids *and* retrieve valuable metals. Kathleen Smith, a research geologist with the US Geological Survey who has since retired, made headlines in 2015 after a conference presentation in which she described finding bits of gold, silver, platinum, and other precious metals in poop. That revelation went viral and led to jokes about how the gold could be fashioned into rings. And the inevitable payoff: "Tell her she's number one with number two!" Smith still cringes a bit at the memory. She can laugh now but having years of scientific work reduced to a punch line was mortifying.

She told me that she and her colleagues had worked through as much of the periodic table as they could. Palladium, copper, zinc, tin, bismuth, lead: Smith found it all. Not all of the metals originate in human biosolids, since wastewater treatment plants process everything that goes down household and business drains—and in the case of combined sewer systems, industrial and storm drains as well. Pinpointing sources is often impossible, but urban infrastructure regularly leaves its mark in the form of proportionately higher levels of specific metals. Under the scanning electron microscope, Smith said, her research team saw little chunks of bismuth, which she suspects came from Pepto-Bismol. Jim Ippolito told me later that the copper piping in houses leaves other telltale signs. "Copper piping, like anything else, decays over time and you see the signature in the biosolids," he said. "And copper piping is joined together by solder and that solder typically has zinc in it and so you see that signature." Lead pipes can leave their own mark in wastewater and warn of water contamination, an environmental justice issue that has disproportionately impacted low-income communities of color.

Gold can come from dental fillings, food decorations, and even supplements and medical treatments. To measure the concentrations, Smith and her colleagues collected biosolids from several wastewater treatment plants, including small towns in Colorado's Mineral Belt, and then irradiated the air-dried samples to sterilize them. After mixing and grinding the bits into powder, the researchers sent them off for two independent analyses. From both measurements, Smith estimated the gold concentrations at one part per million, similar to low-grade deposits in the ground.

Smith repeatedly faced wisecracks and the yuck factor, even from fellow scientists, until she retired. Other efforts, however, suggest she was onto something. In 2015, environmental engineer Paul Westerhoff and colleagues published their own inventory of elements found in sewer sludge. The study used samples from the National Biosolids Repository, a collection sourced from ninety-four wastewater treatment plants for the US EPA's National Sewage Sludge Survey, or the enchantingly named NSSS. Wait, so there's basically a national poop archive? Yes, Smith told me, as I envisioned a *Raiders of the Lost Ark*–type government warehouse filled with a different sort of treasure. Anyway, for a city of one million, Westerhoff and his collaborators estimated that as much as $13 million worth of metals could be plucked from its wastewater every year. A treasure indeed.

As a proof-of-principle, Smith's team used cyanide to leach gold from the biosolid samples and recovered more than 80 percent of it. Using a poison for metal extraction, of course, isn't ideal. A friendlier metal-leaching option that uses a common fertilizer ingredient called thiosulfate (which is also, interestingly enough, an antidote to cyanide poisoning) recovered more than half of the gold from the biosolids. Sulfuric acid yielded a similar recovery rate for metals like copper and zinc. Researchers who specialize in metal extraction, Smith said, could likely devise better methods. She tried to play matchmaker with separate research groups that might be able to learn from each other—"the research yenta," she said with a laugh—but that work would have to be continued by someone else.

The city of Suwa in Japan's Nagano prefecture—with metal plating facilities, precision machining companies, and hot springs all potentially contributing to above-average gold content in its wastewater—has already provided a glimpse of the possibilities. In 2009, a sewage treatment plant there retrieved a bit less than four pounds of gold from every ton of ash. That's a far higher concentration than the Colorado gold rush miners of 1858 or even top modern mines have achieved.

What about extracting the notorious PFAS family of "forever chemicals"? The persistent and ubiquitous chemicals are fairly water-soluble, meaning that they can travel with groundwater and wastewater. They bind to proteins, meaning they can bioaccumulate within us and accompany our poop as well. Environmental and ecological engineer Linda Lee is a big proponent of reusing biosolids as fertilizers but understands the contamination concerns. If PFAS compounds could be prevented from ever entering wastewater treatment plants, she told me, they wouldn't end up in composted biosolids or effluent released to rivers. "We need to get them out of our products, not allow them in our products anymore, and stop point sources right now coming into treatment plants."

Environmental scientist John Lavery pointed out that the nearly ubiquitous family of chemicals is often found in far greater concentrations in cosmetics, clothes, and other consumer products than in our wastewater. "It's like anything in biosolids: it is the fingerprint of how we live, and when we look at PFAS, when we look at micro-constituents like cosmetics, personal care products, these are the fingerprints of decades of better living through chemistry." Maybe it's time to re-think what that means, he said.

Figuring out how best to turn off the spigot will take both time and resources. For the forever chemicals that are already flowing into the environment, Lee and colleagues are working on another application of activated carbon to mop some of them up. By attaching nickel and iron nanoparticles to the etched biochar, they've shown that it can bind up and even destroy some PFAS compounds within twenty-four hours. When used in combination under moderate heat, the two metals can effectively break the tight carbon-fluorine bonds that make PFAS chemicals so persistent in wastewater, biosolids, and the broader environment.

Flush

No single process can degrade everything in the large and highly diverse chemical family, but the experimentation with nickel and iron has raised hopes that a combination of chemical reactions can wrench apart many of the compounds. "We are pretty excited about these particles because they are pretty effective," Lee told me. In one EPA-funded collaboration, she had begun to explore the potential of deploying a two-part PFAS decontamination strategy within a wastewater treatment plant. The method sends wastewater through the tiny pores of a nanofilter and follows with what's known as electrochemical oxidation, a reaction that creates highly reactive chemical species that attack pollutants in the concentrated solution trapped by the filters. If successful, the combination approach might significantly reduce the PFAS levels in biosolids and wastewater effluent and remove another obstacle to expanding their reuse.

Ippolito had branched out into biochar as well. He was less convinced of its utility in improving the physical condition of soil, he told me, and was instead focusing on its potential to remediate heavy metal-tainted soils at former mining sites and clean up other contamination hotspots. Ippolito was working on two EPA Superfund sites—one in Oregon and another in Missouri—in which biochar had successfully adsorbed and locked up metal contaminants. Covering the sites with new plant growth had helped sequester the pollutants in place and keep them from migrating to adjacent land or water. The tiny carbon sponges had essentially pulled the metals out of solution and bound them up, so they weren't taken up by the plants either.

For another collaboration, he was working with scientists in China, where an estimated 7 percent of arable land has been contaminated with cadmium. Chronic exposure to the silvery-blue metal can cause cancer, and eating cadmium-contaminated food and water can weaken bones and damage the kidneys and other organs over time. More traditional reclamation methods don't work because the softer metal resists precipitation out of its water-soluble form. Applying carbon-rich biochar from a range of sources, though, could offer a cleanup alternative. A screen of roughly sixty kinds of biochar had yielded four promising candidates for cadmium remediation, though Ippolito hadn't yet had a chance to test human poop–derived biochar.

What he'd really love to do, he said, is set up a greenhouse study to see whether biochar could sequester metals such as cadmium from contaminated soil while simultaneously adding nutrients back into it. At the two Superfund sites, he said, he and his team had proven that the biochars work well for sequestering heavy metals in the environment. "That's great, but these soils that I'm working with are sort of devoid of nutrients, and they're devoid of microbial activity," he said. At cleanup sites, his team typically adds biochar to precipitate the metals, lime to raise the pH, and manure or biosolids to add back nutrients and spur microbial growth. "But what if we could just eliminate two of those three products and just use biochar that was derived from biosolids?" he mused.

His hypothesis is that biochar made from poop could both sequester the heavy metals *and* supply the nutrients needed to help amend the soil. That's because whatever nutrients are in the biomass to start with tend to stay put after pyrolysis transforms it into biochar. "Woody base materials: there's hardly any nutritive value. It's mostly carbon," Ippolito explained. As we've seen, though, human poop is packed with nutrients. Could it work? Ippolito sounded optimistic. Plenty of work had been done on biochar made from cow manure. The nutrient signatures in manure before and after pyrolysis tend to be very similar, he said, and the same should be true for our own excrement.

He retrieved two Ziploc bags of human poop–derived biochar, a small stash amid the many boxes and bags and bottles of the carbon material from a wide variety of sources in his office and lab. It looked like charcoal briquettes pulverized into little bits. I took several deep breaths. Nothing. All of the volatile organic compounds had been pyrolyzed at a temperature of 500 or 600 degrees Fahrenheit. Vaporized, along with their smells.

● ● ●

At the Fernhill Wetlands in Forest Grove, Oregon, the raging wildfires were too far away to smell any smoke; instead, they appeared like a late-summer storm cloud dominating the southern horizon, set off by the vivid green of the aquatic vegetation around me. I meandered down a shaded walkway in the afternoon heat, listening to the quiet rush of small waterfalls and the wind rustling the sedges. Two small black snakes, likely garters, slid away as I climbed up some rock steps. The graceful space, partly curated and partly wild, joined the essence of a Japanese healing garden with a functional marsh. On one side, landscape architect Hoichi Kurisu had shaped the space with boulders, pines, and blue spruce; two handsome wooden bridges arced gently over the water. On the other, a great egret posed in the shallows and a juvenile green heron sat motionless and well camouflaged on a snag. A sanctuary and oasis for both birds and humans.

Diane Taniguchi-Dennis, CEO of Clean Water Services, said many visitors can't imagine that the regional hotspot for birdwatching and a popular destination for weddings is all part of a natural water purification process at a sewage treatment plant. The constructed treatment wetlands here are an "ecological bridge" between the Forest Grove plant and the Tualatin River, she said. Years ago, nature had transformed old sewage lagoons next to the facility into a prettier expanse of water and cattails. People came to walk and view the birds that gathered. But the cattails had become a monoculture and even they were struggling due to winter inundations. Taniguchi-Dennis wanted to restore the degraded habitat and create a hybrid space where people could understand why their choices matter—what they choose to dump down the drain, what kinds of infrastructure they choose to support. She wanted to extend the Hawaiian concept of ohana—of family and kinship and interconnectedness—to this place. The healing garden could help reconnect us with nature, while the wetlands could extend that healing to nature as well.

The day I visited, roughly five million gallons of water were flowing into the natural treatment system from two resource recovery facilities operated by Clean Water Services. In the first section, large rectangular pools with beds of gravel six feet deep provide habitat for a concentrated population of bacteria that naturally degrade the remaining ammonia. Then the water cascades over multiple waterfalls, helping to add back oxygen. Through a meandering series of wetlands, native plants absorb nitrogen, phosphorus, and other nutrients. The abundant vegetation cools the water by almost four degrees Fahrenheit over its five-day journey to the river, protecting salmon and other wildlife from a warmer discharge.

As part of the extreme makeover, the utility varied the contours and elevations of the old pond beds to create better habitat for species that preferred shallower water. Workers added back 180 logs and snags to provide other spaces for wildlife and more than one million native plants to re-create the marsh ecosystem. Then they let nature "fill in the blanks," Taniguchi-Dennis said. Humans had started the restoration. "We think we're controlling it, right?" she said. But just as I had discovered in my own yard, nature took over and charted a new course.

Artificially cooling and wringing more nitrogen and phosphorus from the water would have required a much costlier course of treatment. "It takes a lot of concrete and steel and energy to be able to run these processes that nature runs," Taniguchi-Dennis said. Without the wetlands, in fact, the utility would have spent roughly twice as much to expand a wastewater treatment facility that still lacked the ability to cool the water. The hybrid approach—treating the water with a nature-inspired reactor and then finishing the process with nature itself—joined an urban system with a natural one and created a mutualism that has benefited both. "It's leveraging science and technology with the power of Mother Nature," she said. Now that she had seen the possibilities, Taniguchi-Dennis had bigger dreams for Fernhill. "What if we could create the right biodiversity within the wetland that actually amplified what the river needs to restore its health and its waters? What if we could create the right type of algae that the river needs?" Seeding the wetlands with some beneficial algae species, for instance, might increase the amount of dissolved oxygen in the river. The by-products of wastewater treatment were being reimagined as tools for not just reducing pollution but actively restoring a wetland, a river, an entire ecosystem.

The first time I talked with David Sedlak, he was visiting another unusual wetland in California's Bay Area. The Oro Loma Sanitary District in San Lorenzo had undertaken an ambitious experiment, with his team's help, to construct a flood control levee in the form of a heavily planted horizontal slope instead of a vertical wall. The horizontal levee, the district hoped, could do multiple things at once: mitigate flooding risks from storm surges as sea levels rise; filter the sanitary district's treated effluent to help improve water quality in the San Francisco Bay; and bring back some key wetland habitat for the region's flora and fauna. And the vegetated slope might accomplish all of that for a fraction of the cost of traditional flood-control alternatives.

In the ocean, coral reefs can help break incoming waves during a storm and reduce the surge. Healthy wetlands can, too, and the Oro Loma project was using their inherent strength to build more resiliency in a coastal community that would be increasingly impacted by global warming. Like the restored wetlands in Forest Grove, Oregon, the Oro Loma slope was also taking advantage of coastal vegetation's natural ability to filter water before it drains into the ocean. From the bay, a tidal marsh and then a brackish marsh slope inland over a distance of two hundred feet to reach a height of five feet. That's the horizontal levee that attenuates incoming waves. Then the land dips down again into a freshwater basin that receives treated wastewater and stormwater from the sanitary district's treatment plant. The water in this basin, or swale, seeps back into the marsh, where microbes in the soil help filter out more nitrogen, sequester more metals, and break down more organic chemicals during the water's journey to the bay. "Call it polishing, if you will," Sedlak said.

As part of the project, researchers set up quadrants with four soil types, plant species, and watering rates to find the best combination for cleaning the effluent and creating sustainable habitat. Willow trees seemed to do a bit better at filtering the water than sedges and other meadow plants; Sedlak said he suspected the trees' aggressive roots were churning up the soil and creating macro-pores that improved the permeability and aided the water flow. To their surprise, though, he and colleagues found that for improving water quality, the type of plants didn't matter nearly as much as where and how quickly the wastewater was flowing back through the marsh. "If you can keep the water flowing underground, it takes longer for it to move across the slope and there are more opportunities for the microbes to do their magic," he said. The system, in fact, had worked wonders at removing pharmaceuticals and antibiotics that are common concerns in coastal waters.

To maintain the underground flow, the researchers regulated the rate at which they added treated wastewater to the swale; on average, it percolated through the levee and reached the bay within three to seven days. The key to maximizing the natural filtering, they found, was to use porous sand and gravel mixed with wood chips, which act as a food source for the microbes. An alternative design, a vegetated creek-like swale that ran through the center of the wetland, provided an attractive feature but a relatively poor filter because most of the flow was faster and confined to the surface. Willows might be less attractive, Sedlak said, but the fast-growing trees could absorb more nutrients as well as more energy from a storm. To meet all of the objectives, he mused, maybe the trees could be combined with other plants in a curated wetland that optimized the levee's strength and filtering abilities while providing an aesthetically pleasing habitat.

In the meantime, his group was testing another strategy that focused not on getting more wastewater through the filtering wetlands but on treating *more concentrated* wastewater, like the coffee-colored brine I saw in the steel sink at the Orange County Groundwater Replenishment System. The concentrate left after the reverse osmosis purification step—maybe 15 percent of the starting amount—is full of salts and nutrients and chemicals. That all needs to go somewhere. In Orange County, it goes back to the adjacent wastewater treatment plant for another round of treatment. But Sedlak thinks the concentrated solution could be filtered through a vegetated horizontal levee instead. It might be saltier than normal wastewater, but the salt levels are still only a fraction of seawater salinity and easily handled by wetland plants well-adapted to coastal living.

The idea of tying horizontal levees to coastal treatment plants was gaining traction with other cities in the Bay Area, Sedlak told me. He had found a kindred spirit in Belgium, too, where an engineer had begun using a willow bed to filter the concentrated nutrients left behind after the reverse osmosis step at a potable water recycling plant. "I'm bullish, personally, on using these kinds of managed natural systems, or nature-based systems, to improve water quality," Sedlak said. Utilities have traditionally focused on gray infrastructure like concrete boxes and treatment plants. But he saw tremendous potential in green infrastructure; it could be less expensive *and* more attractive to the public.

It can also appear where you might least expect it, like the rooftop of a film and television production studio in a heavily industrialized corner of Brooklyn. Greenpoint, once a predominantly Polish neighborhood and my home for eight years, is bordered on its northern and eastern flanks by Newtown Creek, which has the dubious distinction of being one of the most polluted waterways in the US. Multiple oil refineries along the banks of what was once a creek and salt marsh contributed to the country's largest underground oil spill, of which an estimated thirteen million gallons have been removed so far. From the mid-nineteenth century onward, dozens of other refineries for everything from glue and fertilizer to copper and sulfuric acid dumped more toxins and solvents into what had become an industrial canal. In 2010, the EPA designated the creek a Superfund site to remediate some of the accumulated waste.

But chemical contamination is only part of the creek's woes. Whenever New York City's combined sewer system is overwhelmed with storm water, outfall pipes dump the untreated mix of sewage and storm water into city waterways to prevent it from backing up into city streets. Sometimes it's not enough, like when a sewer overflow swamped eight midtown hotels and the Times Square subway station during a storm in July 1913. The regular disasters have led to nearly surreal headlines more than a century later, like this one in 2018: "Please Don't Flush the Toilet. It's Raining." With more intense storms regularly battering the city, sewage discharges from the antiquated system and the resulting spike in bacterial contamination in the creek and other waterways have become necessary evils.

In the fourth-floor office of the Newtown Creek Alliance, executive director Willis Elkins showed me a color-coded map that pinpointed all of the outfalls; the creek, he said, has some of the largest. "And unfortunately, these are placed at the most stagnant points of the waterway because they're sort of these dead-end tributaries," he said. Other outfalls benefit from more natural water movement in the East River. But because nearly every square inch of Newtown Creek had been artificially reshaped to meet the city's industrial needs, it lost its ability to flush and filter and otherwise assist with the city's sanitation needs. After rainstorms, in other words, New York's shit tends to linger in the waterway.

In 2002, the Newtown Creek Alliance formed in an effort to force the polluters and the government to clean up the creek; Elkins told me that the nonprofit tries to make the connection between environmental justice and investments in water quality and infrastructure. The federal Superfund process, though complicated and contentious, had successfully remediated the copper contamination and was moving forward with other projects. But a partial fix proposed by the city and state to address the biological contamination was proving far more controversial. The plan called for a massive concrete tunnel that would act like a temporary storage facility for the untreated mix of storm and sewer water until it could be cleaned by Greenpoint's Newtown Creek Wastewater Treatment Plant. As proposed, the tunnel might reduce overflows by 60 percent or so once completed in 2042. But the alliance and its partners argued that those plans were based on older standards for acceptable bacteria levels and on the "representative" annual rainfall at John F. Kennedy International Airport in 2008, which could badly underestimate the expected precipitation more than thirty years later. New York City was already experiencing more intense rainstorms. The gray infrastructure, although necessary, was unlikely to be sufficient on its own to make up for the decades of environmental abuse and neglect.

Another kind of remediation effort was taking root, though. To prevent some of the storm water runoff from ever reaching the sewer system, the city was heeding a state mandate to add pervious pavement, curbside rain gardens, and bioswales to help soak it up. In a city sometimes called the Concrete Jungle, of course, permeable ground is at a premium. Elkins contended that the city could be more aggressive about making its remaining parking lots pervious and remaking its acres of rooftops into bastions of green. These reimagined roofs, essentially gardens planted atop waterproof membranes, could be a powerful and sustainable part of the city's urban sewer infrastructure—a green infrastructure that wouldn't require another twenty years to build. "Of course you have all these other auxiliary benefits, and many of those are completely tied to climate change as well," Elkins said. Green roofs can increase wildlife habitat while decreasing the energy consumption needed to cool buildings during heat waves. And they can improve air quality while reducing the urban heat island effect that has caused some concrete-rich and canopy-poor neighborhoods to swelter.

Racist policies like redlining, which actively enforced and deepened preexisting segregation in the US, also uprooted heat-lessening greenery and replaced it with heat-absorbing asphalt and concrete in urban islands of disadvantage. Today, the result is clearly visible as deep red blotches on heat maps of cities throughout the country. A 2019 study of the urban heat island effect across Baltimore; Washington, DC; and Richmond, Virginia, found summertime temperature differences of more than 18 degrees Fahrenheit between the hottest and coolest locations in all three cities. The anomalies, the researchers concluded, were largely attributable to land cover patterns: a density of buildings and heat-absorbing surfaces tended to amplify the heat, while parks and open spaces tempered it. Another recent study of 108 urban areas found that 94 percent had consistently hotter surface temperatures in formerly redlined areas. Reversing that trend will require active attention and sustained remediation. Investing in more green roofs, though, could be a start.

Brenda Suchilt, who manages the five green roofs on the Broadway Stages building for the Newtown Creek Alliance, led me to the "upper meadow," the largest of the gardens, and then to the even more dramatic "front yard." A stone walkway curved across an expanse thick with wildflowers, with a glass orb in the center reflecting the afternoon light. Goldenrod and wild strawberry. Milkweed for the caterpillars of Monarch butterflies. Echinacea for bees. A falcon sometimes stops by; Elkins thought it might nest atop one of the giant anaerobic digester "eggs" that turns food scraps and solid waste into biogas at the hulking wastewater treatment plant just to the southwest. I smelled the faint scent of ammonia from the imposing facility and heard the rumble of trucks and excavators moving mounds of scrap metal at a recycling center. The blooming of nature here on the small archipelago of green seemed nearly miraculous.

Settlement money paid to the Greenpoint Community Environmental Fund helped pay for the set of demonstration roofs and provided a precedent for similar efforts. Landscape architect Marni Majorelle worked with the Newtown Creek Alliance to choose native plant species that would be drought tolerant and survive drier periods in the five-inch soil substrate while providing valuable habitat. By the creek, the nonprofit had planted other green corridors with native plants as well. Pollinator pathways, Suchilt called them, that would help knit together a fledgling ecosystem amid the rusting metal and concrete. After maturing, she said, the rooftop gardens arguably required less maintenance than their ground-level counterparts, though she was keeping an eye on more aggressive plants like the goldenrod so they wouldn't crowd out the others. Suchilt had noted a recent decline in insects, and she fretted over the Monarchs that had been more abundant in previous summers. She recalled the giddiness she felt upon seeing fifteen of them flitting around the rooftop front yard in 2018. Seeing them, she said, "was just the most magical experience that I think I've ever had." I walked along the stone pathway and felt a momentary thrill as well: there was a single Monarch, fluttering above the flowers. Some pollinators had made their way back after all.

The same thing was happening, slowly, in Newtown Creek. Elkins had seen some of the changes for himself during periodic outings by canoe and kayak. "The creek is cleaner than it's been in a hundred years," he said. The oxygen levels were rising, making the water more habitable for wildlife. The bacterial contaminant levels were still high, but some life was coming back to the reintroduced buffers, natural crevices, and created spaces. Eels and horseshoe crabs. Clams and mussels in some triangular cutouts made to mimic small tidal pools. Even oysters in a few places. Nature was finding a way.

I wanted to stay longer on the unlikely rooftop oasis, but Elkins told me that the creek was at low tide, and he thought I'd enjoy a recently completed nature walk that follows the creek through part of the neighborhood. I nearly missed the discreet entrance and then found myself in a lushly planted corridor that slices through the heavy industrial zone, crosses the creek, and then parallels it. A small, nearly hidden buffer of green, but a buffer nonetheless. At one point, a long row of concrete steps rises out of the water, with the lowest featuring the triangular tidal pool cutouts. Sure enough, I saw mussels in many of them. A dead crab in one, but living crabs in others. A bit of sea lettuce, a lot of algae. The walk was aspirational, offering a view of what *might* be a more natural setting one day. Life, some seeded and some returning on its own, had been given a toehold.

Three days after my visit to the rooftop gardens and Newtown Creek, the remnants of Hurricane Henri provided a more sobering lesson on the power of nature. Over a thirty-six-hour period, Central Park and parts of Brooklyn received more than eight inches of rain; if extended across the entire city, the volume would have equaled more than forty-two billion gallons of water. As a recap of the deluge in the *New York Times* reminded readers, the city's combined sewer system was designed to handle somewhat less than four billion gallons of water a day, or perhaps six billion for the duration of the storm. Anything beyond that, sewage and all, was destined for the harbor. Less than two weeks later, the remnants of Hurricane Ida proved to be even more destructive. It was as if the skies were adding their own exclamation point to Elkins's complaint about the inadequacy of gray sewer infrastructure.

Seeing life persist in Newtown Creek despite the odds, though, reminded me of the recovery of Cuba's coral reefs. In 2015, a joint Cuban-US research expedition collected the first long core from a Cuban coral. Much like the growth rings of a tree, an expanding coral can chronicle local conditions in successive layers of the colony's calcium carbonate skeleton. Daria Siciliano, who joined the expedition, said corals can essentially capture hundreds of years' worth of environmental conditions including climactic fluctuations. By attaching a handheld pneumatic drill to a scuba tank, the scientists extracted a lengthy core from a massive starlet coral in a patch reef within the Gulf of Ana María, between the Cuban coast and Gardens of the Queen. This time capsule, roughly the diameter of a coffee mug and the length of a broomstick, includes growth layers dating back to the late 1700s.

Siciliano's lab initially extracted information about water temperature and salinity to reconstruct the historical climate. For a follow-up phase, the lab had begun using a state-of-the-art mass spectrometer—like a sensitive scale for molecules—to analyze tiny amounts of nitrogen trapped in the calcium carbonate and capture any trends in the quantity and quality of the nitrogen seeping into the gulf. Based on the ratio of two isotopes, or stable variants of the element, the technique can differentiate among nitrogen sources such as sewage, organic fertilizers, and synthetics. By comparing fluctuations in fertilizer levels with ocean conditions and the coral's annual growth bands, the research might help solidify a suspected correlation between pollution levels and reef conditions that other scientists have documented in samples around the world.

Definitively answering how lower-impact farming methods may have reduced Cuba's flow of chemical fertilizer into the marine environment and improved the health of these ecosystems will require more analysis of corals from other strategic locations, Siciliano emphasized. Stronger US-Cuba relations, though, could help address the need for follow-up collaborations on land and at sea—with the potential to improve coral reef management and restoration throughout the Caribbean and beyond.

Environmental scientist Jane Lubchenco, the former administrator of the US National Oceanic and Atmospheric Administration, talks about the ocean as a singular entity. "The reality is that it's all a single ocean, it's all connected," she said at a virtual conference in 2020. "The ocean connects us, not divides us." At first, she said, we thought the ocean was so immense that it was simply too big to fail. And then as its problems mounted—degraded corals, failing fisheries, dead zones, and increasing pollution among them—a second and equally false narrative suggested that the ocean had become so hopelessly depleted and disrupted that it was simply too big to fix. "It's all about doom and gloom, doom and gloom," Lubchenco said. The ocean as a victim.

But we could pivot to a new narrative that envisions it as a solution instead. Lubchenco had advised a group of fourteen world leaders, called the Ocean Panel, on how to accomplish that. One way to both reduce carbon emissions and improve food insecurity, a panel report suggested, would be to derive more of our protein from the sea. A startling finding from the report suggested that the ocean could sustainably supply sixfold more food than it does today, mostly in the form of bivalves such as mussels and oysters. While the vast amount of money allocated for COVID-19 stimulus plans focused on terrestrial activities and infrastructure, Lubchenco noted, attending to the ocean as well could address a major driver of the global economy. One way to aid coastal communities and protect shellfish aquaculture, to reduce the burden of water-borne diseases and improve the health of coral reef ecosystems, would be to invest in better wastewater treatment infrastructure and stop "thinking of the ocean as a dumping ground," she said.

If you accept Lubchenco's new narrative that says the ocean is so central to our health and well-being and prosperity that it's "too big to ignore," then it makes sense that advocates and utilities are looking for new ways to clean the water that ends up there. That coral experts are considering how to preserve the health of a regional wonder that draws tourists to Cuba and may reseed marine life elsewhere. That commercial shellfish farmers in the US—who are increasingly concerned about global warming, agricultural runoff, and pollution from leaking septic systems and straight pipes into coastal waters—are promoting alternatives like composting toilets. In Washington State, the same record-breaking heat wave in 2021 that scorched Melissa Meyer's Tacoma farm literally cooked shellfish exposed by an extreme low tide, killing bivalves by the hundreds of millions. One witness said it smelled like a clambake.

We can look to seeded shellfish farms and replenished vegetable farms to help feed the planet. To coral reefs and vegetated levees to help break storm waves and reconnect us to nature. To restored forests and revegetated mines to remake the landscape. To green roofs to cool the air while sopping up storm water and preventing flooded sewers. If we look at nature as an ally instead of a competitor, then it makes sense to wisely use a tool we can provide in abundance to help aid our alliance.

Ten weeks after the brutal heat wave, we had an intentional clam bake on Washington's Key Peninsula to celebrate the long Labor Day weekend. The next day, I walked with Geoff and friends at low tide along a beach littered with so many dead Pacific sand dollars that they crunched underfoot. But out in the water of the shallow and surprisingly warm Case Inlet where I waded up to my knees, thousands of healthy black ones still angled out of the sand. A local resident was inspecting a cage of seeded oysters. He figured that he had lost about a quarter of them to the heat. But horse clams and geoducks burrowed deep in the sand spit little geysers of water at us as we passed. We're still here, they signaled. And with a little help from their human neighbors, the populations might return the favor and help ensure our persistence as well.

CHAPTER TWELVE

Momentum

I MAGINE THAT YOU'RE A CITY worker living in an alternate version of today's Anthropocene. Your home and office buildings reach nearly three hundred feet in height—no small feat considering that they're made mostly of engineered wood. The buildings can make their own energy from the sun or wind or underground heat. They can gather their own water from rooftop rain and grow their own leafy greens. And when you shit in them? Well, that's where things get interesting.

A living building, one that works with nature to maximize the potential of natural daylight and solar power and fresh air and sustainable timber and rainwater, may have seemed like utopian fiction not so long ago. But skyscrapers made of renewable resources like laminated timber have sprouted up around the world; in 2019, Norway's eighteen-story Mjøstårnet tower of offices, apartments, a hotel, and a restaurant became the world's tallest wooden building. A plyscraper, some media dubbed it. By 2022, about thirty buildings had been fully certified by a separate movement called the Living Building Challenge that has pushed the construction industry to see how far it can go toward creating fully sustainable spaces that benefit the environment while promoting the health of their inhabitants. The buildings have the life span of a tree, are lit mainly by the sun, and exclude a long "red list" of toxic materials such as mercury, polyvinyl chloride, formaldehyde, and hormone-mimicking compounds. In Seattle, a six-story experiment called the Bullitt Center was billed as the world's most self-sufficient office building when it opened its doors in 2013. And during the many public tours that followed, visitors invariably asked to see a bathroom.

More than almost any other feature, the Bullitt's bathrooms seemed to elicit an overwhelming sense of curiosity among the urbanites. They wanted to know whether its twenty-four composting toilets, well, stank. They didn't, much to the surprise of nearly everyone who has been in an outhouse (and having helped my dad build one, I can confirm that odor control is often an ongoing concern). As one of many features in the $32.5 million living building, the foaming toilets and ten connected aerobic composters in its bowels used the output of office workers as the input for microbes and worms, taking an age-old idea and updating it for the twenty-first century. The final product, at least in principle, isn't just compost that can enrich the soil for gardens and environmental reclamation projects, but an enviable asset that may seem nearly unthinkable for a modern structure in the middle of a major city: independence.

Extensive power grids and sewer connections are normal parts of urban life for much of the Western world, where electricity travels along a convoluted network from power plants to transmission and distribution lines, and where we send our unwanted output hurtling through a miles-long odyssey of pipes. Wastewater treatment plants will continue to be necessary resource recovery plants. But what if, in parallel, we can create a more adaptable and nimble system that makes space for clusters of buildings with their own energy-production and by-product-recycling systems that don't tax an aging infrastructure threatened by increasingly devastating natural disasters?

I know what you're thinking. *Seriously, composting toilets?* Yes, seriously. Beyond the gee-whiz innovations of poop-fueled rocket ships and gold rushes that may create more apparent value, maybe true progress can be measured by the expanding reach of the deceptively simple composting toilet. I like to think of it as the love child of Bartleby the Scrivener and Jane Eyre: declaring that it would "prefer not to" do what's expected by postindustrial capitalism, and then achieving its own independence and sharing the wealth. It may not be a perfect vehicle for recapturing the value of poop, but its modern incarnation doesn't require a radical change in our behavior—only a willingness to reconsider what's normal, what's valuable, and what's possible.

Denis Hayes, president and chief executive of the Bullitt Foundation that built the Bullitt Center, said everyone wanted to know what behavioral changes were necessary to make the building's self-sufficiency work. "It turns out, not much," he said. Sure, the building features a lovely glass-enclosed staircase that offers panoramic views to reward those who climb the stairs instead of riding the elevators (Hayes once called it the "irresistible stairway"). And the tenants know enough to turn off their lights and computers at night to save energy. But mostly it was a matter of the architects, engineers, and builders buying the most efficient lights and computers, and of creating a sustainable environment around them. "You're not changing behavior. You're just causing people to operate within a different medium," Hayes said.

Deborah Sigler, an education and outreach specialist for the University of Washington's Center for Integrated Design, has led countless tours of the Bullitt Center, where the design lab occupies part of the second floor. On the tours, she encouraged her guests to go to a bathroom and inspect and smell what seemed like a reasonably normal toilet. "I think they're always pleasantly surprised, and they're curious about hey, where does it go?" Sigler said. So they'd all head down to the basement to see for themselves. The solids ended up in ten bright blue Phoenix composting tanks, each the size of a small shed, while the urine and water leached from the compost, or the leachate, filled four containers that could handle four hundred gallons apiece.

A properly equipped composting tank is two-thirds full of the same kind of pine shavings you might use for hamster bedding, but large enough to prevent clumping. The carbon-rich bulking agent adds structure to what becomes a three-part compost pile. The tanks contain three stacked chambers, each with an attached hand crank that turns rotating tines to stir the contents and direct them downward. As organic matter accumulates and decomposes, it gradually moves from top to middle to bottom. Some composters, like those in the Bullitt, are seeded with red worms to aid the decomposition (sometimes known as manure worms, *Eisenia fetida* can eat half their body weight every day). Given enough time, the wood shavings and poop and toilet paper break down into a stabilized, earthy compost that can be removed through a bottom door.

A company called McNel Septic Service collected the composted biosolids from tanks as they filled, Sigler said, but only had to do so once every two years, on average. Maybe it shouldn't be so surprising, given that our poop is roughly three-fourths water. Remove most of it, and the remaining mass becomes relatively compact. For their first-ever removal, the McNel workers weren't sure what to expect and wore hazmat suits—"ready for battle," as Sigler put it. The suits came off when they opened the bottom hatch and began shoveling out wet, decomposing wood shavings, and then eventually dark organic matter that smelled like soil.

The periodic removals weren't necessarily pleasant, but they normally required little more than shovels, wheelbarrows, and a good-sized pickup truck that would haul the matter to Sawdust Supply to be intermingled in its own composting piles. Yes, it's the same bark and landscaping supplier that converted King County's Loop biosolids into the coveted GroCo soil amendment that I used on my vegetable gardens. The Bullitt Center was essentially producing the same starting material, just on a smaller scale and over a longer period of time.

Every four months or so, the same septic service pumped the liquid leachate into a tanker and took it to an aerobic processing facility operated by King County. After filtering and treating the nutrient-rich water with ultraviolet light to kill off pathogens, county workers used it for restoration projects in a natural area called Chinook Bend, a fifty-nine-acre tract of woods, wetlands, and former pastureland nestled in a bend of the Snoqualmie River. The wildlife habitat includes a prime spawning site for Chinook salmon. Sigler was thrilled that the liquid didn't end up in Puget Sound, where it would have gone if treated by a wastewater treatment plant.

Critics have worried about the safety and cost and maintenance of composting toilets. But nearly all have agreed that water-based sanitation is far less efficient. Consider that what we normally use to flush is treated to drinking-water standards and then immediately contaminated every time we push the lever. "It's criminal," Sigler said. The Bullitt was using roughly one-fifteenth to one-twentieth of the potable water used by a typical office building, she said, mostly due to its waterless toilets.

There's a quiet rebellion in incorporating tanks full of worms and poop into buildings that represent a vision of the future. But our future, our planet, and our economic survival all depend on our willingness to responsibly deal with our own shit. Instead of dispersers, we have acted as concentrators, draining resources from some places and piling up so much in others that the accumulations have created new problems. Hayes pointed out that we've even concentrated livestock into destructive "cattle cities," turning what were once periodic bursts of nutrients distributed throughout a prairie ecosystem into floods that have given rise to stinking sewage lagoons.

We are concentrating ourselves into megacities as well, requiring creative and flexible approaches like container-based sanitation and reinvented toilets to safely handle all of the humanure. The lessons aren't applicable only to faraway places, or only to poop. Even in Seattle, a long train of boxcars carries the city's trash 320 miles to a landfill in northern Oregon. "Day after day after day. It's just fucking crazy. And all of that has value," Hayes said. Instead of distributing wealth, we are simply piling up our problems in a new location.

Embracing the concept of a composting toilet means accepting that progress will come by working with nature and valuing things that aren't necessarily glamorous or sexy but sturdy and practical. We live on an impermanent planet, one that is always rotting and recombining and reinventing itself. Every bit of life—including the human body—is on loan, and everything goes back into the system for re-use when it's done. Death, decomposition, and even ugliness are necessary for beauty to bloom. Maybe composting toilets aren't your cup of tea. But in the landscapes we've chosen to inhabit, they can improve sanitation, conserve water, create value, restore the environment, and grant independence. So much so, in fact, that regulations and city planning based on traditional sewer hookups may no longer make sense in many parts of the world.

Like many other Western nations, the United States has accumulated plenty of liabilities through its inadequate and crumbling infrastructure. In 2021, the American Society of Civil Engineers gave the country's wastewater infrastructure a D-plus rating. The decline in aging pipes and treatment plants near the end of their normal life spans is being compounded by a population shift toward metropolitan areas that will be forced to accommodate a bigger portion of the nation's wastewater treatment needs.

US schools scored no better in the infrastructure report card, saddled in part with the tens of thousands of portable classrooms that hunker on the margins of overcrowded schoolyards. More than a third of all public schools in the country rely on portable classrooms. Of those "temporary" buildings that are often anything but, the American Society of Civil Engineers estimated that an astounding 45 percent were in poor or fair condition. The "vinyl-wrapped boxes" filled with kids are among the worst offenders of poor indoor air quality, Seattle architect Stacy Smedley told me. Most lack adequate fresh air and daylight and nearly all lack plumbing, meaning that kids have to walk to other buildings or portable toilets to use the bathroom.

Smedley, inspired by the loss of her grandfather's woods and her beloved "green sky" trees in Clackamas, Oregon, when she was a little girl, has helped design healthier and more environmentally friendly schools and learning environments with ample input from the students themselves. At one of those spaces, a science classroom at the private Bertschi School in Seattle, three guides were waiting for me and began talking excitedly, all at once, about the things they loved the most. Fifth-graders Isabel, Isabel, and Jack finished each other's sentences and corrected each other in the exuberant and brutally honest way that only ten- and eleven-year-olds can. My young tour guides met me in an outdoor garden irrigated by reclaimed water, where they had grown a variety of plants like huckleberries for paint and pancakes. Then we moved inside, where they pointed out an eighteen-foot-tall wall covered with four kinds of plants that help purify the classroom's air and treat used gray water. Ooh, and the indoor stream that starts with rainwater collected on the roof, then flows down through an exposed pipe along the way, across a pebble-bottomed concrete channel that crosses the floor, and into two cisterns for storage.

Who wanted to explain the composting toilet? Both Isabels shot up their hands. "Me, me! That's my favorite," Isabel #1 said. "I love the toilet. I did a video on it." We headed to the bathroom, where the girls immediately did the big reveal by opening the louvered closet doors that hid the toilet's two composting waste storage tanks. Isabel #1 launched into presentation mode and gestured first to the toilet, "where you do your business," and then to a button on the wall, which when pressed, sucks everything into a setup within the large closet (each vacuum flush uses about a pint of rainwater). The Isabels did the reveal again, flinging open the closet doors with a flourish. With a little prompting, Isabel #1 told me that it only takes six months for the poop to be converted to compost. So how had their families reacted to the toilet? Jack and Isabel #2 agreed that their parents thought it was cool. Isabel #1, however, had to do a little more convincing. "When I first told my grandmother, she was disgusted and said, 'You are not allowed to eat food from the garden.'" She shrugged. "Bunny poo stinks a lot more."

Julie Blystad, the kids' science teacher, said she was amazed at how her students were integrating the lessons from the living building into their lives. They had become adamant about recycling. Jack had already informed me that the classroom used more energy than it should have the previous year. The class of fifth graders monitoring its usage noticed the gap and convinced school administrators to add more solar panels on two adjacent buildings.

In 2012, Smedley drew inspiration from other schoolchildren at a public school in Jasper, Alberta, who wanted a living classroom of their own. From the lessons she learned designing the Bertschi School's science wing, Smedley and two colleagues reasoned that they could extend the same principles to portable classrooms. So they founded a nonprofit group called The SEED Collaborative (SEED stands for Sustainable Education Every Day) and set out to make their own. When I went to see their prototype for myself, I marveled at how its electricity circuitry and plumbing was deliberately exposed so kids could see how it all worked, like how water flowed from a cistern to a sink operated with a hand-pump, and then to a living wall growing tomatoes and herbs. Smedley pointed out the route above us. "Basically, you can sit here, and you can tell them the whole story," she said.

The nonprofit sold that prototype to the private Perkins School in Seattle for use as its own science classroom—the first of what Smedley and her cofounders hoped would be "hundreds of little green sprouts." When the portable classroom arrived, some kids began referring to the bathroom's centerpiece as a "magic toilet" that converted poop into soil for an on-site garden. Science teacher Zoë Dash later wrote about their enthusiasm: "Not only do they learn about waste treatment and decomposition from the composting toilet, but they find the chance to use it thrilling, and surprisingly many of them attempt to wait to use the bathroom until they are in my class just so they can use the toilet that turns their waste into the nutrient-rich soil!" If there was a yuck factor to overcome, the cool factor had clearly done the trick.

● ● ●

Glenn Nelson, the owner of Advanced Composting Systems in White-fish, Montana, told me that his own interest in environmentally friendly toilets could be traced back to a waterless model from Sweden called the Clivus Multrum. First built in 1939 by inventor Rikard Lindström, the relatively simple setup used two chutes to effectively compost poop and kitchen scraps at the Lindström family's beachfront property on the Baltic Sea. The inventor named his basement system after clivus, the Latin word for "inclination," which referred to the slanting bottom of the concrete collecting chamber (later made of fiberglass). There, a layer of peat moss, grass, and soil starts the aerobic composting process while a natural ventilation system directs air down through the toilet and into the chamber, diverting any odors through a vent pipe. A friend later convinced Lindström to add multrum, a composite Swedish word that means a "composting room" or "place of decay," to the name of his invention.

After Nelson's parents emigrated from Sweden to the US, his mother read an article about the invention in a Swedish magazine. The article was apparently so gripping that she decided she wanted a Clivus Multrum of her own and began exchanging letters with Lindström. They became pen pals and when Nelson and his wife set off for a backpacking trip around Europe in the early 1970s, his mother urged them to visit the Lindströms. They did, and the gracious hosts let them stay at their beachfront home for a week. Nelson was soon hooked on the Swedish pot as well and upon his return home, became a dealer and later a designer and manufacturer of the composting toilet. Nelson's mother got one, and more than 20,000 have now been installed around the world.

But it wasn't enough. Nelson wanted to resolve flaws in the tank's design that interfered with how the compost moved and settled and that made removing the organic matter cumbersome. So he designed the Phoenix Composting Toilet and manufactured tanks in three sizes. Seattle's Bullitt Center received ten of the largest models, while he installed one in his own home as well. When we talked, Nelson was growing tomatoes and peppers—rarities for a northern Montana garden—in a two-story greenhouse attached to his home. It helps that his plants have been well fed. In a forty-foot-long planting bed that runs the length of the greenhouse, initially seeded with wood chips, he regularly deposits the compost from his ultra-low flush toilets and then allows the mixture to continue decomposing into a rich soil. For residential composting, Nelson cautioned that the homemade soil shouldn't be considered pathogen-free, meaning that it's not ideal for growing onions and other root crops. In essence, it's the same thing as the Class B biosolids that the Bullitt Center and some wastewater treatment plants produce.

About one-fourth of Nelson's customers are homeowners. His company had just installed the toilets in two affordable housing units in Moab, Utah, when we talked and he suggested that if properly designed, the $6,300 system could be significantly more affordable than a septic system. Even so, his sales hadn't increased much beyond an annual range of fifty to a hundred units. I asked him why. Lack of awareness? Regulatory hurdles? Disgust? No, he said. In part it seemed to be a matter of cultural momentum. "With a composting toilet, you really have to design a home around it—it's fairly large."

In the US, at least, most architects and home builders haven't had it on their radar and few homeowners have clamored for what's still something of a novelty, even if the underlying concept extends back at least as far as the fertile Amazonian dark earths, or the Terra Preta de Índio compost of feces and manure and food and biochar. In 1860, the Reverend Henry Moule moved the concept indoors with his patented "earth closet" that was an early forerunner of the Clivus Multrum, the Phoenix, and other models. The setup was little more than a tall metal bucket positioned beneath a wooden toilet seat and a hopper filled with dry dirt or peat. Pulling a handle released a bit of the earth or peat to cover the fresh deposits below. Moule, a parish vicar in Fordington, England, had discovered that the natural covering reduced the odor and helped the poop decompose. Full buckets could be deposited in a garden to continue the breakdown process.

After it was introduced to the US in the late 1860s, one enthusiastic reviewer wrote that the system needed but a "trifling" quantity of earth or failing that, coal ashes. "The writer has now used the Moule system for over three years, and with four closets in constant use—three in the house and one in an adjacent street—he has never the least occasion for the preparation of earth. The ashes of the house fires furnish, without cost, all the material required for perfect disinfection." The reverend was reportedly motivated by both public health *and* environmental concerns. After the same cholera epidemics of 1849 and 1854 that prompted physician John Snow's medical sleuthing in London's Soho neighborhood and the Great Stink of 1858 that prodded Parliament to unify the city sewer system, Moule became convinced that cesspools were a health hazard. He also disapproved of the water closets used by wealthier homeowners. According to one account, "He felt it polluted God's rivers and seas and was a waste of God's nutrients contained in excrement, which should be returned to the soil."

In that regard, Moule would have found rapport with French novelist Victor Hugo. The flowery language of *Les Misérables* notwithstanding, Nelson told me he had read the book three times and seen the play twice. One passage, in particular, stuck with him. In 1862, a few years after Moule's invention, Hugo expounded upon the tragedy of Paris throwing its own manure into the sea despite the considerable value of human fertilizer that had long been known to Chinese farmers.

> There is no guano comparable in fertility to the detritus of a capital. A great city is the most powerful of dung producers. To employ the city to enrich the plain would be a sure success. If our gold is manure, on the other hand, our manure is gold. What is done with this gold, manure? It is swept into the abyss. At great expense, we send out convoys of ships, to gather up at the South Pole the droppings of petrels and penguins, and the incalculable element of wealth that we have at hand we send to the sea. All the human and animal manure that the world loses, if restored to the land instead of being thrown into the water, would suffice to nourish the world. This garbage heaped up beside the stone blocks, the tumbrils of mire jolting through the streets at night, the awful scavengers' carts, the fetid streams of subterranean slime that the pavement hides from you, do you know what all this is? It is the flowering meadow, it is the green grass, it is marjoram and thyme and sage, it is game, it is cattle, it is the satisfied lowing of huge oxen in the evening, it is perfumed hay, it is golden wheat, it is bread on your table, it is joy, it is life. So wills that mysterious creation, transformation on earth and transfiguration in heaven.

For the Pueblo of Acoma in New Mexico, in a city built to bring its people closer to the heavens and rain clouds, a transformation on Earth has meant alfalfa and environmental restoration and cultural preservation. About an hour west of Albuquerque, Sky City sits atop a 367-foot-high sandstone mesa and is believed to be the oldest continuously inhabited settlement in North America. The small village of adobe homes, one of four in the Pueblo of Acoma, is the ceremonial heart of the Acoma people and a popular tourist destination. Sky City's seventeenth-century, fortress-like San Estévan del Rey Mission Church, built by Acoma men, women, and children amid a brutal occupation by Spanish conquistadors and priests, now hosts religious ceremonies and holidays that incorporate both Catholic and Indigenous influences.

The village has always been off-grid and lacks even running water except for a few natural cisterns that still collect rainwater. For bathrooms, villagers and visitors used dozens of ancient outhouses and newer portable toilets that clustered atop the mesa and dotted the slope in front of the mission. Jose Antonio, a waterless composting systems operator for the Pueblo of Acoma Utility Authority, told me that over the centuries, eighty-two personal restrooms had been built next to the church alone, creating both an environmental and aesthetic challenge for the tribe's effort to preserve its cultural heritage. He explained that some of the outhouse contents had been seeping into the sandstone for centuries. The rock, when saturated, can begin to flake and lose its strength, and the tribe was determined to keep its mesa from crumbling.

As a solution, the tribe's environment department decided to build public bathrooms equipped with composting toilets instead. Antonio had met Nelson at a conference and heard about his Phoenix composting toilets. Would he be interested in a new project? And so, as part of a $3 million venture, funded with state and federal grants, Advanced Composting Systems built twelve solar-powered restrooms on the mesa's periphery to match the village's stucco look. Altogether, the two-story structures house sixty-two toilets that empty into thirty-one composting tanks beneath them. The sloping roofs hold the necessary solar panels and channel rainwater into other tanks for reuse in washbasins.

The elegant solution still required outreach and education to get everyone on board. At first, some residents didn't want to give up their personal bathrooms in favor of more communal restrooms. "It took a while for our community to get used to something new," Antonio said; the change, in essence, required the same normalization and legitimization that has been essential for water reclamation efforts. Eventually, he and other advocates won over fellow tribal members. Residents liked that they didn't have to maintain the new restrooms themselves and that the ventilated structures weren't nearly as hot as the enclosed outhouses and portable bathrooms. "The reason why we're putting them here is because we're thinking about our future, our kids, our grandkids," Antonio reminded them. "You know, we don't want Acoma to fall because we saturated our mesa." Seven years after the new restrooms arrived, he had noticed a change in the sandstone where some of the outhouses had been: a regrowth of vegetation, a drying-out of the underlying rock. The colors of the mesa, he said, were returning to a more familiar sandstone brown.

Every year, around April or May, Antonio removed some of the composted matter, maybe 450 gallons at most. Much of it initially went to a nearby wastewater treatment plant, but a few years ago he worked with a tribal member to use some to fill in the bald spots on his alfalfa field instead. It worked perfectly, the farmer told him. The farmer had become a believer and was trying it on a cornfield next. Maybe more field trials could be conducted with other fruits and vegetables, Antonio mused. But it was clear that the compost was a valuable resource that should stay in the community. Based on the pueblo's success, he went to other tribal communities in the Southwest to teach them how to take care of their own composting toilets. It took him three years to figure it out, he said; he was still learning all the time but had become a go-to expert in the region and was passing down his accumulated knowledge. Even so, he worried about who would take over when he retired.

• • •

If they haven't yet gone mainstream, composting toilets have long had a strong undercurrent of support. Consider that the 1994 self-published bestseller, *The Humanure Handbook*, by composting toilet pioneer Joseph Jenkins, has been translated into about twenty languages. Jenkins sells his own handcrafted spin on the commode, something he calls a "Loveable Loo." Its environmentally friendly nature is a core attraction for some; for others, it's an incidental feature of a simplified waste disposal process that enables off-the-grid living. There's a reason why composting toilets have been embraced across the political spectrum: they have become an unexpected rallying point for independence by a diverse community of modern pioneers. Just as some models have been showcased at international toilet fairs, others have been major attractions at survivalist and doomsday expos. Below an ad for a twelve-pack of B&M canned brown bread on the HappyPreppers.com website, a thorough discussion of the toilet's merits ended with this upbeat conclusion: "A composting toilet is a happy ending! You'll feel better about the environment and you'll be better prepared for when the stuff hits the fan."

Curiously, a cause célèbre for rugged individualists has also become a catalyst for new communities. In 2014, the UK's five-day Glastonbury Festival began using composting loos at its huge outdoor venue. By 2021, the popular music festival was using more than 1,300 of them to systematically replace their widely reviled plastic portable counterparts. The loos' contents, sprinkled with sawdust after every number two, are combined and composted on a nearby farm, and then redistributed to other farmers. In London, about 10 percent of the roughly 5,000 people who live on canal boats and lack access to the sewer system have likewise embraced composting loos. For that more permanent community, the loos have provided a clever answer to the question of how to properly dispose of all the poop—a particularly apt solution given the fouling of the Thames over the centuries.

In northern Amsterdam, the temporary neighborhood of de Ceuvel has gone one step further. A team of collaborators painstakingly converted a heavily polluted former shipyard into a residential, commercial, and arts community of salvaged houseboats equipped with roof-mounted solar panels and dry composting toilets. The "playground for circular cities," as its architects dubbed it, has become a testbed for new ideas on closing the loop between urban production and consumption. Because the revamped houseboats don't require a foundation or pipes connecting them to the city sewer system, they haven't disturbed the polluted soil and are giving the land a chance to heal.

The pop-up neighborhood, which debuted in the summer of 2014 and is slated to last until January 1, 2024 before being dismantled, is connected by a winding bamboo jetty that gives visitors the impression of being in a drained and revegetated harbor. Even the plants have played a role in the transformation: Amsterdam-based DELVA Landscape Architecture & Urbanism and researchers at the University of Ghent in Belgium used plant-based detoxification, or phytoremediation, to help clean contaminants like heavy metals and polycyclic aromatic hydrocarbons from the soil. Compost collected from the houseboats, further processed in a portable tumbling composter to ensure its safety, has helped to nourish the plants. Another collaborator on the project, Amsterdam's space&matter design studio, described how the transformation of a former wasteland would create a source of value long after the project's conclusion. As largely self-sufficient elements, one of the firm's architects noted, the boats would leave little trace after their departure, "leaving the land more valuable, biodiverse, and cleaner from pollutants."

• • •

The same concept of lasting value has figured prominently in attempts to make better use of the toilet in addressing the global sanitation crisis. In 2020, according to a joint report by UNICEF and the World Health Organization, nearly half of the world's population lacked safe access to sanitation. In parts of the world without a functional sewer system, safely extracting the wastewater from rudimentary toilets or pit latrines has been a major hurdle, especially with limited access to electricity or water for flushing (Kenya's Sanergy and Sanivation have developed two strategies for accomplishing it). The same report estimated that 5 percent of people worldwide still lack even basic toilets and defecate in outdoor spaces like fields, bushes, and bodies of water, though that number has dropped by more than half over the past two decades.

The Reinvent the Toilet Challenge, launched by the Bill & Melinda Gates Foundation in 2011, approached the sanitation problem with an engineering competition to design off-grid receptacles that can safely handle wastewater without electricity or running water. Many of the designs *also* provide a renewable resource—fertilizer, clean water, or a source of power, for example—that could encourage more widespread adoption in communities without basic restrooms. They have harnessed solar and microwave energy, acted like pressure cookers, or used smaller versions of the filtering membranes that water reclamation plants rely upon. One enterprising group of engineers figured out how to burn poop by turning it into pellets with a commercial meat grinder. The Gates-funded prototype featured a semi-gasifier combustion process that's similar to how pellet-burning stoves work. The team nicknamed it the Assifier.

Environmental engineer Karl Linden and colleagues entered the

Reinvent the Toilet Challenge with a prototype they called the Sol-Char. The high-heat, low-oxygen pyrolysis technology produces the same kind of biochar fuel that has been created at waste processing centers. But in this case, the transformation occurs after every deposit. The Sol-Char uses a solar panel and concentrator to collect and focus sunlight into high-intensity energy and then transfer it through fiberoptic cables to heat up a chamber away from the toilet. The heat boils off the water. "Once you get the water out, then you can start to char the organic matter and create basically poop briquettes," Linden said. After the pyrolysis step, adding a sticky binding material like molasses to the bits of biochar creates the fuel source.

To test the biochar's energy, Linden found the perfect collaborator in Lupita Montoya, a mechanical engineer and air quality expert who works at the intersection of science and poverty. She was initially excited about the energy potential and the ability to safely reuse something still widely associated with disease. But she still had concerns, namely the potential for air pollution. The Montoya lab first pyrolyzed feces from twenty-five anonymous volunteers at 570, 840, and 1,380 degrees Fahrenheit to compare the energy content in each batch. Fecal char pyrolyzed at 570 degrees was the clear winner, with an energy content comparable to charcoal made from wood or bituminous coal.

There was also the practical matter of whether the process could be replicated in both a city and a small village. "If you are trying to make sure that this is done locally, then you need to look at the resources locally. You don't want to depend on bringing resources from far away because then that has an energy cost as well," Montoya said. The researchers tested binders that could be easily found, including a sticky plant starch and several combinations of molasses and lime. The fecal briquettes bound together with 5 percent starch retained the highest energy content but were too brittle and broke into little bits when dropped from a few feet. Briquettes made with 20 percent molasses and 7 percent lime were the sturdiest but scored lowest on energy content. Briquettes with 10 percent starch provided the best combination of impact resistance and energy, again comparable to charcoal briquettes.

There's plenty left to learn but Montoya said the study was a necessary step toward showing potential users that the fuel source wouldn't require acclimating to a completely different product. "We come out with these great technological ideas, but then they're so removed from what people are used to." Numerous interventions have failed because they took a technologist approach and not necessarily a holistic approach, she said; to be adopted, a new resource must meet a community's needs *and* its cultural context. She hadn't yet been able to directly address the air pollution question. But she hoped that people might pay attention to her lab's initial findings. The United Nations University Institute for Water, Environment and Health already had, and used the study's conclusions in its number crunching to calculate the theoretical value of converting the world's wastewater into biogas and biochar as alternatives to natural gas and traditional charcoal. Turning at least part of that potential into reality might help the world deal with both its sanitation problem and its energy problem at the same time.

In the end, Linden's team proved that their Sol-Char toilet worked (at a 2014 toilet fair in New Delhi, India, the researchers demonstrated its potential by using the poop-char briquettes to roast peanuts). But concerns over the maintenance and use of a solar panel and solar concentrator in a densely populated urban environment prevented it from being selected for the next round of funding toward the eventual goal of mass production. Even so, Linden said his group's innovation of capturing, concentrating, and transferring high-intensity solar energy could be useful in separate applications such as reducing the energy costs associated with water disinfection and desalination.

Most composting toilets, of course, use a far simpler means to the same end. Is there room for both strategies? The ultimate test may be whether communities want them, can afford them, and keep using them over the long term. The real-world utility will take time to resolve, though other contenders in the toilet sweepstakes have appeared in field trials around the world and proven that their own high-tech concepts work as well. Even with the investment of commercial partners, critics have worried whether the reinvented off-grid models will be affordable enough for widespread use. One strategy has been to combine the best innovations into a single, low-cost version the Gates Foundation is calling the Generation 2 Reinvented Toilet. Bill Gates has compared the toilet's heat, energy, and pressure-based technology to the slightly less complicated mechanics of a large espresso maker. The output from the input: recycled water for flushing, ash, and dried cakes of pasteurized poop that could be emptied every few days and composted.

Another Gates Foundation–funded strategy has circled back to a more familiar design that works directly with nature: a composting system created at the London School of Hygiene & Tropical Medicine and commercialized by TBF Environmental Solutions in India. The US$350 Tiger Toilet, as it's known, is designed to be a "leapfrog" sanitation alternative. The "flush-and-forget" setup uses a familiar squat toilet above. Dug into the ground below, a barrel or tank contains tiger worms, the same manure-loving *Eisenia fetida* species as in the Bullitt Center's composters, nestled in a bedding layer of wood chips or coconut husks atop drainage layers of soil and gravel. The natural filter cleans the water as it percolates through the layers, while the worms feast on the poop and convert it into castings, and then into valuable vermicompost, as it's known. By the end of 2021, more than 4,500 of the toilets had been installed throughout India.

Rural communities, in turns out, may be critical to creating composting toilets that are cheap, practical, and widely used. In June 2021, I went to the northern Minnesota town of Pine River to visit Hunt Utilities Group, a seventy-acre lab of sustainable living. The key to a composting toilet designed for residential use, co-owner Paul Hunt told me over a lunch of mixed greens and mariner's pie (basically a seafood pot pie), is that it has to be both ecological and economical; not just easy to install, but also easy to maintain. He had studied *The Humanure Handbook* closely and the lab had initially tested out a simple bucket method. "All the things I doubted, we tested, and he's right: the thing works," Hunt concluded.

But he was keen to see whether the lab could design a more user-friendly version that was so low maintenance it could be used in affordable housing units. So about six years ago, employee Simon Goble took the lead in designing a prototype within a plywood room in the center of the lab's main workspace, the Manifesting Shop. "It's been torture tested in a number of ways," Hunt said, noting that it had endured both feast and famine. After an on-site concert a few years back, employees pitched about a hundred pounds of vegetables into the rotating composting bin (which to my eyes looked suspiciously like a repurposed kiddie pool). It was smelly for the next two weeks, Hunt recalled, but only really bothered you if you lifted the lid of the plywood box that enclosed the bin. It had been seeded with wood chips and red worms (as in the Tiger Toilets and the Bullitt Center's composting tanks) and smelled only faintly woody and earthy when Goble lifted the lid to point out its features. If the system dried out too much, it would attract moths, but regular sprays of water brought the worms back to the surface, where they ate the moth eggs. "Most of the time, when the worms are happy, it's like a meat loaf in there; it's just *bazillions* of them," Hunt said. More recently, the COVID-19 pandemic had starved the little wrigglers since few workers remained on-site to use the john, and neither he nor Goble had seen recent signs of movement.

Goble had automated the system so that spreaders evened out the piles and the tub rotated every half-hour. In six years of use, they still hadn't had to empty it, and both agreed that it had required very little maintenance. Incorporating a version into homes, though, might still require a community effort, Hunt surmised, especially if the system began acting badly: neighbors could look after each other's systems or hire someone to do it for them.

● ● ●

Normalizing the equitable distribution of community resources in ways that also benefit the environment may sound like a lofty discussion held at a think tank. In reality, designing systems that will be widely adopted and sustainable over time sometimes comes down to accounting for our messy peculiarities and differences when it comes to sitting on the pot. The Bullitt building had gravity on its side to convey its poop to the composters. But the system in its low-ceilinged basement lacked a way to divide up the waste evenly, meaning that some heavily used toilets—mostly in the men's bathrooms on a few floors—filled their connected composting tanks far sooner than other lightly used toilets and didn't always allow the tanks to fully decompose the organic matter. As we've seen, psychology can play a big role in determining who defecates in public restrooms, and Sigler said women in the building have tended to use the toilets much less than the men. That means offices that skew toward a male workforce will have much busier toilets.

When the Bullitt was fully staffed, Sigler said the building's second floor served as a kind of lobby where visitors would often use the bathroom while waiting for friends or colleagues. Workers on other floors would too. That meant the second-floor toilets were especially busy. But the bathroom-goers wouldn't use just any toilet. No, they gravitated to a fully enclosed, handicapped-accessible stall. Why? Unlike the neighboring stall, its dividers extend all the way down to the floor, meaning that no one can see its occupant's feet. In other words, poop shame effectively funneled a significant amount of the building's deposits through specific toilets on a specific floor.

The phenomenon made such an impression that I heard about it from four people. Building engineer Mark Rogers confirmed that men like to go in the enclosed stalls. "And they tend to sit for a *long, long time*," he said, taking his own time on the last three words so I understood the sheer monumentality of it all. Nelson said he had heard that some workers used the stalls to play on their cell phones during extended bathroom breaks.

The problem is that the toilets relied on a photosensor to dispense foam as a flushing agent for as long as they were occupied. When one worker called it the "latte effect," Sigler asked him to clarify. "And he goes, 'Well you know, if you sit on the toilet long enough, that foaming action percolates up and hits you on the butt.' And I'm thinking, how long do you sit on it? How long does it take you to do your business?" Apparently, a *long, long time*. Hayes told me how wet toilet paper could create a dam until the foam reached the undercarriage of startled lingerers. As a retrofit, Nelson's team adjusted the occupancy sensor to stop the foaming after five minutes, but the mechanism still failed to efficiently flush everything it should have.

There were other issues. Nitrogen-rich urine can sometimes jumpstart the composting process; some enthusiastic advocates have added "pee bales" in gardens for that very purpose. Yes, it's exactly what you think it is. But urine can be too much of a good thing, and busier toilets in the men's bathrooms in the Bullitt were emptying too much urea and ammonia into the composting tanks, which can interfere with microbial growth. Because the building lacked a mop sink on every floor, some janitors were apparently dumping the water they used to wash the floors down the toilets as well. The disruptions may have had something to do with one particularly ill-fated emptying of a full composting tank: it released an unmistakably foul odor and effectively cleared the building. On another occasion, small flies variously known as humpbacked, scuttle, sewer, and coffin flies colonized one of the composters. "Oh my God, it's a nightmare," Rogers told me, clearly traumatized by the incident. The tanks, Sigler said, essentially created their own little ecosystems. And just like others, they could become unbalanced and create openings for interlopers.

So it wasn't a perfectly smooth process. But, Sigler maintained, the Bullitt Center's experiment demonstrated to public health officials that composting toilets could be managed safely and responsibly. Ironically, the experiment's final straw was the inefficiency of disposal at the scale of a single building. With no composting toilets in other nearby office buildings, the Bullitt paid a high rate to pump out and haul the liquids nearly thirty miles. When Seattle's Sawdust Supply closed forever, the composted solids had to be hauled nearly seventy miles for responsible reuse. In the end, Hayes and his colleagues calculated that the environmental costs outweighed the benefits and they reluctantly killed the alternative toilets. "This building can be thought of as about a hundred different science fair experiments, many of them nested within one another," he told me. "And by and large, they all succeeded, but this one didn't succeed. And as I at least tried to sell it to my board, if everything succeeded, we just weren't stretching far enough."

But it wasn't a complete failure. The Bullitt Center's experience suggested that progressive policies can mesh well with smarter designs, for instance. One way to correct for differences in whether and how occupants use public toilets, Sigler said, would be to have gender-neutral bathrooms. The designers of other large living buildings also learned from the Bullitt's mistakes and redesigned their composting toilet systems to increase the efficiency and minimize the headaches. For the one-story Arch Nexus SAC office in Sacramento, California, a Norwegian-built system vacuums the waste from the toilets, macerates it, and then pumps it to a manifold, or a fitting that evenly distributes the flow to eight Phoenix composting tanks. A valve opens sequentially for each tank, Nelson explained, "so every time a toilet flushes, it doses the next tank." Portland's PAE Living Building, meant to last five hundred years, has adopted a similar system with more leeway: eighteen vacuum-flushed toilets evenly distribute their output to twenty Phoenix composting tanks in the bowels of the five-story building. Design teams from another living building on the Georgia Tech University campus in Atlanta went with an updated model of Nelson's inspiration: the Clivus Multrum.

Resolving the larger issue of critical mass from multiple buildings may require a more sustained effort. In urban areas, updated regulations and incentives that allow for a coordinated approach could effectively create cooperative eco-districts of homes and businesses that together produce a steady supply of nutrient-rich compost. The districts, maybe comprised of a few city blocks, could provide the necessary scale. Haulers like those at McNel Septic Service could deliver each building's Class B output to a local processing center akin to Seattle's Sawdust Supply. There, the mix could be further composted with wood chips and heated with solar-powered pasteurizers to convert it into a pathogen-free soil that could be certified by local authorities as safe, Class A compost. And then it could be redistributed locally. As we've seen, most major cities in the US have a pressing need for regreening and restoration close to home.

Sigler told me that she had a similar fantasy, that other nearby buildings would adopt composting toilets as well and create a more cost-effective route for what could be another disposal service not unlike the city's curbside recycling and garden composting pickups. I asked Nelson whether it could work, and he seemed optimistic. While traveling around northern Europe in the 1970s, he noted how many high-rise apartment buildings had adjacent community gardens; making space for something similar in the US—including dedicated facilities to provide the compost for those gardens—would help improve our city designs, he suggested.

One key to making such eco-districts work, according to environmental engineer David Sedlak, will be finding the right scale for efficiently separating gray water from black water for treatment and reuse. Or maybe the districts could treat it all together in a more compact system like an automated microfiltration membrane bioreactor, which is essentially a mini–sewage treatment plant. More research and development will be required before water reuse systems become user-friendly enough to feature in residential areas, Sedlak cautioned. But a lack of existing infrastructure in new developments may provide the flexibility to re-think sanitation systems from the ground up. "When you have a built-up city, the retrofit problem is considerable, and the resources that we've sunk into the sewer system make it really hard to abandon," he said. "So the real opportunities are in edge cities and new developments where these kinds of eco-districts or eco-blocks can be reconceived from the beginning."

No one solution will work everywhere, just as no single product will be needed or wanted everywhere. As one of the planet's most prolific poopers, though, we're nowhere near our potential. With a growing diversity of possible solutions, towns and cities might adopt a hybrid approach: a container-based sanitation service and the Omni Processor or microfiltration membrane bioreactors to handle larger volumes transported to hubs in some densely populated areas, say, and composting toilets in linked residential or commercial districts where the resulting biomass can be efficiently gathered up and processed closer to the source and redistributed to nearby parks, farms, or forests.

Sedlak and other engineers have imagined an even more efficient circular economy: after the wastewater is cleaned and reused, the remaining impurities could be incorporated into existing curbside recycling programs. "Instead of an expensive sewer network, the salts and nutrients in wastewater that cannot be converted to carbon dioxide and water during treatment would be dried and left at the curb for recycling," he wrote in 2018. In essence, it's a scaled-down version of recapturing the phosphorus in wastewater for reuse as fertilizer; communities like Amsterdam's de Ceuvel neighborhood have already experimented with using struvite reactors to recapture the nutrients from urine.

● ● ●

The end of the Bullitt's own composting experiment arrived in two installments, or rather, un-installments at the end of 2020. Early in the morning on the first Saturday, a crew from McNel Septic Service arrived to clean out six of the basement tanks. Owner Ken Carlton told me he hated to see the building switch from the composting toilets to more standard models: it represented the loss of a very lucrative account. Even so, he said, business was booming from the estimated 85,000 septic systems in King County alone; he was so busy that he routinely turned down prospective clients. Sigler was there to witness the event as well and took me down to the basement, where Carlton's crew was struggling with two clogged composters that smelled a bit like soiled hamster cages. Soon after we arrived, a worker named Kirk shouted to Sigler that another composting tank was chock full of red worms.

"Aaah!" she yelled in excitement, and we rushed over to take a look. He was right—there were so many that I could see them writhing in the compost and a few dropped through the bottom door. The worms normally exist in a gradient: the most up top and the least down below; they tend to go where the poop is. These had been well fed. "They multiply like crazy," Sigler said, a bit in awe. More and more kept coming out with the shit-compost, which meant to Sigler that the process was working the way it should, even if it hadn't entirely finished. She seemed enormously pleased but confided that she was sad as well. This room was always one of her favorite parts of the building tour.

The workers continued to struggle with a stubborn agglomeration of shavings and poop in the midsection of a digester. COVID-19 had disrupted the building's normal rhythms and the tanks' regular input, and the mixture hadn't been stirred as often as it should have. Without the building's tenants, Sigler told me later, the composting process struggled and became unbalanced. "The shit is like concrete," Carlton said. He moved on to the tank where the worms had been thriving and sprayed it with a hose to loosen the remaining matter. He grinned as he turned to Sigler, sprayer in hand. "It's like a colonoscopy!" She laughed.

The water had made the smell noticeably worse. Sharper, more pungent. Kirk held a flashlight for Carlton. "This is where all the magic happens!" he shouted. And then, to no one in particular, "It's just poop." Another worker was still grappling with a stopped-up composter. "Someone needs more fiber in their diet!" he yelled, and Carlton grinned at me. The matter finally broke free in big chunks and the crew bundled them off in blue wheelbarrows. Another worker was on a ladder, spraying a hose into a composter from the top. "Boss, I'm trying to clean the ice cream machine!" he joked, and Carlton laughed again: "I never thought I'd be doing this kind of shit for a living," he told me. Back up in the alleyway, while his workers secured a tarp over the trailer bed, Carlton expanded on the thought. "This is friggin' disgusting!" But it was a highly profitable business, and he was paying his workers $50 an hour. "I shower in the morning, I shower at night, and laugh all the way to the bank," he said.

Six weeks later, a crew of four returned to empty the last of the composters. This time, everything went smoothly and Kirk hauled away the contents, which all fit into a single trailer bed hooked up to a white Ford F-350. Less than three hours after they had arrived, they were all gone, with Sigler and I looking on as the Ford turned right and headed toward the freeway.

It could have ended there: an expensive one-off curiosity by a do-gooder foundation in a liberal city. Only it hasn't. Not yet. The lessons of the Bullitt Center will help revise and refine the composting systems for other living buildings in other cities, where other visitors will ask to see the bathrooms and wonder at how, well, normal they seem. It won't all go perfectly, but other septic services will help haul away other worm-filled piles of matter that smell and look nothing like the starting material.

In 2018, the Perkins School's science wing became the first portable classroom to earn a full designation as a living building. The recognition was bittersweet for the SEED Collaborative; after four of its classrooms had sprouted on both coasts of the US, Stacy Smedley and her colleagues announced that they had made the difficult decision to put the project on an indefinite hiatus. They didn't have the capacity to continue on their own but made their plans open-access, and in those classrooms with the life span of a tree, other children will clamor to use the "magic toilet" that helps feed a garden.

In the Pueblo of Acoma, Jose Antonio moved on from his position in the utility authority, but he and other caretakers have passed down their knowledge to the next generation, one of whom will answer the questions asked by another group of curious visitors.

In adobe towns and modern plyscrapers, in refurbished houseboats and living schools, our collective power has worth. How poetic it would be if more neighborhoods and cities and rural communities could tap into that surprising source of overlooked energy to reassert their dignity, their independence, their health, and their environment. A shittier future is a joyful one, simpler and homelier than the alternative but far more meaningful and innovative and optimistic.

As the planet's dominant megafauna, we have both the ability and the responsibility to restore and widen circles of value that sync with natural cycles instead of supplanting or suppressing them. Poop isn't everything we need, but it's more than enough to start. Shimogoe. Night soil. Humanure. Terra preta. Black gold. Sometimes hope arrives in unexpected packages. We are the couriers now, distributing them across the landscapes that will define our future.

ACKNOWLEDGMENTS

F OR A BOOK THAT CELEBRATES A COLLABORATIVE ACT OF production, it's only fitting that I start by acknowledging my dear friends and family. Without their support and encouragement, *Flush* would have remained a lonely kernel of an idea. They have listened to me blather on about it for years without flinching, made donations of pig manure in my honor, shared their own stories about the voluminous output of babies and dogs, and sent touching "Thinking of you" messages that have included links to strange stories about poop. Thank you—you know me well.

My husband has been my human thesaurus, part-time editor, fellow gardener, and beloved chef and travel companion who patiently allowed me to pop into wastewater treatment plants or water reclamation facilities while on vacation and let me discuss the finer points of biosolids over dinner. Geoff instilled in me a love of bird-watching and befriended our pandemic crow companions. He has been there every step of the way and given me the space and support to take on a project like this in an unsettled and uncertain time. I also owe an immense debt of gratitude to Russ and Susan Goedde, our dear friends and neighbors who have been enthusiastic cheerleaders from the start. They gamely read early versions of multiple chapters, helped me taste-test beer made from recycled water, and tried out composted biosolids in their own lovely gardens at my request.

The ecosystem that made this all possible extends to my many science and English teachers and professors, who saw my intense curiosity as a feature instead of a bug and who encouraged me to continue exploring both the natural world and the written word. It includes my supportive thesis adviser at the University of Washington, Beth Traxler, who gave me the freedom to be an unconventional graduate student and who rightly understood that I wasn't throwing my career into the crapper by wanting to be a writer instead of a researcher. At *Newsday*, my editors and colleagues mentored and nurtured me and gave me incredible opportunities to develop my skills at what was a dream job for a green journalist. Chrissie Giles, my friend and editor at *Mosaic*, inspired my deep dive into this world of poo with my very first feature story for her, on the rise of fecal transplants.

The science writing community has been another glorious source of encouragement ever since my days at the University of California at Santa Cruz, and I feel immensely lucky to be surrounded by such generous, gifted, and inspiring colleagues. Support groups like SciLance have sustained me over more than a few spells of darkness and self-doubt. A special thanks goes out to Virginia Gewin and Liza Gross, friends and talented writers who offered astute suggestions and advice on key chapters. And here in Seattle, I am deeply indebted to my friend and fellow science writer Michael Bradbury, who went above and beyond with his help on this project. For more than three years, we met regularly as accountability buddies and worked on every step, from finding an agent and publisher to outlining chapters and honing tricky passages.

Flush never would have seen the light of day, of course, had it not been for Anna Sproul-Latimer at Neon Literary, my hilarious and whip-smart agent, who saw something that others didn't and who helped me shape a half-formed idea into a coherent whole. And I will be forever thankful for Maddie Caldwell, my wonderful editor at Grand Central Publishing, who consistently made everything better with her funny and perceptive guidance and her remarkable ability to be simultaneously kind and razor sharp with needed edits. Jacqueline Young helped me navigate the many twists and turns of the publishing process, and Sarah Congdon designed a gorgeous book cover that perfectly captures the beauty of a swirl of hope and transformation. My top-notch fact-checkers, Lowri Daniels, Hannah Furfaro, and Yvonne McGreevy, were as gracious as they were thorough in questioning every fact and assertion and in pushing me to be as accurate as possible (though I alone am responsible for any remaining errors).

Finally, I want to acknowledge the writers who have helped bring so many stories to light, and the scientists, doctors, engineers, patients,

ACKNOWLEDGMENTS

advocates, and others who so generously shared their own stories with me. I couldn't include them all in this book, but their words and wisdom were crucial in allowing me to weave together a coherent narrative. The many revelations about our inner treasure and collective output have come about only because curious and determined people never stopped looking or asking questions, even when others thought it was gross or pointless or useless. This book is a tribute to all of them.

CREDITS

Parts of chapters 2 and 3 were adapted from "Medicine's Dirty Secret," first published April 28, 2014, by Wellcome on mosaicscience.com and reproduced here under a Creative Commons license.

Parts of chapter 2 were adapted from "Desperate Love in a Time of Cholera," first published April 28, 2014, by Wellcome on mosaicscience .com and reproduced here under a Creative Commons license.

Parts of chapter 11 were adapted from "Farm to Reef," first published May 8, 2018, in *bioGraphic* (biographic.com).

Parts of chapter 12 were adapted from "Inside the Green Schools Revolution," first published November 4, 2014, by Wellcome on mosaic science.com and reproduced here under a Creative Commons license.

FURTHER READING

My book has been but a drop in the bucket of scatological studies. Happily, those who want to take a deeper dive can read other works that have examined complementary themes of sanitation, sustainability, and shame-free poop production. Here are a few that deserve attention.

George, Rose. *The Big Necessity: The Unmentionable World of Human Waste and Why It Matters*. New York: Picador, 2008.

Shafner, Shawn. *Know Your Shit: What Your Crap Is Telling You*. New York: Cider Mill Press, 2022.

Wald, Chelsea. *Pipe Dreams: The Urgent Global Quest to Transform the Toilet*. New York: Avid Reader, 2021.

Zeldovich, Lina. *The Other Dark Matter: The Science and Business of Turning Waste into Wealth and Health*. Chicago: University of Chicago Press, 2021.

Introduction

Balkawade, Nilesh Unmesh, and Mangala Ashok Shinde. "Study of Length of Umbilical Cord and Fetal Outcome: A Study of 1,000 Deliveries." *The Journal of Obstetrics and Gynecology of India* 62, no. 5 (2012): 520–525.

Berendes, David M., Patricia J. Yang, Amanda Lai, David Hu, and Joe Brown. "Estimation of Global Recoverable Human and Animal Faecal Biomass." *Nature Sustainability* 1, no. 11 (2018): 679–685.

Chaisson, Clara. "When It Rains, It Pours Raw Sewage into New York City's Waterways." *National Resources Defense Council.* December 12, 2017.

Daisley, Hubert, Arlene Rampersad, and Dawn Lisa Meyers. "Pulmonary Embolism Associated with the Act of Defecation. 'The Bed Pan Syndrome.'" *Journal of Lung, Pulmonary, & Respiratory Research* 5, no. 2 (2018): 74–75.

Doughty, Caitlin. *From Here to Eternity: Traveling the World to Find the Good Death.* New York: W.W. Norton, 2017.

FBI. "Unearthing Stories for 20 Years at the 'Body Farm.'" March 20, 2019.

Gomi, Tarō. *Everyone Poops.* Translated by Amanda Mayer Stinchecum. Brooklyn, New York: Kane/Miller, 1993.

Gupta, Ashish O., and John E. Wagner. "Umbilical Cord Blood Transplants: Current Status and Evolving Therapies." *Frontiers in Pediatrics* (2020): 629.

Hu, Winnie. "Please Don't Flush the Toilet. It's Raining." *New York Times.* March 2, 2018.

Ishiyama, Yusuke, Satoshi Hoshide, Hiroyuki Mizuno, and Kazuomi Kario. "Constipation-Induced Pressor Effects as Triggers for Cardiovascular Events." *The Journal of Clinical Hypertension* 21, no. 3 (2019): 421–425.

Laporte, Dominique. *History of Shit.* Translated by Nadia Benadbid and Rodolphe el-Khoury. Cambridge, Massachusetts: MIT Press, 2002.

Markel, Howard. "Elvis' Addiction Was the Perfect Prescription for an Early Death." *PBS News Hour.* August 16, 2018.

Meissner, Dirk. "Victoria No Longer Flushes Raw Sewage into Ocean After Area Opens Treatment Plant." *The Canadian Press.* January 9, 2021.

Mufson, Steven, and Brady Dennis. "In Irma's Wake, Millions of Gallons of Sewage and Wastewater Are Bubbling up across Florida." *The Washington Post.* September 15, 2017.

Nelson, Bryn. "Cord Blood Banking: What You Need to Know." *Mosaic.* March 27, 2017.

Nelson, Bryn. "Death Down to a Science/Experiments at 'Body Farm,'" *Newsday,* November 24, 2003.

Nelson, Bryn. "The Life-Saving Treatment That's Being Thrown in the Trash." *Mosaic*. March 27, 2017.

Niziolomski, J., J. Rickson, N. Marquez-Grant, and M. Pawlett. "Soil Science Related to Human Body After Death." School of Energy, Environment and Agrifood, Cranfield University [ebook], available at: http://www. thecorpseproject. net /wp-content/uploads/2016/06/Corpseand-Soils-literature-review-March-2016 .pdf (2016).

Odell, Jenny. *How to Do Nothing: Resisting the Attention Economy*. Brooklyn: Melville House, 2020.

Roach, Mary. *Gulp: Adventures on the Alimentary Canal*. New York: W.W. Norton, 2013.

Rose, C., Alison Parker, Bruce Jefferson, and Elise Cartmell. "The Characterization of Feces and Urine: A Review of the Literature to Inform Advanced Treatment Technology." *Critical Reviews in Environmental Science and Technology* 45, no. 17 (2015): 1827–1879.

Rytkheu, Yuri. *The Chukchi Bible*. Translated by Ilona Yazhbin Chavasse. Brooklyn, New York: Archipelago, 2011.

Smallwood, Karl. "Do People Really Defecate Directly after Death and, If So, How Often Does It Occur?" *TodayIFoundOut.com*. June 3, 2019.

Stuckey, Alex. "Harvey Caused Sewage Spills." *Houston Chronicle*. September 19, 2017.

Zeng, Qing, Lishan Lv, and Xifu Zheng. "Is Acquired Disgust More Difficult to Extinguish Than Acquired Fear? An Event-Related Potential Study." *Frontiers in Psychology* 12 (2021): 687779.

Chapter One

Achour, L., S. Nancey, D. Moussata, I. Graber, B. Messing, and B. Flourie. "Faecal Bacterial Mass and Energetic Losses in Healthy Humans and Patients with a Short Bowel Syndrome." *European Journal of Clinical Nutrition* 61, no. 2 (2007): 233–238.

Almeida, Alexandre, Alex L. Mitchell, Miguel Boland, Samuel C. Forster, Gregory B. Gloor, Aleksandra Tarkowska, Trevor D. Lawley, and Robert D. Finn. "A New Genomic Blueprint of the Human Gut Microbiota." *Nature* 568, no. 7753 (2019): 499–504.

Anderson, James W., Pat Baird, Richard H. Davis, Stefanie Ferreri, Mary Knudtson, Ashraf Koraym, Valerie Waters, and Christine L. Williams. "Health Benefits of Dietary Fiber." *Nutrition Reviews* 67, no. 4 (2009): 188–205.

ARTIS Micropia. "Sustainability with Microbes." Accessed April 20, 2022, https:// www.micropia.nl/en/discover/stories/blog-lab-technician/sustainability -microbes/.

Bandaletova, Tatiana, Nina Bailey, Sheila A. Bingham, and Alexandre Loktionov. "Isolation of Exfoliated Colonocytes from Human Stool as a New Technique for Colonic Cytology." *Apmis* 110, no. 3 (2002): 239–246.

Banskota, Suhrid, Jean-Eric Ghia, and Waliul I. Khan. "Serotonin in the Gut: Blessing or a Curse." *Biochimie* 161 (2019): 56–64.

BIBLIOGRAPHY

Barr, Wendy, and Andrew Smith. "Acute Diarrhea in Adults." *American Family Physician* 89, no. 3 (2014): 180–189.

Beaumont, William. *Experiments and Observations on the Gastric Juice, and the Physiology of Digestion.* Plattsburgh: F.P. Allen, 1833.

Ben-Amor, Kaouther, Hans Heilig, Hauke Smidt, Elaine E. Vaughan, Tjakko Abee, and Willem M. de Vos. "Genetic Diversity of Viable, Injured, and Dead Fecal Bacteria Assessed by Fluorescence-Activated Cell Sorting and 16S rRNA Gene Analysis." *Applied and Environmental Microbiology* 71, no. 8 (2005): 4679–4689.

Berendes, David M., Patricia J. Yang, Amanda Lai, David Hu, and Joe Brown. "Estimation of Global Recoverable Human and Animal Faecal Biomass." *Nature Sustainability* 1, no. 11 (2018): 679–685.

Berstad, Arnold, Jan Raa, and Jørgen Valeur. "Indole–the Scent of a Healthy 'Inner Soil.'" *Microbial Ecology in Health and Disease* 26, no. 1 (2015): 27997.

Berstad, Arnold, Jan Raa, Tore Midtvedt, and Jørgen Valeur. "Probiotic Lactic Acid Bacteria–the Fledgling Cuckoos of the Gut?" *Microbial Ecology in Health and Disease* 27, no. 1 (2016): 31557.

Betts, J. Gordon, Kelly A. Young, James A. Wise, Eddie Johnson, Brandon Poe, Dean H. Kruse, Oksana Korol, Jody E. Johnson, Mark Womble, and Peter DeSaix. "Chemical Digestion and Absorption: A Closer Look." In *Anatomy and Physiology.* OpenStax, 2013.

Bhattacharya, Sudip, Vijay Kumar Chattu, and Amarjeet Singh. "Health Promotion and Prevention of Bowel Disorders through Toilet Designs: A Myth or Reality?" *Journal of Education and Health Promotion* 8 (2019).

Boback, Scott M., Christian L. Cox, Brian D. Ott, Rachel Carmody, Richard W. Wrangham, and Stephen M. Secor. "Cooking and Grinding Reduces the Cost of Meat Digestion." *Comparative Biochemistry and Physiology Part A: Molecular & Integrative Physiology* 148, no. 3 (2007): 651–656.

Bohlin, Johan, Erik Dahlin, Julia Dreja, Bodil Roth, Olle Ekberg, and Bodil Ohlsson. "Longer Colonic Transit Time Is Associated with Laxative and Drug Use, Lifestyle Factors, and Symptoms of Constipation." *Acta Radiologica Open* 7, no. 10 (2018): 2058460118807232.

Breidt, Fred, Roger F. McFeeters, Ilenys Perez-Diaz, and Cherl-Ho Lee. "Fermented Vegetables." In *Food Microbiology: Fundamentals and Frontiers.* 841–855. ASM Press, 2012.

Carding, Simon R., Nadine Davis, and L. J. A. P. Hoyles. "The Human Intestinal Virome in Health and Disease." *Alimentary Pharmacology & Therapeutics* 46, no. 9 (2017): 800–815.

Carpenter, Siri. "That Gut Feeling." *Monitor on Psychology,* 43, no. 8 (2012): 50.

Chandel, Dinesh S., Gheorghe T. Braileanu, June-Home J. Chen, Hegang H. Chen, and Pinaki Panigrahi. "Live Colonocytes in Newborn Stool: Surrogates for Evaluation of Gut Physiology and Disease Pathogenesis." *Pediatric Research* 70, no. 2 (2011): 153–158.

Chapkin, Robert S., Chen Zhao, Ivan Ivanov, Laurie A. Davidson, Jennifer S. Goldsby, Joanne R. Lupton, Rose Ann Mathai et al. "Noninvasive Stool-Based Detection of

Infant Gastrointestinal Development Using Gene Expression Profiles from Exfoliated Epithelial Cells." *American Journal of Physiology-Gastrointestinal and Liver Physiology* 298, no. 5 (2010): G582-G589.

Chen, Tingting, Wenmin Long, Chenhong Zhang, Shuang Liu, Liping Zhao, and Bruce R. Hamaker. "Fiber-Utilizing Capacity Varies in *Prevotella*-vVersus *Bacteroides*-Dominated Gut Microbiota." *Scientific Reports* 7, no. 1 (2017): 1–7.

Compound Chemistry. "The Chemistry of the Odour of Decomposition." October 30, 2014. https://www.compoundchem.com/2014/10/30/decompositionodour/.

Cummings, J. H., W. Branch, D. J. A. Jenkins, D. A. T. Southgate, Helen Houston, and W. P. T. James. "Colonic Response to Dietary Fibre from Carrot, Cabbage, Apple, Bran, and Guar Gum." *The Lancet* 311, no. 8054 (1978): 5–9.

Dalrymple, George H., and Oron L. Bass. "The Diet of the Florida Panther in Everglades National Park, Florida." *Bulletin—Florida Museum of Natural History*. 39, No. 5 (1996): 173–193.

DeGruttola, Arianna K., Daren Low, Atsushi Mizoguchi, and Emiko Mizoguchi. "Current Understanding of Dysbiosis in Disease in Human and Animal Models." *Inflammatory Bowel Diseases* 22, no. 5 (2016): 1137–1150.

Degen, L. P., and S. F. Phillips. "Variability of Gastrointestinal Transit in Healthy Women and Men." *Gut* 39, no. 2 (1996): 299–305.

Doughty, Christopher E., Joe Roman, Søren Faurby, Adam Wolf, Alifa Haque, Elisabeth S. Bakker, Yadvinder Malhi, John B. Dunning, and Jens-Christian Svenning. "Global Nutrient Transport in a World of Giants." *Proceedings of the National Academy of Sciences* 113, no. 4 (2016): 868–873.

Elias-Oliveira, Jefferson, Jefferson Antônio Leite, Ítalo Sousa Pereira, Jhefferson Barbosa Guimarães, Gabriel Martins da Costa Manso, João Santana Silva, Rita Cássia Tostes, and Daniela Carlos. "NLR and Intestinal Dysbiosis-Associated Inflammatory Illness: Drivers or Dampers?" *Frontiers in Immunology* 11 (2020): 1810.

Enders, Giulia. *Gut: The Inside Story of Our Body's Most Underrated Organ (Revised Edition)*. Vancouver: Greystone Books Ltd, 2018.

Eschner, Kat. "This Man's Gunshot Wound Gave Scientists a Window into Digestion." *Smithsonian*. June 6, 2017.

Faith, J. Tyler, and Todd A. Surovell. "Synchronous Extinction of North America's Pleistocene Mammals." *Proceedings of the National Academy of Sciences* 106, no. 49 (2009): 20641–20645.

Ferreira, Becky. "Another Thing a Triceratops Shares with an Elephant." *New York Times*. January 8, 2021.

Figueirido, Borja, Juan A. Pérez-Claros, Vanessa Torregrosa, Alberto Martín-Serra, and Paul Palmqvist. "Demythologizing *Arctodus Simus*, the 'Short-Faced' Long-Legged and Predaceous Bear That Never Was." *Journal of Vertebrate Paleontology* 30, no. 1 (2010): 262–75.

Flint, Harry J., Karen P. Scott, Sylvia H. Duncan, Petra Louis, and Evelyne Forano. "Microbial Degradation of Complex Carbohydrates in the Gut." *Gut Microbes* 3, no. 4 (2012): 289–306.

BIBLIOGRAPHY

Forget, Ph, Maarten Sinaasappel, Jan Bouquet, N. E. P. Deutz, and C. Smeets. "Fecal Polyamine Concentration in Children with and without Nutrient Malabsorption." *Journal of Pediatric Gastroenterology and Nutrition* 24, no. 3 (1997): 285–288.

Garner, Catherine E., Stephen Smith, Ben de Lacy Costello, Paul White, Robert Spencer, Chris SJ Probert, and Norman M. Ratcliffem. "Volatile Organic Compounds from Feces and Their Potential for Diagnosis of Gastrointestinal Disease." *The FASEB Journal* 21, no. 8 (2007): 1675–1688.

Gensollen, Thomas, Shankar S. Iyer, Dennis L. Kasper, and Richard S. Blumberg. "How Colonization by Microbiota in Early Life Shapes the Immune System." *Science* 352, no. 6285 (2016): 539–544.

Giridharadas, Anand. "The American Dream is Now in Denmark." *The.Ink.* February 23, 2021.

Gonzalez, Liara M., Adam J. Moeser, and Anthony T. Blikslager. "Porcine Models of Digestive Disease: The Future of Large Animal Translational Research." *Translational Research* 166, no. 1 (2015): 12–27.

Grant, Bethan. "How Fast Are Your Bowels? Take the Sweetcorn Test to Find out!" *ERIC*, 2012. Accessed April 20, 2022, https://www.eric.org.uk/blog/how-fast-are -your-bowels-take-the-sweetcorn-test-to-find-out.

Guinane, Caitriona M., and Paul D. Cotter. "Role of the Gut Microbiota in Health and Chronic Gastrointestinal Disease: Understanding a Hidden Metabolic Organ." *Therapeutic Advances in Gastroenterology* 6, no. 4 (2013): 295–308.

Hartley, Louise, Michael D. May, Emma Loveman, Jill L. Colquitt, and Karen Rees. "Dietary Fibre for the Primary Prevention of Cardiovascular Disease." *Cochrane Database of Systematic Reviews* 1 (2016).

Hu, Xiu, Xiangying Wei, Jie Ling, and Jianjun Chen. "Cobalt: An Essential Micronutrient for Plant Growth?" *Frontiers in Plant Science.* 12 (2021): 768523.

Hylla, Silke, Andrea Gostner, Gerda Dusel, Horst Anger, Hans-P. Bartram, Stefan U. Christl, Heinrich Kasper, and Wolfgang Scheppach. "Effects of Resistant Starch on the Colon in Healthy Volunteers: Possible Implications for Cancer Prevention." *The American Journal of Clinical Nutrition* 67, no. 1 (1998): 136–142.

Iqbal, Jahangir, and M. Mahmood Hussain. "Intestinal Lipid Absorption." *American Journal of Physiology-Endocrinology and Metabolism* 296, no. 6 (2009): E1183-E1194.

Iyengar, Vasantha, George P. Albaugh, Althaf Lohani, and Padmanabhan P. Nair. "Human Stools as a Source of Viable Colonic Epithelial Cells." *The FASEB Journal* 5, no. 13 (1991): 2856–2859.

Johnson, Jon. "Why is Pooping so Pleasurable?" *Medical News Today.* February 26, 2021.

Khan, Shahnawaz Umer, Mohammad Ashraf Pal, Sarfaraz Ahmad Wani, and Mir Salahuddin. "Effect of Different Coagulants at Varying Strengths on the Quality of Paneer Made from Reconstituted Milk." *Journal of Food Science and Technology* 51, no. 3 (2014): 565–570.

Kiela, Pawel R., and Fayez K. Ghishan. "Physiology of Intestinal Absorption and Secretion." *Best Practice & Research Clinical Gastroenterology* 30, no. 2 (2016): 145–159.

BIBLIOGRAPHY

Kim, Young Sun, and Nayoung Kim. "Sex-Gender Differences in Irritable Bowel Syndrome." *Journal of Neurogastroenterology and Motility* 24, no. 4 (2018): 544.

Kimmerer, Robin Wall. *Braiding Sweetgrass: Indigenous Wisdom, Scientific Knowledge and the Teachings of Plants.* Minneapolis: Milkweed Editions, 2013.

Kwon, Diana. "Scientists Question Discovery of New Human Salivary Gland." *The Scientist.* January 12, 2021.

Lamont, Richard J., and Howard F. Jenkinson. *Oral Microbiology at a Glance.* Vol. 38. John Wiley & Sons, 2010.

Lee, Jae Soung, Seok-Young Kim, Yoon Shik Chun, Young-Jin Chun, Seung Yong Shin, Chang Hwan Choi, and Hyung-Kyoon Choi. "Characteristics of Fecal Metabolic Profiles in Patients with Irritable Bowel Syndrome with Predominant Diarrhea Investigated Using 1H-NMR Coupled with Multivariate Statistical Analysis." *Neurogastroenterology & Motility* 32, no. 6 (2020): e13830.

Leffingwell, John C. "Olfaction—Update No. 5." *Leffingwell Reports* 2, No. 1 (2002): 1–34.

Levitt, Michael D., and William C. Duane. "Floating Stools—Flatus Versus Fat." *New England Journal of Medicine* 286, no. 18 (1972): 973–975.

Liang, Guanxiang, and Frederic D. Bushman. "The Human Virome: Assembly, Composition and Host Interactions." *Nature Reviews Microbiology* 19, no. 8 (2021): 514–527.

Lineback, Paul E. "The Development of the Spiral Coil in the Large Intestine of the Pig." *American Journal of Anatomy* 20, no. 3 (1916): 483–503.

Lurie-Weinberger, Mor N., and Uri Gophna. "Archaea in and on the Human Body: Health Implications and Future Directions." *PLoS Pathogens* 11, no. 6 (2015): e1004833.

Magnúsdóttir, Stefanía, Dmitry Ravcheev, Valérie de Crécy-Lagard, and Ines Thiele. "Systematic Genome Assessment of B-Vitamin Biosynthesis Suggests Co-Operation Among Gut Microbes." *Frontiers in Genetics* 6 (2015): 148.

Malhi, Yadvinder, Christopher E. Doughty, Mauro Galetti, Felisa A. Smith, Jens-Christian Svenning, and John W. Terborgh. "Megafauna and Ecosystem Function from the Pleistocene to the Anthropocene." *Proceedings of the National Academy of Sciences* 113, no. 4 (2016): 838–846.

Marco, Maria L., Dustin Heeney, Sylvie Binda, Christopher J. Cifelli, Paul D. Cotter, Benoit Foligné, Michael Gänzle et al. "Health Benefits of Fermented Foods: Microbiota and Beyond." *Current Opinion in Biotechnology* 44 (2017): 94–102.

Matheus, Paul Edward. *Paleoecology and Ecomorphology of the Giant Short-Faced Bear in Eastern Beringia.* Doctoral thesis, University of Alaska Fairbanks, 1997.

Mayo Clinic. "Stool DNA Test." Accessed April 20, 2022, https://www.mayoclinic .org/tests-procedures/stool-dna-test/about/pac-20385153.

Meek, Walter. "The Beginnings of American Physiology." *Annals of Medical History* 10, no. 2 (1928): 111–125.

Moore, J. G., L. D. Jessop, and D. N. Osborne. "Gas-Chromatographic and Mass-Spectrometric Analysis of the Odor of Human Feces." *Gastroenterology* 93, no. 6 (1987): 1321–1329.

Mukhopadhya, Indrani, Jonathan P. Segal, Simon R. Carding, Ailsa L. Hart, and Georgina L. Hold. "The Gut Virome: the 'Missing Link' between Gut Bacteria and Host Immunity?" *Therapeutic Advances in Gastroenterology* 12 (2019): 1756284819836620.

Muñoz-Esparza, Nelly C., M. Luz Latorre-Moratalla, Oriol Comas-Basté, Natalia Toro-Funes, M. Teresa Veciana-Nogués, and M. Carmen Vidal-Carou. "Polyamines in Food." *Frontiers in Nutrition* 6 (2019): 108.

Mushegian, A. R. "Are There 10^{31} Virus Particles on Earth, or More, or Fewer?" *Journal of Bacteriology* 202, no. 9 (2020): e00052-20.

Nakamura, Atsuo, Takushi Ooga, and Mitsuharu Matsumoto. "Intestinal Luminal Putrescine is Produced by Collective Biosynthetic Pathways of the Commensal Microbiome." *Gut Microbes* 10, no. 2 (2019): 159–171.

Nelson, Bryn. "Life System That Relies on Guano." *Newsday*. November 27, 2001.

Nightingale, J., and Jeremy M. Woodward. "Guidelines for Management of Patients with a Short Bowel." *Gut* 55, no. suppl 4 (2006): iv1-iv12.

Nijhuis, Michelle. *Beloved Beasts: Fighting for Life in an Age of Extinction*. New York: W.W. Norton, 2021.

Niziolomski, J., J. Rickson, N. Marquez-Grant, and M. Pawlett. "Soil Science Related to Human Body after Death." The Corpse Project, 2016.

Nkamga, Vanessa Demonfort, Bernard Henrissat, and Michel Drancourt. "Archaea: Essential Inhabitants of the Human Digestive Microbiota." *Human Microbiome Journal* 3 (2017): 1–8.

Oliphant, Kaitlyn, and Emma Allen-Vercoe. "Macronutrient metabolism by the human gut microbiome: major fermentation by-products and their impact on host health." *Microbiome* 7, no. 1 (2019): 1–15.

Parvez, S., Karim A. Malik, S. Ah Kang, and H-Y. Kim. "Probiotics and their Fermented Food Products Are Beneficial for Health." *Journal of Applied Microbiology* 100, no. 6 (2006): 1171–1185.

Paytan, Adina, and Karen McLaughlin. "The Oceanic Phosphorus Cycle." *Chemical Reviews* 107, no. 2 (2007): 563–576.

Peñuelas, Josep, and Jordi Sardans. "The Global Nitrogen-Phosphorus Imbalance." *Science* 375, no. 6578 (2022): 266–267.

Phillips, Jodi, Jane G. Muir, Anne Birkett, Zhong X. Lu, Gwyn P. Jones, Kerin O'Dea, and Graeme P. Young. "Effect of Resistant Starch on Fecal Bulk and Fermentation-Dependent Events in Humans." *The American Journal of Clinical Nutrition* 62, no. 1 (1995): 121–130.

Pokusaeva, Karina, Gerald F. Fitzgerald, and Douwe van Sinderen. "Carbohydrate Metabolism in Bifidobacteria." *Genes & Nutrition* 6, no. 3 (2011): 285–306.

Prasad, Kedar N., and Stephen C. Bondy. "Dietary Fibers and their Fermented Short-Chain Fatty Acids in Prevention of Human Diseases." *Bioactive Carbohydrates and Dietary Fibre* 17 (2019): 100170.

Price, Catherine. "Probing the Mysteries of Human Digestion." *Distillations*. August 13, 2018.

Prout, William. "III. On the Nature of the Acid and Saline Matters Usually Existing in the Stomachs of Animals." *Philosophical Transactions of the Royal Society of London* 114 (1824): 45–49.

Purwantini, Endang, Trudy Torto-Alalibo, Jane Lomax, João C. Setubal, Brett M. Tyler, and Biswarup Mukhopadhyay. "Genetic Resources for Methane Production from Biomass Described with the Gene Ontology." *Frontiers in Microbiology* 5 (2014): 634.

Raimondi, Stefano, Alberto Amaretti, Caterina Gozzoli, Marta Simone, Lucia Righini, Francesco Candeliere, Paola Brun et al. "Longitudinal Survey of Fungi in the Human Gut: ITS Profiling, Phenotyping, and Colonization." *Frontiers in Microbiology* (2019): 1575.

Rao, S. S., Kimberley Welcher, Bridget Zimmerman, and Phyllis Stumbo. "Is Coffee a Colonic Stimulant?" *European Journal of Gastroenterology & Hepatology* 10, no. 2 (1998): 113–118.

Ratnarajah, Lavenia, Andrew Bowie, and Indi Hodgson-Johnston. "Bottoms up: How Whale Poop Helps Feed the Ocean." *The Conversation.* August 4, 2014.

Ray, C. Claiborne. "The Toughest Seed." *New York Times*, Dec. 26, 2011.

Reynolds, Andrew, Jim Mann, John Cummings, Nicola Winter, Evelyn Mete, and Lisa Te Morenga. "Carbohydrate Quality and Human Health: A Series of Systematic Reviews and Meta-Analyses." *The Lancet* 393, no. 10170 (2019): 434–445.

Richman, Josh, and Anish Sheth. *What's Your Poo Telling You?* San Francisco: Chronicle Books, 2007.

Roager, Henrik M., and Tine R. Licht. "Microbial Tryptophan Catabolites in Health and Disease." *Nature Communications* 9, no. 1 (2018): 1–10.

Roman, Joe, and James J. McCarthy. "The Whale Pump: Marine Mammals Enhance Primary Productivity in a Coastal Basin." *PloS One* 5, no. 10 (2010): e13255.

Rosario, Karyna, Erin M. Symonds, Christopher Sinigalliano, Jill Stewart, and Mya Breitbart. "Pepper Mild Mottle Virus as an Indicator of Fecal Pollution." *Applied and Environmental Microbiology* 75, no. 22 (2009): 7261–7267.

Rose, C., Alison Parker, Bruce Jefferson, and Elise Cartmell. "The Characterization of Feces and Urine: A Review of the Literature to Inform Advanced Treatment Technology." *Critical Reviews in Environmental Science and Technology* 45, no. 17 (2015): 1827–1879.

Rosenfeld, Louis. "William Prout: Early 19th Century Physician-Chemist." *Clinical Chemistry* 49, no. 4 (2003): 699-705.

Sandom, Christopher, Søren Faurby, Brody Sandel, and Jens-Christian Svenning. "Global Late Quaternary Megafauna Extinctions Linked to Humans, not Climate Change." *Proceedings of the Royal Society B: Biological Sciences* 281, no. 1787 (2014): 20133254.

Sanford, Kiki. "Spermine and Spermidine." *Chemistry World.* March 15, 2017.

Savoca, Matthew S., Max F. Czapanskiy, Shirel R. Kahane-Rapport, William T. Gough, James A. Fahlbusch, K. C. Bierlich, Paolo S. Segre et al. "Baleen Whale Prey Consumption Based on High-Resolution Foraging Measurements." *Nature* 599, no. 7883 (2021): 85–90.

BIBLIOGRAPHY

Schubert, Blaine W., Richard C. Hulbert, Bruce J. Macfadden, Michael Searle, and Seina Searle. "Giant Short-Faced Bears (*Arctodus simus*) in Pleistocene Florida USA, a Substantial Range Extension." *Journal of Paleontology* 84, no. 1 (2010): 79–87.

Sender, Ron, Shai Fuchs, and Ron Milo. "Revised Estimates for the Number of Human and Bacteria Cells in the Body." *PLoS Biology* 14, no. 8 (2016): e1002533.

Soto, Ana, Virginia Martín, Esther Jiménez, Isabelle Mader, Juan M. Rodríguez, and Leonides Fernández. "Lactobacilli and bifidobacteria in Human Breast Milk: Influence of Antibiotherapy and Other Host and Clinical Factors." *Journal of Pediatric Gastroenterology and Nutrition* 59, no. 1 (2014): 78.

Stephen, Alison M., and J. H. Cummings. "The Microbial Contribution to Human Faecal Mass." *Journal of Medical Microbiology* 13, no. 1 (1980): 45–56.

Stevenson, L. E. O., Frankie Phillips, Kathryn O'Sullivan, and Jenny Walton. "Wheat Bran: Its Composition and Benefits to Health, a European Perspective." *International Journal of Food Sciences and Nutrition* 63, no. 8 (2012): 1001–1013.

Stewart, Mathew, W. Christopher Carleton, and Huw S. Groucutt. "Climate Change, not Human Population Growth, Correlates with Late Quaternary Megafauna Declines in North America." *Nature Communications* 12, no. 1 (2021): 1–15.

Stokstad, Erik. "Rootin', Poopin' African Elephants Help Keep Soil Fertile." *Science.* April 1, 2020.

Suau, Antonia, Régis Bonnet, Malène Sutren, Jean-Jacques Godon, Glenn R. Gibson, Matthew D. Collins, and Joel Doré. "Direct Analysis of Genes Encoding 16S rRNA from Complex Communities Reveals Many Novel Molecular Species within the Human Gut." *Applied and Environmental Microbiology* 65, no. 11 (1999): 4799–4807.

Symonds, Erin M., Karyna Rosario, and Mya Breitbart. "Pepper Mild Mottle Virus: Agricultural Menace Turned Effective Tool for Microbial Water Quality Monitoring and Assessing (Waste) Water Treatment Technologies." *PLoS Pathogens* 15, no. 4 (2019): e1007639.

Szarka, Lawrence A., and Michael Camilleri. "Methods for the Assessment of Small-Bowel and Colonic Transit." In *Seminars in Nuclear Medicine*, vol. 42, no. 2, pp. 113–123. WB Saunders, 2012.

Tamime, A. Y. "Fermented Milks: A Historical Food with Modern Applications–a Review." *European Journal of Clinical Nutrition* 56, no. 4 (2002): S2–S15.

Terry, Natalie, and Kara Gross Margolis. "Serotonergic Mechanisms Regulating the GI Tract: Experimental Evidence and Therapeutic Relevance." *Gastrointestinal Pharmacology* (2016): 319–342.

Tesfaye, W., J. A. Suarez-Lepe, I. Loira, F. Palomero, and A. Morata. "Dairy and Nondairy-Based Beverages as a Vehicle for Probiotics, Prebiotics, and Symbiotics: Alternatives to Health Versus Disease Binomial Approach through Food." In *Milk-Based Beverages*, pp. 473-520. Cambridge, United Kingdom: Woodhead Publishing, 2019.

Valstar, Matthijs H., Bernadette S. de Bakker, Roel JHM Steenbakkers, Kees H. de Jong, Laura A. Smit, Thomas JW Klein Nulent, Robert JJ van Es et al. "The Tubarial

Salivary Glands: A Potential New Organ at Risk for Radiotherapy." *Radiotherapy and Oncology* 154 (2021): 292–298.

Van Valkenburgh, Blaire, Matthew W. Hayward, William J. Ripple, Carlo Meloro, and V. Louise Roth. "The Impact of Large Terrestrial Carnivores on Pleistocene Ecosystems." *Proceedings of the National Academy of Sciences* 113, no. 4 (2016): 862–867.

Vodusek, David B., and François Boller, eds. *Neurology of Sexual and Bladder Disorders.* Amsterdam: Elsevier, 2015.

Wastyk, Hannah C., Gabriela K. Fragiadakis, Dalia Perelman, Dylan Dahan, Bryan D. Merrill, B. Yu Feiqiao, Madeline Topf et al. "Gut-Microbiota-Targeted Diets Modulate Human Immune Status." *Cell* 184, no. 16 (2021): 4137–4153.

Wexler, Hannah M. "Bacteroides: The Good, the Bad, and the Nitty-Gritty." *Clinical Microbiology Reviews* 20, no. 4 (2007): 593–621.

Wolf, Adam, Christopher E. Doughty, and Yadvinder Malhi. "Lateral Diffusion of Nutrients by Mammalian Herbivores in Terrestrial Ecosystems." *PloS One* 8, no. 8 (2013): e71352.

Yu, Siegfried W. B., and Satish SC Rao. "Anorectal Physiology and Pathophysiology in the Elderly." *Clinics in Geriatric Medicine* 30, no. 1 (2014): 95–106.

Zafar, Hassan, and Milton H. Saier Jr. "Gut Bacteroides Species in Health and Disease." *Gut Microbes* 13, no. 1 (2021): 1848158.

Ziegler, Amanda, Liara Gonzalez, and Anthony Blikslager. "Large Animal Models: The Key to Translational Discovery in Digestive Disease Research." *Cellular and Molecular Gastroenterology and Hepatology* 2, no. 6 (2016): 716–724.

Zylberberg, Nadine. "Fermenting Your Compost." *Medium.* Aug. 2, 2020.

Chapter Two

Allen, David J., and Terry Oleson. "Shame and Internalized Homophobia in Gay Men." *Journal of Homosexuality* 37, no. 3 (1999): 33–43.

Al-Shawaf, Laith, David MG Lewis, and David M. Buss. "Disgust and Mating Strategy." *Evolution and Human Behavior* 36, no. 3 (2015): 199–205.

Applebaum, Anne. "Trump is Turning America into the 'Shithole Country' He Fears." *The Atlantic.* July 3, 2020.

Bennett, Brian, and Tessa Berenson. "How Donald Trump Lost the Election." *Time.* November 7, 2020.

Campanile, Carl, and Yaron Steinbuch. "Rioters Left Feces, Urine in Hallways and Offices during Mobbing of US Capitol." *New York Post.* January 8, 2021.

Case, Trevor I., Betty M. Repacholi, and Richard J. Stevenson. "My Baby Doesn't Smell as Bad as Yours: The Plasticity of Disgust." *Evolution and Human Behavior* 27, no. 5 (2006): 357–365.

Cepon-Robins, Tara J., Aaron D. Blackwell, Theresa E. Gildner, Melissa A. Liebert, Samuel S. Urlacher, Felicia C. Madimenos, Geeta N. Eick, J. Josh Snodgrass, and Lawrence S. Sugiyama. "Pathogen Disgust Sensitivity Protects against Infection in a High Pathogen Environment." *Proceedings of the National Academy of Sciences* 118, no. 8 (2021): e2018552118.

Clifford, Scott, Cengiz Erisen, Dane Wendell, and Francisco Cantu. "Disgust Sensitivity and Support for Immigration across Five Nations." *Politics and the Life Sciences*. Published online March 4, 2022.

Costello, Kimberly, and Gordon Hodson. "Explaining Dehumanization among Children: The Interspecies Model of Prejudice." *British Journal of Social Psychology* 53, no. 1 (2014): 175–197.

Curtis, Val. *Don't Look, Don't Touch, Don't Eat: The Science Behind Revulsion.* University of Chicago Press, 2013.

Darling-Hammond, Sean, Eli K. Michaels, Amani M. Allen, David H. Chae, Marilyn D. Thomas, Thu T. Nguyen, Mahasin M. Mujahid, and Rucker C. Johnson. "After 'The China Virus' Went Viral: Racially Charged Coronavirus Coverage and Trends in Bias against Asian Americans." *Health Education & Behavior* 47, no. 6 (2020): 870–879.

Davey, Graham CL. "Disgust: The Disease-Avoidance Emotion and its Dysfunctions." *Philosophical Transactions of the Royal Society B: Biological Sciences* 366, no. 1583 (2011): 3453–3465.

Doughty, Caitlin. *From Here to Eternity: Traveling the World to Find the Good Death.* New York: W.W. Norton, 2017.

Dozo, Nerisa. "Gender Differences in Prejudice: A Biological and Social Psychological Analysis." Doctoral thesis, University of Queensland, 2015.

"Elephants Get a Big Thank You." *New York Times*. February 27, 2002.

Fessler, Daniel, and Kevin Haley. "Guarding the Perimeter: The Outside-Inside Dichotomy in Disgust and Bodily Experience." *Cognition & Emotion* 20, no. 1 (2006): 3–19.

Foggatt, Tyler. "Giuliani Vs. the Virgin." *The New Yorker*. May 21, 2018.

Gabriel, Trip. "In Statehouse Races, Suburban Voters' Disgust with Trump Didn't Translate into a Rebuke of Other Republicans." *New York Times*. November 29, 2020.

Gerba, Charles P. "Environmentally Transmitted Pathogens." In *Environmental Microbiology*, pp. 445-484. Oxford: Academic Press, 2009.

Goff, Phillip Atiba, Jennifer L. Eberhardt, Melissa J. Williams, and Matthew Christian Jackson. "Not Yet Human: Implicit Knowledge, Historical Dehumanization, and Contemporary Consequences." *Journal of Personality and Social Psychology* 94, no. 2 (2008): 292.

Gomi, Tarō. *Everyone Poops.* Translated by Amanda Mayer Stinchecum. Brooklyn, New York: Kane/Miller, 1993.

Hodson, Gordon, Becky L. Choma, Jacqueline Boisvert, Carolyn L. Hafer, Cara C. MacInnis, and Kimberly Costello. "The Role of Intergroup Disgust in Predicting Negative Outgroup Evaluations." *Journal of Experimental Social Psychology* 49, no. 2 (2013): 195-205.

Hodson, Gordon, Blaire Dube, and Becky L. Choma. "Can (Elaborated) Imagined Contact Interventions Reduce Prejudice among Those Higher in Intergroup

Disgust Sensitivity (ITG-DS)?" *Journal of Applied Social Psychology* 45, no. 3 (2015): 123-131.

Hodson, Gordon, Nour Kteily, and Mark Hoffarth. "Of Filthy Pigs and Subhuman Mongrels: Dehumanization, Disgust, and Intergroup Prejudice." *TPM: Testing, Psychometrics, Methodology in Applied Psychology* 21, no. 3 (2014).

Hu, Jane C. "The Panic over Chinese People Doesn't Come from the Coronavirus." *Slate.* February 4, 2020.

Igielnik, Ruth. "Men and Women in the U.S. Continue to Differ in Voter Turnout Rate, Party Identification." *Pew Research Center.* August 18, 2020.

Jack, Rachael E., Oliver GB Garrod, Hui Yu, Roberto Caldara, and Philippe G. Schyns. "Facial Expressions of Emotion Are Not Culturally Universal." *Proceedings of the National Academy of Sciences* 109, no. 19 (2012): 7241-7244.

Jacobson, Gary C. "Extreme Referendum: Donald Trump and the 2018 Midterm Elections." *Political Science Quarterly* 134, no. 1 (2019): 9-38.

Johnson, Steven. *The Ghost Map: The Story of London's Most Terrifying Epidemic—and How it Changed Science, Cities, and the Modern World.* New York: Penguin, 2006.

Kiss, Mark J., Melanie A. Morrison, and Todd G. Morrison. "A Meta-Analytic Review of the Association between Disgust and Prejudice toward Gay Men." *Journal of Homosexuality* 67, no. 5 (2020): 674-696.

Klein, Charlotte. "Watch Giuliani Demand 'Trial by Combat' to Settle the Election." *New York.* January 6, 2021.

Laporte, Dominique. *History of Shit.* Translated by Nadia Benadbid and Rodolphe el-Khoury. Cambridge, Massachusetts: The MIT Press, 2002.

Levin, Brian. "Report to the Nation: Anti-Asian Prejudice & Hate Crime. New 2020–21 First Quarter Comparison Data." California State University–San Bernardino, 2021.

"Louis Pasteur." Science History Institute, accessed April 20, 2022, https://www.sciencehistory.org/historical-profile/louis-pasteur.

Lowrey, Annie. "The One Issue That's Really Driving the Midterm Elections." *The Atlantic.* November 2, 2018.

Martin, Jonathan. "Despite Big House Losses, G.O.P. Shows No Signs of Course Correction." *New York Times.* December 2, 2018.

Mayor, Adrienne. *Greek Fire, Poison Arrows, & Scorpion Bombs: Biological and Chemical Warfare in the Ancient World.* New York: Abrams Press, 2003.

McCarrick, Christopher, and Tim Ziaukas. "Still Scary After All These Years: Mr. Yuk Nears 40." *Western Pennsylvania History: 1918-2018* (2009): 18-31.

McCrystal, Laura, and Erin McCarthy. "'Disgusted' Voters in the Philly Suburbs Could Help Biden Offset Trump's Gains in Pennsylvania." *The Philadelphia Inquirer.* September 20, 2020.

Michalak, Nicholas M., Oliver Sng, Iris M. Wang, and Joshua Ackerman. "Sounds of Sickness: Can People Identify Infectious Disease Using Sounds of Coughs and Sneezes?" *Proceedings of the Royal Society B: Biological Sciences*, 2020: 287 (1928): 20200944.

Migdon, Brooke. "Gov. DeSantis Spokesperson Says 'Don't Say Gay' Opponents Are 'Groomers.'" *The Hill.* March 7, 2022.

BIBLIOGRAPHY

Milligan, Susan. "Bipartisan Disgust Could Save the Republic." *U.S. News & World Report.* January 8, 2021.

Morris Jr, J. Glenn. "Cholera—Modern Pandemic Disease of Ancient Lineage." *Emerging Infectious Diseases* 17, no. 11 (2011): 2099-2104.

Morrison, Todd G., Mark J. Kiss, C. J. Bishop, and Melanie A. Morrison. "'We're Disgusted with Queers, Not Fearful of Them': The Interrelationships among Disgust, Gay Men's Sexual Behavior, and Homonegativity." *Journal of Homosexuality* 66, no. 7 (2019): 1014-1033.

Newcomb, Steven. "On Historical Narratives and Dehumanization." *Indian Country Today.* June 20, 2012.

Nilsen, Ella. "Suburban Women Have Had Their Lives Upended by Covid-19. Trump Might Pay the Price." *Vox.* October 27, 2020.

O'Shea, Brian A., Derrick G. Watson, Gordon DA Brown, and Corey L. Fincher. "Infectious Disease Prevalence, Not Race Exposure, Predicts Both Implicit and Explicit Racial Prejudice across the United States." *Social Psychological and Personality Science* 11, no. 3 (2020): 345-355.

Pajak, Rosanna, Christine Langhoff, Sue Watson, and Sunjeev K. Kamboj. "Phenomenology and Thematic Content of Intrusive Imagery in Bowel and Bladder Obsession." *Journal of Obsessive-Compulsive and Related Disorders* 2, no. 3 (2013): 233-240.

Pollitzer, Robert. "Cholera Studies. 1. History of the Disease." *Bulletin of the World Health Organization* 10, no. 3 (1954): 421-461.

Richardson, Michael. "The Disgust of Donald Trump." *Continuum* 31, no. 6 (2017): 747-756.

Rose-Stockwell, Tobias. "This is How Your Fear and Outrage Are Being Sold for Profit." *Quartz.* July 28, 2017.

Rottman, Joshua. "Evolution, Development, and the Emergence of Disgust." *Evolutionary Psychology* 12, no. 2 (2014): 147470491401200209.

Rozin, Paul. "Disgust, Psychology of." In *International Encyclopedia of the Social & Behavioral Sciences, 2nd edition* Vol 6. 546–549. Oxford: Elsevier, 2015.

Rozin, Paul, Larry Hammer, Harriet Oster, Talia Horowitz, and Veronica Marmora. "The Child's Conception of Food: Differentiation of Categories of Rejected Substances in the 16 Months to 5 Year Age Range." *Appetite* 7, no. 2 (1986): 141-151.

Rubenking, Bridget, and Annie Lang. "Captivated and Grossed out: An Examination of Processing Core and Sociomoral Disgusts in Entertainment Media." *Journal of Communication* 64, no. 3 (2014): 543-565.

Santucci, John. "Trump Makes Sexually Derogatory Remark about Hillary Clinton, Calls Bathroom Break 'Disgusting.'" *ABCNews.com.* December 21, 2015.

Schaller, Mark, and L. A. Duncan. "The Behavioral Immune System." In *The Handbook of Evolutionary Psychology, Second Edition, Vol. 1.* 206-224. New York: Wiley, 2015.

Schlatter, Evelyn, and Robert Steinback. "10 Anti-Gay Myths Debunked." *Intelligence Report.* February 27, 2011.

Shear, Michael D., Katie Benner, and Michael S. Schmidt. "'We Need to Take away Children,' No Matter How Young, Justice Dept. Officials Said." *New York Times.* October 6, 2020.

Shorrocks, Rosalind. "Gender Gaps in the 2019 General Election." *UK in a Changing Europe*. March 8, 2021.

Skinner, Allison L., and Caitlin M. Hudac. "'Yuck, You Disgust Me!' Affective Bias Against Interracial Couples." *Journal of Experimental Social Psychology* 68 (2017): 68-77.

Spinelli, Marcelo. "Decorative Beauty Was a Taboo Thing." *Brilliant! New Art from London*, exhibit catalogue, 67. Minneapolis, Walker Art Center, 1995.

Terrizzi Jr, John A., Natalie J. Shook, and Michael A. McDaniel. "The Behavioral Immune System and Social Conservatism: A Meta-Analysis." *Evolution and Human Behavior* 34, no. 2 (2013): 99-108.

Thompson, Derek. "Why Men Vote for Republicans, and Women Vote for Democrats." *The Atlantic*. February 10, 2020.

"#ToiletPaperApocalypse: Australia's Toilet Paper Problem and the Subsequent Explosion." *Asiaville*. March 4, 2020.

Tuite, Ashleigh R., Christina H. Chan, and David N. Fisman. "Cholera, Canals, and Contagion: Rediscovering Dr Beck's Report." *Journal of Public Health Policy* 32, no. 3 (2011): 320-333.

Tulchinsky, Theodore H., and Elena A. Varavikova. "A History of Public Health." In *The New Public Health* 1-42. Cambridge, Massachusetts: Academic Press, 2014.

Turnbull, Stephen. *Siege Weapons of the Far East (1): AD 612–1300*. Oxford: Osprey Publishing, 2012.

Tybur, Joshua M., Debra Lieberman, and Vladas Griskevicius. "Microbes, Mating, and Morality: Individual Differences in Three Functional Domains of Disgust." *Journal of Personality and Social Psychology* 97, no. 1 (2009): 103-122.

Vitali, Ali, Kasie Hunt, and Frank Thorp V. "Trump Referred to Haiti and African Nations as 'Shithole' Countries." *NBCNews.com*. January 11, 2018.

Vogel, Carol. "An Artist Who's Grateful for Elephants." *New York Times*. February 21, 2002.

Young, Allison. "Chris Ofili, The Holy Virgin Mary." *Smarthistory*, August 9, 2015.

Zakrzewska, Marta, Jonas K. Olofsson, Torun Lindholm, Anna Blomkvist, and Marco Tullio Liuzza. "Body Odor Disgust Sensitivity Is Associated with Prejudice Towards a Fictive Group of Immigrants." *Physiology & Behavior* 201 (2019): 221-227.

Zint, Bradley. "Costa Mesa Restaurant's Special: Spend $20 on Takeout, Get a Free Roll of Toilet Paper." *The Los Angeles Times*. March 18, 2020.

Chapter Three

Allen-Vercoe, Emma, and Elaine O. Petrof. "Artificial Stool Transplantation: Progress Towards a Safer, More Effective and Acceptable Alternative." *Expert Review of Gastroenterology & Hepatology* 7, no. 4 (2013): 291-293.

Aroniadis, Olga C., and Lawrence J. Brandt. "Fecal Microbiota Transplantation: Past, Present and Future." *Current Opinion in Gastroenterology* 29, no. 1 (2013): 79-84.

BIBLIOGRAPHY

Bassler, Anthony. "A New Method of Treatment for Chronic Intestinal Putrefactions by Means of Rectal Instillations of Autogenous Bacteria and Strains of Human Bacillus coli communis." *Medical Record (1866-1922)* 78, no. 13 (1910): 519.

Baunwall, Simon Mark Dahl, Mads Ming Lee, Marcel Kjærsgaard Eriksen, Benjamin H. Mullish, Julian R. Marchesi, Jens Frederik Dahlerup, and Christian Lodberg Hvas. "Faecal Microbiota Transplantation for Recurrent Clostridioides difficile Infection: An Updated Systematic Review and Meta-Analysis." *EClinicalMedicine* 29 (2020): 100642.

Boneca, Ivo G., and Gabriela Chiosis. "Vancomycin Resistance: Occurrence, Mechanisms and Strategies to Combat It." *Expert Opinion on Therapeutic Targets* 7, no. 3 (2003): 311-328.

Borody, Thomas J., Eloise F. Warren, Sharyn Leis, Rosa Surace, and Ori Ashman. "Treatment of Ulcerative Colitis Using Fecal Bacteriotherapy." *Journal of Clinical Gastroenterology* 37, no. 1 (2003): 42-47.

Borody, Thomas J., Eloise F. Warren, Sharyn M. Leis, Rosa Surace, Ori Ashman, and Steven Siarakas. "Bacteriotherapy Using Fecal Flora: Toying with Human Motions." *Journal of Clinical Gastroenterology* 38, no. 6 (2004): 475-483.

Bourke, John Gregory. *Scatalogic Rites of All Nations*. Washington, DC: Lowdermilk & Company, 1891.

Brandt, Lawrence J. "Editorial Commentary: Fecal Microbiota Transplantation: Patient and Physician Attitudes." *Clinical Infectious Diseases* 55, no. 12 (2012): 1659-1660.

Bryce, E., T. Zurberg, M. Zurberg, S. Shajari, and D. Roscoe. "Identifying Environmental Reservoirs of Clostridium difficile with a Scent Detection Dog: Preliminary Evaluation." *Journal of Hospital Infection* 97, no. 2 (2017): 140-145.

Cammarota, Giovanni, Gianluca Ianiro, Colleen R. Kelly, Benjamin H. Mullish, Jessica R. Allegretti, Zain Kassam, Lorenza Putignani et al. "International Consensus Conference on Stool Banking for Faecal Microbiota Transplantation in Clinical Practice." *Gut* 68, no. 12 (2019): 2111-2121.

Craven, Laura J., Seema Nair Parvathy, Justin Tat-Ko, Jeremy P. Burton, and Michael S. Silverman. "Extended Screening Costs Associated with Selecting Donors for Fecal Microbiota Transplantation for Treatment of Metabolic Syndrome-Associated Diseases." *Open Forum Infectious Diseases* 4, no. 4 (2017): ofx243.

Dahlhamer, James M., Emily P. Zammitti, Brian W. Ward, Anne G. Wheaton, and Janet B. Croft. "Prevalence of Inflammatory Bowel Disease among Adults Aged ≥ 18 Years—United States, 2015." *Morbidity and Mortality Weekly Report* 65, no. 42 (2016): 1166-1169.

DeFilipp, Zachariah, Patricia P. Bloom, Mariam Torres Soto, Michael K. Mansour, Mohamad RA Sater, Miriam H. Huntley, Sarah Turbett, Raymond T. Chung, Yi-Bin Chen, and Elizabeth L. Hohmann. "Drug-Resistant E. coli Bacteremia Transmitted by Fecal Microbiota Transplant." *New England Journal of Medicine* 381, no. 21 (2019): 2043-2050.

DePeters, E. J., and L. W. George. "Rumen Transfaunation." *Immunology Letters* 162, no. 2 (2014): 69-76.

BIBLIOGRAPHY

Du, Huan, Ting-ting Kuang, Shuang Qiu, Tong Xu, Chen-Lei Gang Huan, Gang Fan, and Yi Zhang. "Fecal Medicines Used in Traditional Medical System of China: A Systematic Review of Their Names, Original Species, Traditional Uses, and Modern Investigations." *Chinese Medicine* 14, no. 1 (2019): 1-16.

Eiseman, Ben, W. Silen, G. S. Bascom, and A. J. Kauvar. "Fecal Enema as an Adjunct in the Treatment of Pseudomembranous Enterocolitis." *Surgery* 44, no. 5 (1958): 854-859.

Falkow, Stanley. "Fecal Transplants in the 'Good Old Days.'" *Small Things Considered.* May 13, 2013.

Freeman, J., M. P. Bauer, Simon D. Baines, J. Corver, W. N. Fawley, B. Goorhuis, E. J. Kuijper, and M. H. Wilcox. "The Changing Epidemiology of Clostridium difficile Infections." *Clinical Microbiology Reviews* 23, no. 3 (2010): 529-549.

Grady, Denise. "Fecal Transplant Is Linked to a Patient's Death, the F.D.A. Warns." *New York Times.* June 13, 2019.

Guh, Alice Y., Yi Mu, Lisa G. Winston, Helen Johnston, Danyel Olson, Monica M. Farley, Lucy E. Wilson et al. "Trends in US Burden of Clostridioides difficile Infection and Outcomes." *New England Journal of Medicine* 382, no. 14 (2020): 1320-1330.

HomeFMT. "Fecal Transplant (FMT)" *YouTube.* May 13, 2013.

Hopkins, Roy J., and Robert B. Wilson. "Treatment of Recurrent *Clostridium difficile* Colitis: A Narrative Review." *Gastroenterology Report* 6, no. 1 (2018): 21-28.

Jacobs, Andrew. "Drug Companies and Doctors Battle over the Future of Fecal Transplants." *New York Times.* March 3, 2019.

Jacobs, Andrew. "How Contaminated Stool Stored in a Freezer Left a Fecal Transplant Patient Dead." *New York Times.* October 30, 2019.

Kao, Dina, Karen Wong, Rose Franz, Kyla Cochrane, Keith Sherriff, Linda Chui, Colin Lloyd et al. "The Effect of a Microbial Ecosystem Therapeutic (MET-2) on Recurrent Clostridioides difficile Infection: A Phase 1, Open-Label, Single-Group Trial." *The Lancet Gastroenterology & Hepatology* 6, no. 4 (2021): 282-291.

Katz, Kevin C., George R. Golding, Kelly Baekyung Choi, Linda Pelude, Kanchana R. Amaratunga, Monica Taljaard, Stephanie Alexandre et al. "The Evolving Epidemiology of Clostridium difficile Infection in Canadian Hospitals during a Postepidemic Period (2009–2015)." *CMAJ* 190, no. 25 (2018): E758-E765.

Kelly, Colleen R., Sachin S. Kunde, and Alexander Khoruts. "Guidance on Preparing an Investigational New Drug Application for Fecal Microbiota Transplantation Studies." *Clinical Gastroenterology and Hepatology* 12, no. 2 (2014): 283-288.

Khoruts, Alexander, Johan Dicksved, Janet K. Jansson, and Michael J. Sadowsky. "Changes in the Composition of the Human Fecal Microbiome after Bacteriotherapy for Recurrent Clostridium difficile-Associated Diarrhea." *Journal of Clinical Gastroenterology* 44, no. 5 (2010): 354-360.

Li, Cheng, Teresa Zurberg, Jaime Kinna, Kushal Acharya, Jack Warren, Salomeh Shajari, Leslie Forrester, and Elizabeth Bryce. "Using Scent Detection Dogs to Identify Environmental Reservoirs of Clostridium difficile: Lessons from the Field." *Canadian Journal of Infection Control* 34, no. 2 (2019): 93-95.

Li, Simone S., Ana Zhu, Vladimir Benes, Paul I. Costea, Rajna Hercog, Falk Hildebrand, Jaime Huerta-Cepas et al. "Durable Coexistence of Donor and Recipient

BIBLIOGRAPHY

Strains after Fecal Microbiota Transplantation." *Science* 352, no. 6285 (2016): 586-589.

Marchione, Marilyn. "Pills Made from Poop Cure Serious Gut Infections." *Associated Press.* October 3, 2013.

"OpenBiome Announces New Collaboration with the University of Minnesota to Treat Patients with Recurrent C. difficile Infections." News release, OpenBiome, January 20, 2022.

"OpenBiome Announces New Direct Testing for SARS-CoV-2 in Fecal Microbiota Transplantation (FMT) Preparations and Release of New Inventory." News release, OpenBiome, February 23, 2021.

Ratner, Mark. "Microbial Cocktails Raise Bar for C. *diff.* Treatments." *Nature Biotechnology.* December 3, 2020.

Sachs, Rachel E., and Carolyn A. Edelstein. "Ensuring the Safe and Effective FDA Regulation of Fecal Microbiota Transplantation." *Journal of Law and the Biosciences* 2, no. 2 (2015): 396-415.

Scudellari, Megan. "News Feature: Cleaning up the Hygiene Hypothesis." *Proceedings of the National Academy of Sciences* 114, no.7 (2017): 1433-1436.

Sheridan, Kate. "Months of Limbo at OpenBiome Put Fecal Matter Transplants on Hold Across the Country." *STAT+.* December 8, 2020.

Sholeh, Mohammad, Marcela Krutova, Mehdi Forouzesh, Sergey Mironov, Nourkhoda Sadeghifard, Leila Molaeipour, Abbas Maleki, and Ebrahim Kouhsari. "Antimicrobial Resistance in Clostridioides (Clostridium) difficile Derived from Humans: A Systematic Review and Meta-Analysis." *Antimicrobial Resistance & Infection Control* 9, no. 1 (2020): 1-11.

Smith, Sean B., Veronica Macchi, Anna Parenti, and Raffaele De Caro. "Hieronymus Fabricius Ab Acquapendente (1533–1619)." *Clinical Anatomy* 17, no. 7 (2004): 540-543.

Stein, Rob. "FDA Backs off on Regulation of Fecal Transplants." *NPR.* June 18, 2013.

Turner, Nicholas A., Steven C. Grambow, Christopher W. Woods, Vance G. Fowler, Rebekah W. Moehring, Deverick J. Anderson, and Sarah S. Lewis. "Epidemiologic Trends in Clostridioides difficile Infections in a Regional Community Hospital Network." *JAMA Network Open* 2, no. 10 (2019): e1914149-e1914149.

U.S. Food and Drug Administration. "Important Safety Alert Regarding Use of Fecal Microbiota for Transplantation and Risk of Serious Adverse Reactions Due to Transmission of Multi-Drug Resistant Organisms." Press release, June 13, 2019.

U.S. Food and Drug Administration. "Update to March 12, 2020 Safety Alert Regarding Use of Fecal Microbiota for Transplantation and Risk of Serious Adverse Events Likely Due to Transmission of Pathogenic Organisms." Press release, March 13, 2020.

Van Nood, Els, Anne Vrieze, Max Nieuwdorp, Susana Fuentes, Erwin G. Zoetendal, Willem M. de Vos, Caroline E. Visser et al. "Duodenal Infusion of Donor Feces for Recurrent Clostridium difficile." *New England Journal of Medicine* 368, no. 5 (2013): 407-415.

Williams, Shawna. "Fecal Microbiota Transplantation Is Poised for a Makeover." *The Scientist.* June 1, 2021.

Woodworth, Michael H., Cynthia Carpentieri, Kaitlin L. Sitchenko, and Colleen S. Kraft. "Challenges in Fecal Donor Selection and Screening for Fecal Microbiota Transplantation: A Review." *Gut Microbes* 8, no. 3 (2017): 225-237.

Worcester, Sharon. "FDA Eases Some Fecal Transplant Restrictions." *MDedge News*. June 19, 2013.

Yatsunenko, Tanya, Federico E. Rey, Mark J. Manary, Indi Trehan, Maria Gloria Dominguez-Bello, Monica Contreras, Magda Magris et al. "Human Gut Microbiome Viewed across Age and Geography." *Nature* 486, no. 7402 (2012): 222-227.

Yong, Ed. "Sham Poo Washes Out." *The Atlantic*. Aug. 1, 2016.

Zhang, Faming, Wensheng Luo, Yan Shi, Zhining Fan, and Guozhong Ji. "Should We Standardize the 1,700-Year-Old Fecal Microbiota Transplantation?" *The American Journal of Gastroenterology* 107, no. 11 (2012): 1755.

Chapter Four

Amann, Anton, Ben de Lacy Costello, Wolfram Miekisch, Jochen Schubert, Bogusław Buszewski, Joachim Pleil, Norman Ratcliffe, and Terence Risby. "The Human Volatilome: Volatile Organic Compounds (VOCs) In Exhaled Breath, Skin Emanations, Urine, Feces and Saliva." *Journal of Breath Research* 8, no. 3 (2014): 034001.

Angle, Craig, Lowell Paul Waggoner, Arny Ferrando, Pamela Haney, and Thomas Passler. "Canine Detection of the Volatilome: A Review of Implications for Pathogen and Disease Detection." *Frontiers in Veterinary Science* 3 (2016): 47.

Appelt, Sandra, Fabrice Armougom, Matthieu Le Bailly, Catherine Robert, and Michel Drancourt. "Polyphasic Analysis of a Middle Ages Coprolite Microbiota, Belgium." *PloS One* 9, no. 2 (2014): e88376.

Appelt, Sandra, Laura Fancello, Matthieu Le Bailly, Didier Raoult, Michel Drancourt, and Christelle Desnues. "Viruses in a 14th-Century Coprolite." *Applied and Environmental Microbiology* 80, no. 9 (2014): 2648-2655.

Benecke, Mark. "Arthropods and Corpses." In *Forensic Pathology Reviews*. 207-240. Totowa, New Jersey: Humana Press, 2005.

Benecke, Mark, Eberhard Josephi, and Ralf Zweihoff. "Neglect of the Elderly: Forensic Entomology Cases and Considerations." *Forensic Science International* 146 (2004): S195-S199.

Benecke, Mark, and Rüdiger Lessig. "Child Neglect and Forensic Entomology." *Forensic Science International* 120, no. 1-2 (2001): 155-159.

Bennett, Matthew R., David Bustos, Jeffrey S. Pigati, Kathleen B. Springer, Thomas M. Urban, Vance T. Holliday, Sally C. Reynolds et al. "Evidence of Humans in North America during the Last Glacial Maximum." *Science* 373, no. 6562 (2021): 1528-1531.

Berstad, Arnold, Jan Raa, and Jørgen Valeur. "Indole–the Scent of a Healthy 'Inner Soil.'" *Microbial Ecology in Health and Disease* 26, no. 1 (2015): 27997.

Bol, Peter Kees. "The Washing Away of Wrongs [Hsi yuan chi lu, by Sung Tz'u (1186–1249)]: Forensic Medicine in Thirteenth-Century China. Translated and introduced by Brian E. McKnight. Ann Arbor: University of Michigan Center for

BIBLIOGRAPHY

Chinese Studies, Science, Medicine, and Technology in East Asia no. 1, 1981. xv, 181 pp. Illustrations, Bibliography, Index. 6." *The Journal of Asian Studies* 42, no. 3 (1983): 643-644.

Bonacci, Teresa, Vannio Vercillo, and Mark Benecke. "Flies and Ants: A Forensic Entomological Neglect Case of an Elderly Man in Calabria, Southern Italy." *Romanian Journal of Legal Medicine* 25 (2017): 283-286.

Bowers, C. Michael. "Review of a Forensic Pseudoscience: Identification of Criminals from Bitemark Patterns." *Journal of Forensic and Legal Medicine* 61 (2019): 34-39.

Brewer, Kirstie. "Paleoscatologists Dig up Stools 'as Precious as the Crown Jewels.'" *The Guardian*. May 12, 2016.

Bryant, Vaughn M. "The Eric O. Callen Collection." *American Antiquity* 39, no. 3 (1974): 497-498.

Bryant, Vaughn M., and Glenna W. Dean. "Archaeological Coprolite Science: the Legacy of Eric O. Callen (1912–1970)." *Palaeogeography, Palaeoclimatology, Palaeoecology* 237, no. 1 (2006): 51-66.

Catts, E. Paul, and M. Lee Goff. "Forensic Entomology in Criminal Investigations." *Annual Review of Entomology* 37, no. 1 (1992): 253-272.

Curran, Allison M., Scott I. Rabin, Paola A. Prada, and Kenneth G. Furton. "Comparison of the Volatile Organic Compounds Present in Human Odor Using SPME-GC/MS." *Journal of Chemical Ecology* 31, no. 7 (2005): 1607-1619.

D'Anjou, Robert M., Raymond S. Bradley, Nicholas L. Balascio, and David B. Finkelstein. "Climate Impacts on Human Settlement and Agricultural Activities in Northern Norway Revealed through Sediment Biogeochemistry." *Proceedings of the National Academy of Sciences* 109, no. 50 (2012): 20332-20337.

Daswick, Tyler. "How the Ultimate Men's Health Dog Tracks down Missing Persons." *Men's Health*. June 27, 2017.

Drabinska, Natalia, Cheryl Flynn, Norman Ratcliffe, Ilaria Belluomo, Antonis Myridakis, Oliver Gould, Matteo Fois, Amy Smart, Terry Devine, and Ben PJ de Lacy Costello. "A Literature Survey of Volatiles from the Healthy Human Breath and Bodily Fluids: The Human Volatilome." *Journal of Breath Research* (2021).

Duggan, W. Dennis. "A History of the Bench and Bar of Albany County." Historical Society of the New York Courts, 2021.

Ensminger, John J., and Tadeusz Jezierski. "Scent Lineups in Criminal Investigations and Prosecutions." In *Police and Military Dogs*. 101-116. Boca Raton, Florida: CRC Press, 2011.

Ferry, Barbara, John J. Ensminger, Adee Schoon, Zbignev Bobrovskij, David Cant, Maciej Gawkowski, Illkka Hormila et al. "Scent Lineups Compared across Eleven Countries: Looking for the Future of a Controversial Forensic Technique." *Forensic Science International* 302 (2019): 109895.

Foley, Denis. *Lemuel Smith and the Compulsion to Kill: The Forensic Story of a Multiple Personality Serial Killer*. Delmar, New York: New Leitrim House, 2003.

Friedmaan, Albert B. "The Scatological Rites of Burglars." *Western Folklore*. 27, No. 3 (1968): 171-179.

Gerritsen, Resi, and Ruud Haak. "History of the Police Dog." In *K9 Working Breeds: Characteristics and Capabilities*. Calgary: Detselig, 2007.

Gilbert, M. Thomas P., Dennis L. Jenkins, Anders Gotherstrom, Nuria Naveran, Juan J. Sanchez, Michael Hofreiter, Philip Francis Thomsen et al. "DNA from Pre-Clovis Human Coprolites in Oregon, North America." *Science* 320, no. 5877 (2008): 786-789.

Gopalakrishnan, S., VM Anantha Eashwar, M. Muthulakshmi, and A. Geetha. "Intestinal Parasitic Infestations and Anemia among Urban Female School Children in Kancheepuram District, Tamil Nadu." *Journal of Family Medicine and Primary Care* 7, no. 6 (2018): 1395-1400.

Hald, Mette Marie, Betina Magnussen, Liv Appel, Jakob Tue Christensen, Camilla Haarby Hansen, Peter Steen Henriksen, Jesper Langkilde, Kristoffer Buck Pedersen, Allan Dørup Knudsen, and Morten Fischer Mortensen. "Fragments of Meals in Eastern Denmark from the Viking Age to the Renaissance: New Evidence from Organic Remains in Latrines." *Journal of Archaeological Science: Reports* 31 (2020): 102361.

Hald, Mette Marie, Jacob Mosekilde, Betina Magnussen, Martin Jensen Søe, Camilla Haarby Hansen, and Morten Fischer Mortensen. "Tales from the Barrels: Results from a Multi-Proxy Analysis of a Latrine from Renaissance Copenhagen, Denmark." *Journal of Archaeological Science: Reports* 20 (2018): 602-610.

Hald, Mette Marie, Morten Fischer Mortensen, and Andreas Tolstrup. "Lortemorgen! Forskning og Formidling af Latriner." *Nationalmuseets Arbejdsmark* (2019): 124-133.

Harrault, Loïc, Karen Milek, Emilie Jardé, Laurent Jeanneau, Morgane Derrien, and David G. Anderson. "Faecal Biomarkers Can Distinguish Specific Mammalian Species in Modern and Past Environments." *PLoS One* 14, no. 2 (2019): e0211119.

Horowitz, Alexandra, and Becca Franks. "What Smells? Gauging Attention to Olfaction in Canine Cognition Research." *Animal Cognition* 23, no. 1 (2020): 11-18.

Hunter, Andrea A., James Munkres, and Barker Fariss. "Osage Nation NAGPRA Claim for Human Remains Removed from the Clarksville Mound Group (23PI6), Pike County, Missouri," Osage Nation Historic Preservation Office (2013): 1–60.

Jeffrey, Simon. "Museum's Broken Treasure Not Just Any Old Shit." *The Guardian*. June 6, 2003.

Jensen, Peter Mose, Christian Vrængmose Jensen, Jette Linaa, and Jakob Ørnbjerg. "Biskoppernes Latrin. En Tværvidenskabelig Undersøgelse af 1700-Tals Latrin fra Aalborg." *Kulturstudier* 7, no. 2 (2016): 41-76.

Krichbaum, Sarah, Bart Rogers, Emma Cox, L. Paul Waggoner, and Jeffrey S. Katz. "Odor Span Task in Dogs (*Canis familiaris*)." *Animal Cognition* 23, no. 3 (2020): 571–580.

Kudo, Keiko, Chiaki Miyazaki, Ryo Kadoya, Tohru Imamura, Narumi Jitsufuchi, and Noriaki Ikeda. "Laxative Poisoning: Toxicological Analysis of Bisacodyl and its Metabolite in Urine, Serum, and Stool." *Journal of Analytical Toxicology* 22, no. 4 (1998): 274-278.

Landry, Alyssa. "Native history: Osage Forced to Abandon Lands in Missouri and Arkansas." *Indian Country Today*. November 10, 2013.

BIBLIOGRAPHY

Lanska, Douglas J. "Optograms and Criminology: Science, News Reporting, and Fanciful Novels." *Progress in Brain Research*. 205 (2013): 55-84.

Levenson, Eric. "How Cadaver Dogs Found a Missing Pennsylvania Man Deep Underground." *CNN*. July 13, 2017.

"Lloyds Bank Coprolite." *Atlas Obscura*. December 26, 2018.

Long, Robert A., Therese M. Donovan, Paula Mackay, William J. Zielinski, and Jeffrey S. Buzas. "Effectiveness of Scat Detection Dogs for Detecting Forest Carnivores." *The Journal of Wildlife Management* 71, no. 6 (2007): 2007-2017.

Lozano, Alicia Victoria. "In Their Own Words: Admitted Killer Cosmo DiNardo, Accused Accomplice Sean Kratz Detail Bucks County Farm Murders in Confession Recordings." *NBCPhiladelphia.com*. May 15, 2018.

Marchal, Sophie, Olivier Bregeras, Didier Puaux, Rémi Gervais, and Barbara Ferry. "Rigorous Training of Dogs Leads to High Accuracy in Human Scent Matching-To-Sample Performance." *Plos One* 11, no. 2 (2016): e0146963.

Meier, Allison C. "Finding a Murderer in a Victim's Eye." *JSTOR Daily*. October 31, 2018.

Mitchell, Piers D. "Human Parasites in Medieval Europe: Lifestyle, Sanitation and Medical Treatment." In *Advances in Parasitology*, Vol. 90. 389-420. Oxford: Academic Press, 2015.

Mitchell, Piers D. "Human Parasites in the Roman World: Health Consequences of Conquering an Empire." *Parasitology* 144, no. 1 (2017): 48-58.

Mitchell, Piers D. "The Origins of Human Parasites: Exploring the evidence for Endoparasitism throughout Human Evolution." *International Journal of Paleopathology* 3, no. 3 (2013): 191-198.

Mitchell, Piers D., Hui-Yuan Yeh, Jo Appleby, and Richard Buckley. "The Intestinal Parasites of King Richard III." *The Lancet* 382, no. 9895 (2013): 888.

Nicholson, Rebecca, Jennifer Robinson, Mark Robinson, and Erica Rowan. "From the Waters to the Plate to the Latrine: Fish and Seafood from the Cardo V Sewer, Herculaneum." *Journal of Maritime Archaeology* 13, no. 3 (2018): 263-284.

Norris, David O., and Jane H. Bock. "Use of Fecal Material to Associate a sSuspect with a Crime Scene: Report of Two Cases." *Journal of Forensic Science* 45, no. 1 (2000): 184-187.

Pearce, Jemah. "Copenhagen Burnt Down 3 Times in 80 Years. It Was Not All Bad." *Uniavisen*. May 14, 2019.

"Peoria Tribe of Indians of Oklahoma." Accessed April 20, 2022, https://peoriatribe.com/culture/.

Pinc, Ludvík, Luděk Bartoš, Alice Reslova, and Radim Kotrba. "Dogs Discriminate Identical Twins." *PLoS One* 6, no. 6 (2011): e20704.

Rampelli, Simone, Silvia Turroni, Carolina Mallol, Cristo Hernandez, Bertila Galván, Ainara Sistiaga, Elena Biagi et al. "Components of a Neanderthal gut microbiome recovered from fecal sediments from El Salt." *Communications Biology* 4, no. 1 (2021): 1-10.

Rankin, Caitlin G., Casey R. Barrier, and Timothy J. Horsley. "Evaluating Narratives of Ecocide with the Stratigraphic Record at Cahokia Mounds State Historic Site, Illinois, USA." *Geoarchaeology* 36, no. 3 (2021): 369-387.

BIBLIOGRAPHY

Robinson, Mark, and Erica Rowan. "Roman Food Remains in Archaeology and the Contents of a Roman Sewer at Herculaneum." In *A Companion to Food in the Ancient World*, First Edition. 105-115. Hoboken, New Jersey: Wiley, 2015.

Robinson, Nathan J. "Forensic Pseudoscience." *Boston Review*. November 16, 2015.

Sakr, Rania, Cedra Ghsoub, Celine Rbeiz, Vanessa Lattouf, Rachelle Riachy, Chadia Haddad, and Marouan Zoghbi. "COVID-19 Detection by Dogs: From Physiology to Field Application—a Review Article." *Postgraduate Medical Journal* 98, no. 1157 (2022): 212-218.

Saks, Michael J., Thomas Albright, Thomas L. Bohan, Barbara E. Bierer, C. Michael Bowers, Mary A. Bush, Peter J. Bush et al. "Forensic Bitemark Identification: Weak Foundations, Exaggerated Claims." *Journal of Law and the Biosciences* 3, no. 3 (2016): 538-575.

Schneider, Judith, Eduard Mas-Carrió, Catherine Jan, Christian Miquel, Pierre Taberlet, Katarzyna Michaud, and Luca Fumagalli. "Comprehensive Coverage of Human Last Meal Components Revealed by a Forensic DNA Metabarcoding Approach." *Scientific Reports* 11, no. 1 (2021): 1-8.

Sistiaga, Ainara, Carolina Mallol, Bertila Galván, and Roger Everett Summons. "The Neanderthal Meal: A New Perspective Using Faecal Biomarkers." *PloS One* 9, no. 6 (2014): e101045.

Sistiaga, Ainara, Francesco Berna, Richard Laursen, and Paul Goldberg. "Steroidal Biomarker Analysis of a 14,000 Years Old Putative Human Coprolite from Paisley Cave, Oregon." *Journal of Archaeological Science* 41 (2014): 813-817.

Verheggen, François, Katelynn A. Perrault, Rudy Caparros Megido, Lena M. Dubois, Frédéric Francis, Eric Haubruge, Shari L. Forbes, Jean-François Focant, and Pierre-Hugues Stefanuto. "The Odor of Death: An Overview of Current Knowledge on Characterization and Applications." *Bioscience* 67, no. 7 (2017): 600-613.

Vynne, Carly, John R. Skalski, Ricardo B. Machado, Martha J. Groom, Anah TA Jácomo, J. A. D. E. R. Marinho-Filho, Mario B. Ramos Neto et al. "Effectiveness of Scat-Detection Dogs in Determining Species Presence in a Tropical Savanna Landscape." *Conservation Biology* 25, no. 1 (2011): 154-162.

White, A. J., Lora R. Stevens, Varenka Lorenzi, Samuel E. Munoz, Carl P. Lipo, and Sissel Schroeder. "An Evaluation of Fecal Stanols as Indicators of Population Change at Cahokia, Illinois." *Journal of Archaeological Science* 93 (2018): 129-134.

White, A. J., Lora R. Stevens, Varenka Lorenzi, Samuel E. Munoz, Sissel Schroeder, Angelica Cao, and Taylor Bogdanovich. "Fecal Stanols Show Simultaneous Flooding and Seasonal Precipitation Change Correlate with Cahokia's Population Decline." *Proceedings of the National Academy of Sciences* 116, no. 12 (2019): 5461-5466.

White, A. J., Samuel E. Munoz, Sissel Schroeder, and Lora R. Stevens. "After Cahokia: Indigenous Repopulation and Depopulation of the Horseshoe Lake Watershed AD 1400–1900." *American Antiquity* 85, no. 2 (2020): 263-278.

Wilke, Philip J., and Henry Johnson Hall. *Analysis of Ancient Feces: A Discussion and Annotated Bibliography*. Berkeley: Archaeological Research Facility, Department of Anthropology, University of California, 1975.

BIBLIOGRAPHY

Yeh, Hui-Yuan, and Piers D. Mitchell. "Ancient Human Parasites in Ethnic Chinese Populations." *The Korean Journal of Parasitology* 54, no. 5 (2016): 565.

Chapter Five

Ahmed, Iftikhar, Rosemary Greenwood, Ben de Lacy Costello, Norman M. Ratcliffe, and Chris S. Probert. "An Investigation of Fecal Volatile Organic Metabolites in Irritable Bowel Syndrome." *PloS One* 8, no. 3 (2013): e58204.

Ahmed, Imtiaz, Muhammad Najmuddin Shabbir, Mohammad Ali Iqbal, and Muhammad Shahzeb. "Role of Defecation Postures on the Outcome of Chronic Anal Fissure." *Pakistan Journal of Surgery* 29, no. 4 (2013): 269-271.

Allen, Thomas. *Plain Directions for the Prevention and Treatment of Cholera.* Oxford: J. Vincent, 1848.

Amann, Anton, Ben de Lacy Costello, Wolfram Miekisch, Jochen Schubert, Bogusław Buszewski, Joachim Pleil, Norman Ratcliffe, and Terence Risby. "The Human Volatilome: Volatile Organic Compounds (VOCs) In Exhaled Breath, Skin Emanations, Urine, Feces and Saliva." *Journal of Breath Research* 8, no. 3 (2014): 034001.

Antoniou, Georgios P., Giovanni De Feo, Franz Fardin, Aldo Tamburrino, Saifullah Khan, Fang Tie, Ieva Reklaityte et al. "Evolution of Toilets Worldwide through the Millennia." *Sustainability* 8, no. 8 (2016): 779.

Asbjørnsen, Peter Christen & Jørgen Engebretsen Moe. *The Complete Norwegian Folktales and Legends of Asbjørnsen & Moe.* Translated by Simon Roy Hughes. 2020.

Asnicar, Francesco, Emily R. Leeming, Eirini Dimidi, Mohsen Mazidi, Paul W. Franks, Haya Al Khatib, Ana M. Valdes et al. "Blue Poo: Impact of Gut Transit Time on the Gut Microbiome Using a Novel Marker." *Gut* 70, no. 9 (2021): 1665-1674.

Bala, Manju, Asha Sharma, and Gaurav Sharma. "Assessment of Heavy Metals in Faecal Pellets of Blue Rock Pigeon from Rural and Industrial Environment in India." *Environmental Science and Pollution Research* 27, no. 35 (2020): 43646-43655.

Barbieri, Annalisa. "The Truth about Poo: We're Doing It Wrong." *The Guardian.* May 18, 2015.

Barclay, Eliza. "For Best Toilet Health: Squat or Sit?" *NPR.* September 28, 2012.

Baron, Ruth, Meron Taye, Isolde Besseling-van der Vaart, Joanne Ujčič-Voortman, Hania Szajewska, Jacob C. Seidell, and Arnoud Verhoeff. "The Relationship of Prenatal Antibiotic Exposure and Infant Antibiotic Administration with Childhood Allergies: A Systematic Review." *BMC Pediatrics* 20, no. 1 (2020): 1-14.

Bekkali, Noor, Sofie L. Hamers, Johannes B. Reitsma, Letty Van Toledo, and Marc A. Benninga. "Infant Stool form Scale: Development and Results." *The Journal of Pediatrics* 154, no. 4 (2009): 521-526.

Bharucha, Adil E., John H. Pemberton, and G. Richard Locke. "American Gastroenterological Association Technical Review on Constipation." *Gastroenterology* 144, no. 1 (2013): 218-238.

BIBLIOGRAPHY

Blasdel, Alex. "Bowel Movement: The Push to Change the Way You Poo." *The Guardian*. November 30, 2018.

BMJ. "Cliff and C. diff—Smelling the Diagnosis." *YouTube*. December 14, 2012.

Bomers, Marije K., Michiel A. Van Agtmael, Hotsche Luik, Merk C. Van Veen, Christina MJE Vandenbroucke-Grauls, and Yvo M. Smulders. "Using a Dog's Superior Olfactory Sensitivity to Identify Clostridium difficile in Stools and Patients: Proof of Principle Study." *BMJ* 345 (2012): e7396.

Bond, Allison. "A 'Shark Tank'-Funded Test for Food Sensitivity Is Medically Dubious, Experts Say." *STAT*. January 23, 2018.

Branswell, Helen. "The Dogs Were Supposed to Be Experts at Sniffing out C. diff. Then They Smelled Breakfast." *STAT*. Aug. 22, 2018.

Brown, S. R., P. A. Cann, and N. W. Read. "Effect of Coffee on Distal Colon Function." *Gut* 31, no. 4 (1990): 450-453.

Bryce, E., T. Zurberg, M. Zurberg, S. Shajari, and D. Roscoe. "Identifying Environmental Reservoirs of Clostridium difficile with a Scent Detection Dog: Preliminary Evaluation." *Journal of Hospital Infection* 97, no. 2 (2017): 140-145.

Carlson, Alexander L., Kai Xia, M. Andrea Azcarate-Peril, Barbara D. Goldman, Mihye Ahn, Martin A. Styner, Amanda L. Thompson, Xiujuan Geng, John H. Gilmore, and Rebecca C. Knickmeyer. "Infant Gut Microbiome Associated with Cognitive Development." *Biological Psychiatry* 83, no. 2 (2018): 148-159.

Chakrabarti, S. D., R. Ganguly, S. K. Chatterjee, and A. Chakravarty. "Is Squatting a Triggering Factor for Stroke in Indians?" *Acta Neurologica Scandinavica* 105, no. 2 (2002): 124-127.

Czepiel, Jacek, Mirosław Dróźdź, Hanna Pituch, Ed J. Kuijper, William Perucki, Aleksandra Mielimonka, Sarah Goldman, Dorota Wultańska, Aleksander Garlicki, and Grażyna Biesiada. "Clostridium difficile Infection." *European Journal of Clinical Microbiology & Infectious Diseases* 38, no. 7 (2019): 1211-1221.

David, Lawrence A., Arne C. Materna, Jonathan Friedman, Maria I. Campos-Baptista, Matthew C. Blackburn, Allison Perrotta, Susan E. Erdman, and Eric J. Alm. "Host Lifestyle Affects Human Microbiota on Daily Timescales." *Genome Biology* 15, no. 7 (2014): 1-15.

David, Lawrence A., Corinne F. Maurice, Rachel N. Carmody, David B. Gootenberg, Julie E. Button, Benjamin E. Wolfe, Alisha V. Ling et al. "Diet Rapidly and Reproducibly Alters the Human Gut Microbiome." *Nature* 505, no. 7484 (2014): 559-563.

Davis, Jasmine CC, Sarah M. Totten, Julie O. Huang, Sadaf Nagshbandi, Nina Kirmiz, Daniel A. Garrido, Zachery T. Lewis et al. "Identification of Oligosaccharides in Feces of Breast-Fed Infants and their Correlation with the Gut Microbial Community." *Molecular & Cellular Proteomics* 15, no. 9 (2016): 2987-3002.

De Leoz, Maria Lorna A., Shuai Wu, John S. Strum, Milady R. Niñonuevo, Stephanie C. Gaerlan, Majid Mirmiran, J. Bruce German, David A. Mills, Carlito B. Lebrilla, and Mark A. Underwood. "A Quantitative and Comprehensive Method to Analyze Human Milk Oligosaccharide Structures in the Urine and Feces of Infants." *Analytical and Bioanalytical Chemistry* 405, no. 12 (2013): 4089-4105.

BIBLIOGRAPHY

Deweerdt, Sarah. "Estimate of Autism's Sex Ratio Reaches New Low." *Spectrum*. April 27, 2017.

Douglas, Bruce R., J. B. Jansen, R. T. Tham, and C. B. Lamers. "Coffee Stimulation of Cholecystokinin Release and Gallbladder Contraction in Humans." *The American Journal of Clinical Nutrition* 52, no. 3 (1990): 553-556.

Ebert, Vince. "Deutsche Thoroughness." *Journal*. July 7, 2020.

Enders, Giulia. *Gut: The Inside Story of Our Body's Most Underrated Organ (Revised Edition)*. Vancouver: Greystone Books Ltd, 2018.

Essler, Jennifer L., Sarah A. Kane, Pat Nolan, Elikplim H. Akaho, Amalia Z. Berna, Annemarie DeAngelo, Richard A. Berk et al. "Discrimination of SARS-CoV-2 Infected Patient Samples by Detection Dogs: A Proof of Concept Study." *PLoS One* 16, no. 4 (2021): e0250158.

"Fact Check-No Evidence that 'Urine Therapy' Cures COVID-19." *Reuters*. January 12, 2022.

Foreman, Judy. "Beware of Colon Cleansing Claims." *Los Angeles Times*. June 30, 2008.

Frew, John W. "The Hygiene Hypothesis, Old Friends, and New Genes." *Frontiers in Immunology* 10 (2019): 388.

Frias, Bárbara, and Adalberto Merighi. "Capsaicin, Nociception and Pain." *Molecules* 21, no. 6 (2016): 797.

Fujimura, Kei E., Alexandra R. Sitarik, Suzanne Havstad, Din L. Lin, Sophia Levan, Douglas Fadrosh, Ariane R. Panzer et al. "Neonatal Gut Microbiota Associates with Childhood Multisensitized Atopy and T Cell Differentiation." *Nature Medicine* 22, no. 10 (2016): 1187-1191.

Gil-Riaño, Sebastián, and Sarah E. Tracy. "Developing Constipation: Dietary Fiber, Western Disease, and Industrial Carbohydrates." *Global Food History* 2, no. 2 (2016): 179-209.

Hecht, Jen, Travis Sanchez, Patrick S. Sullivan, Elizabeth A. DiNenno, Natalie Cramer, and Kevin P. Delaney. "Increasing Access to HIV Testing Through Direct-to-Consumer HIV Self-Test Distribution—United States, March 31, 2020–March 30, 2021." *Morbidity and Mortality Weekly Report* 70, no. 38 (2021): 1322-1325.

Ho, Vincent. "What's the Best Way to Go to the Toilet—Squatting or Sitting?" *The Conversation*. August 16, 2016.

Huang, Pien. "How Ivermectin Became the New Focus of the Anti-Vaccine Movement." *NPR*. September 19, 2021.

Hussain, Ghulam, Jing Wang, Azhar Rasul, Haseeb Anwar, Ali Imran, Muhammad Qasim, Shamaila Zafar et al. "Role of Cholesterol and sSphingolipids in Brain Development and Neurological Diseases." *Lipids in Health and Disease* 18, no. 1 (2019): 1-12.

Huysentruyt, Koen, Ilan Koppen, Marc Benninga, Tom Cattaert, Jiqiu Cheng, Charlotte De Geyter, Christophe Faure et al. "The Brussels Infant and Toddler Stool Scale: A Study on Interobserver Reliability." *Journal of Pediatric Gastroenterology and Nutrition* 68, no. 2 (2019): 207-213.

Ishihara, Nobuo, and Takashi Matsushiro. "Biliary and Urinary Excretion of Metals in Humans." *Archives of Environmental Health: An International Journal* 41, no. 5 (1986): 324-330.

BIBLIOGRAPHY

Jairoun, Ammar A., Sabaa Saleh Al-Hemyari, Moyad Shahwan, and Sa'ed H. Zyoud. "Adulteration of Weight Loss Supplements by the Illegal Addition of Synthetic Pharmaceuticals." *Molecules* 26, no. 22 (2021): 6903.

Johnson, Steven. *The Ghost Map: The Story of London's Most Terrifying Epidemic—and How it Changed Science, Cities, and the Modern World.* New York: Penguin, 2006.

Jun-yong, Ahn, and Lee Kil-seong. "Kim Jong-un's Flight to Singapore a Precision Maneuver." *The Chosun Ilbo.* June 11, 2018.

Kim, Byoung-Ju, So-Yeon Lee, Hyo-Bin Kim, Eun Lee, and Soo-Jong Hong. "Environmental Changes, Microbiota, and Allergic Diseases." *Allergy, Asthma & Immunology Research* 6, no. 5 (2014): 389-400.

Knapp, Alex. "SEC Charges Microbiome Startup uBiome's Cofounders with Defrauding Investors for $60 Million." *Forbes*. March 18, 2021.

Kobayashi, T. "Studies on *Clostridium difficile* and Antimicrobial Associated Diarrhea or Colitis." *The Japanese Journal of Antibiotics* 36, no. 2 (1983): 464-476.

Korownyk, Christina, Michael R. Kolber, James McCormack, Vanessa Lam, Kate Overbo, Candra Cotton, Caitlin Finley et al. "Televised Medical Talk Shows—What They Recommend and the Evidence to Support Their Recommendations: A Prospective Observational Study." *BMJ* 349 (2014): g7346.

Korpela, Katja, Marjo Renko, Petri Vänni, Niko Paalanne, Jarmo Salo, Mysore V. Tejesvi, Pirjo Koivusaari et al. "Microbiome of the First Stool and Overweight at Age 3 Years: A Prospective Cohort Study." *Pediatric Obesity* 15, no. 11 (2020): e12680.

Krisberg, Kim. "Is Everlywell for Real?" *Austin Monthly.* February 2021.

Kybert, Nicholas, Katharine Prokop-Prigge, Cynthia M. Otto, Lorenzo Ramirez, EmmaRose Joffe, Janos Tanyi, Jody Piltz-Seymour, AT Charlie Johnson, and George Preti. "Exploring Ovarian Cancer Screening Using a Combined Sensor Approach: A Pilot Study." *AIP Advances* 10, no. 3 (2020): 035213.

Levin, Albert M., Alexandra R. Sitarik, Suzanne L. Havstad, Kei E. Fujimura, Ganesa Wegienka, Andrea E. Cassidy-Bushrow, Haejin Kim et al. "Joint Effects of Pregnancy, Sociocultural, and Environmental Factors on Early Life Gut Microbiome Structure and Diversity." *Scientific Reports* 6, no. 1 (2016): 1-16.

Li, Cheng, Teresa Zurberg, Jaime Kinna, Kushal Acharya, Jack Warren, Salomeh Shajari, Leslie Forrester, and Elizabeth Bryce. "Using Scent Detection Dogs to Identify Environmental Reservoirs of Clostridium difficile: Lessons from the Field." *Canadian Journal of Infection Control* 34, no. 2 (2019): 93-95.

Marris, Emma. *Rambunctious Garden: Saving Nature in a Post-Wild World.* New York: Bloomsbury, 2011.

Masood, R., and M. Miraftab. "Psyllium: Current and Future Applications." In *Medical and Healthcare Textiles.* 244-253, New Delhi: Woodhead Publishing, 2010.

Mayo Clinic. "Dietary Fiber: Essential for a Healthy Diet." January 6, 2021, https://www.mayoclinic.org/healthy-lifestyle/nutrition-and-healthy-eating/in-depth/fiber/art-20043983.

Mayo Clinic. "Over-the-Counter Laxatives for Constipation: Use with Caution." March 3, 2022, https://www.mayoclinic.org/diseases-conditions/constipation/in-depth/laxatives/art-20045906.

BIBLIOGRAPHY

McDonald, Daniel, Embriette Hyde, Justine W. Debelius, James T. Morton, Antonio Gonzalez, Gail Ackermann, Alexander A. Aksenov et al. "American Gut: An Open Platform for Citizen Science Microbiome Research." *Msystems* 3, no. 3 (2018): e00031-18.

Melendez, Johan H., Matthew M. Hamill, Gretchen S. Armington, Charlotte A. Gaydos, and Yukari C. Manabe. "Home-Based Testing for Sexually Transmitted Infections: Leveraging Online Resources during the COVID-19 Pandemic." *Sexually Transmitted Diseases* 48, no. 1 (2021): e8-e10.

Mitchell, Piers D. "Human Parasites in the Roman World: Health Consequences of Conquering an Empire." *Parasitology* 144, no. 1 (2017): 48-58.

Modi, Rohan M., Alice Hinton, Daniel Pinkhas, Royce Groce, Marty M. Meyer, Gokulakrishnan Balasubramanian, Edward Levine, and Peter P. Stanich. "Implementation of a Defecation Posture Modification Device: Impact on Bowel Movement Patterns in Healthy Subjects." *Journal of Clinical Gastroenterology* 53, no. 3 (2019): 216.

National Library of Medicine. "Stools—Foul Smelling." *MedlinePlus.* July 16, 2020, https://medlineplus.gov/ency/article/003132.html.

National Library of Medicine. "White Blood Cell (WBC) in Stool." *MedlinePlus.* Accessed April 20, 2022, https://medlineplus.gov/lab-tests/white-blood-cell-wbc-in-stool/.

Ng, Siew C., Michael A. Kamm, Yun Kit Yeoh, Paul KS Chan, Tao Zuo, Whitney Tang, Ajit Sood et al. "Scientific Frontiers in Faecal Microbiota Transplantation: Joint Document of Asia-Pacific Association of Gastroenterology (APAGE) and Asia-Pacific Society for Digestive Endoscopy (APSDE)." *Gut* 69, no. 1 (2020): 83-91.

Oz, Mehmet. "Everybody Poops." *Oprah.com.* January 1, 2006.

Ozaki, Eijiro, Haru Kato, Hiroyuki Kita, Tadahiro Karasawa, Tsuneo Maegawa, Youko Koino, Kazumasa Matsumoto et al. "Clostridium difficile Colonization in Healthy Adults: Transient Colonization and Correlation with Enterococcal Colonization." *Journal of Medical Microbiology* 53, no. 2 (2004): 167-172.

Palm, Noah W., Rachel K. Rosenstein, and Ruslan Medzhitov. "Allergic Host Defences." *Nature* 484, no. 7395 (2012): 465-472.

Park, Seung-min, Daeyoun D. Won, Brian J. Lee, Diego Escobedo, Andre Esteva, Amin Aalipour, T. Jessie Ge et al. "A Mountable Toilet System for Personalized Health Monitoring via the Analysis of Excreta." *Nature Biomedical Engineering* 4, no. 6 (2020): 624-635.

Philpott, Hamish L., Sanjay Nandurkar, John Lubel, and Peter Raymond Gibson. "Drug-Induced Gastrointestinal Disorders." *Frontline Gastroenterology* 5, no. 1 (2014): 49-57.

Picco, Michael F. "Stool Color: When to Worry." Mayo Clinic, October 10, 2020, https://www.mayoclinic.org/stool-color/expert-answers/faq-20058080.

Prinsenberg, Tamara, Sjoerd Rebers, Anders Boyd, Freke Zuure, Maria Prins, Marc van der Valk, and Janke Schinkel. "Dried Blood Spot Self-Sampling at Home Is a Feasible Technique for Hepatitis C RNA detection." *PLoS One* 15, no. 4 (2020): e0231385.

BIBLIOGRAPHY

Rao, Satish S.C., Kimberly Welcher, Bridget Zimmerman, and Phyllis Stumbo. "Is Coffee a Colonic Stimulant?" *European Journal of Gastroenterology & Hepatology* 10, no. 2 (1998): 113-118.

Rosenberg, Steven. "Stalin 'Used Secret Laboratory to Analyse Mao's Excrement.'" *BBC News.* January 28, 2016.

Saeidnia, Soodabeh and Azadeh Manayi. "Phenolphthalein." In *Encyclopedia of Toxicology* (Third Edition). 877-880. Cambridge, Massachusetts: Academic Press, 2014.

Sakakibara, Ryuji, Kuniko Tsunoyama, Hiroyasu Hosoi, Osamu Takahashi, Megumi Sugiyama, Masahiko Kishi, Emina Ogawa, Hitoshi Terada, Tomoyuki Uchiyama, and Tomonori Yamanishi. "Influence of Body Position on Defecation in Humans." *LUTS: Lower Urinary Tract Symptoms* 2, no. 1 (2010): 16-21.

Sberro, Hila, Brayon J. Fremin, Soumaya Zlitni, Fredrik Edfors, Nicholas Greenfield, Michael P. Snyder, Georgios A. Pavlopoulos, Nikos C. Kyrpides, and Ami S. Bhatt. "Large-Scale Analyses of Human Microbiomes Reveal Thousands of Small, Novel Genes." *Cell* 178, no. 5 (2019): 1245-1259.

Scudellari, Megan. "Cleaning up the Hygiene Hypothesis." *Proceedings of the National Academy of Sciences of the United States of America.* 114, no. 7 (2017): 1433-1436.

Sethi, Saurabh. "Squatting: A Forgotten Natural Instinct to Prevent Hemorrhoids!" *American Journal of Gastroenterology* 105 (2010): S142.

Sheth, Anish, and Josh Richman. *What's Your Baby's Poo Telling You?: A Bottoms-Up Guide to Your Baby's Health.* New York: Avery, 2014.

Shirasu, Mika, and Kazushige Touhara. "The Scent of Disease: Volatile Organic Compounds of the Human Body Related to Disease and Disorder." *The Journal of Biochemistry* 150, no. 3 (2011): 257-266.

Sikirov, Berko A. "Etiology and Pathogenesis of Diverticulosis Coli: A New Approach." *Medical Hypotheses* 26, no. 1 (1988): 17-20.

Sikirov, Dov. "Comparison of Straining During Defecation in Three Positions: Results and Implications for Human Health." *Digestive Diseases and Sciences* 48, no. 7 (2003): 1201-1205.

Specter, Michael. "The Operator." *The New Yorker.* January 27, 2013.

Squatty Potty. "This Unicorn Changed the Way I Pooped." *YouTube.* October 6, 2015.

Stanford Medicine. "Bristol Stool Form Scale." Accessed April 20, 2022, https://pediatricsurgery.stanford.edu/Conditions/BowelManagement/bristol-stool-form-scale.html.

Stempel, Jonathan. "Co-Founders of San Francisco Biotech Startup uBiome Charged with Fraud." *Reuters.* March 18, 2021.

Taft, Diana H., Jinxin Liu, Maria X. Maldonado-Gomez, Samir Akre, M. Nazmul Huda, S. M. Ahmad, Charles B. Stephensen, and David A. Mills. "Bifidobacterial Dominance of the Gut in Early Life and Acquisition of Antimicrobial Resistance." *MSphere* 3, no. 5 (2018): e00441-18.

Tamana, Sukhpreet K., Hein M. Tun, Theodore Konya, Radha S. Chari, Catherine J. Field, David S. Guttman, Allan B. Becker et al. "Bacteroides-Dominant Gut Microbiome of Late Infancy Is Associated with Enhanced Neurodevelopment." *Gut Microbes* 13, no. 1 (2021): 1930875.

BIBLIOGRAPHY

Taylor, Maureen T., Janine McCready, George Broukhanski, Sakshi Kirpalaney, Haydon Lutz, and Jeff Powis. "Using Dog Scent Detection as a Point-of-Care Tool to Identify Toxigenic Clostridium difficile in Stool." *Open Forum Infectious Diseases* 5, no. 8 (2018): ofy179.

"The Myth of IgG Food Panel Testing." American Academy of Allergy, Asthma & Immunology, September 28, 2020, https://www.aaaai.org/tools-for-the-public/conditions-library/allergies/igg-food-test.

Thompson, Henry J., and Mark A. Brick. "Perspective: Closing the Dietary Fiber Gap: An Ancient Solution for a 21st Century Problem." *Advances in Nutrition* 7, no. 4 (2016): 623-626.

Tsai, Pei-Yun, Bingkun Zhang, Wei-Qi He, Juan-Min Zha, Matthew A. Odenwald, Gurminder Singh, Atsushi Tamura et al. "IL-22 Upregulates Epithelial Claudin-2 to Drive Diarrhea and Enteric Pathogen Clearance." *Cell Host & Microbe* 21, no. 6 (2017): 671-681.

Tucker, Jenna, Tessa Fischer, Laurence Upjohn, David Mazzera, and Madhur Kumar. "Unapproved Pharmaceutical Ingredients Included in Dietary Supplements Associated with US Food and Drug Administration Warnings." *JAMA Network Open* 1, no. 6 (2018): e183337-e183337.

U.S. Food and Drug Administration. "Direct-to-Consumer Tests." Accessed April 20, 2022, https://www.fda.gov/medical-devices/in-vitro-diagnostics/direct-consumer-tests.

U.S. Food and Drug Administration. "Questions and Answers about FDA's Initiative against Contaminated Weight Loss Products." Accessed April 20, 2022, https://www.fda.gov/drugs/questions-answers/questions-and-answers-about-fdas-initiative-against-contaminated-weight-loss-products.

U.S. Preventive Services Task Force. "Final Recommendation Statement. Thyroid Dysfunction: Screening." Last modified March 24, 2015, https://www.uspreventiveservicestaskforce.org/uspstf/recommendation/thyroid-dysfunction-screening.

U.S. Preventive Services Task Force. "Final Recommendation Statement. Vitamin D Deficiency in Adults: Screening." Last modified April 13, 2021, https://www.uspreventiveservicestaskforce.org/uspstf/recommendation/vitamin-d-deficiency-screening.

Vandenplas, Yvan, Hania Szajewska, Marc Benninga, Carlo Di Lorenzo, Christophe Dupont, Christophe Faure, Mohamed Miqdadi et al. "Development of the Brussels Infant and Toddler Stool Scale ('BITSS'): Protocol of the Study." *BMJ Open* 7, no. 3 (2017): e014620.

Wastyk, Hannah C., Gabriela K. Fragiadakis, Dalia Perelman, Dylan Dahan, Bryan D. Merrill, B. Yu Feiqiao, Madeline Topf et al. "Gut-Microbiota-Targeted Diets Modulate Human Immune Status." *Cell* 184, no. 16 (2021): 4137-4153.

Whorton, James. "Civilisation and the Colon: Constipation as the 'Disease of Diseases.'" *BMJ* 321, no. 7276 (2000): 1586-1589.

World Health Organization. "Coronavirus Disease (COVID-19) Advice for the Public: Mythbusters." January 19, 2022, https://www.who.int/emergencies/diseases/novel-coronavirus-2019/advice-for-public/myth-busters.

Wypych, Tomasz P., Céline Pattaroni, Olaf Perdijk, Carmen Yap, Aurélien Trompette, Dovile Anderson, Darren J. Creek, Nicola L. Harris, and Benjamin J. Marsland. "Microbial Metabolism of L-Tyrosine Protects against Allergic Airway Inflammation." *Nature Immunology* 22, no. 3 (2021): 279-286.

Yabe, John, Shouta MM Nakayama, Yoshinori Ikenaka, Yared B. Yohannes, Nesta Bortey-Sam, Abel Nketani Kabalo, John Ntapisha, Hazuki Mizukawa, Takashi Umemura, and Mayumi Ishizuka. "Lead and Cadmium Excretion in Feces and Urine of Children from Polluted Townships near a Lead-Zinc Mine in Kabwe, Zambia." *Chemosphere* 202 (2018): 48-55.

Zarrell, Rachel. "People Who Ate Burger King's Black Whopper Said It Turned Their Poop Green." *BuzzFeed*. October 5, 2015.

Chapter Six

Aburto, José Manuel, Jonas Schöley, Ilya Kashnitsky, Luyin Zhang, Charles Rahal, Trifon I. Missov, Melinda C. Mills, Jennifer B. Dowd, and Ridhi Kashyap. "Quantifying Impacts of the COVID-19 Pandemic Through Life-Expectancy Losses: A Population-Level Study of 29 Countries." *International Journal of Epidemiology* 51, no. 1 (2022): 63-74.

Albert, Sandra, Alba Ruíz, Javier Pemán, Miguel Salavert, and Pilar Domingo-Calap. "Lack of Evidence for Infectious SARS-CoV-2 in Feces and Sewage." *European Journal of Clinical Microbiology & Infectious Diseases* 40, no. 12 (2021): 2665-2667.

Aghamohammadi, Asghar, Hassan Abolhassani, Necil Kutukculer, Steve G. Wassilak, Mark A. Pallansch, Samantha Kluglein, Jessica Quinn et al. "Patients with Primary Immunodeficiencies Are a Reservoir of Poliovirus and a Risk to Polio Eradication." *Frontiers in Immunology* 8 (2017): 685.

Ahmed, Warish, Nicola Angel, Janette Edson, Kyle Bibby, Aaron Bivins, Jake W. O'Brien, Phil M. Choi et al. "First Confirmed Detection of SARS-CoV-2 in Untreated Wastewater in Australia: a Proof of Concept for the Wastewater Surveillance of COVID-19 in the Community." *Science of the Total Environment* 728 (2020): 138764.

Ahmed, Warish, Paul M. Bertsch, Nicola Angel, Kyle Bibby, Aaron Bivins, Leanne Dierens, Janette Edson et al. "Detection of SARS-CoV-2 RNA in Commercial Passenger Aircraft and Cruise Ship Wastewater: A Surveillance Tool for Assessing the Presence of COVID-19 Infected Travellers." *Journal of Travel Medicine* 27, no. 5 (2020): taaa116.

Avadhanula, Vasanthi, Erin G. Nicholson, Laura Ferlic-Stark, Felipe-Andres Piedra, Brittani N. Blunck, Sonia Fragoso, Nanette L. Bond et al. "Viral Load of Severe Acute Respiratory Syndrome Coronavirus 2 in Adults during the First and Second Wave of Coronavirus Disease 2019 Pandemic in Houston, Texas: The Potential of the Superspreader." *The Journal of Infectious Diseases* 223, no. 9 (2021): 1528-1537.

Azzoni, Tales, and Andrew Dampf. "Game Zero: Spread of Virus Linked to Champions League Match." *Associated Press*. March 25, 2020.

BIBLIOGRAPHY

Bedford, Trevor, Alexander L. Greninger, Pavitra Roychoudhury, Lea M. Starita, Michael Famulare, Meei-Li Huang, Arun Nalla et al. "Cryptic Transmission of SARS-CoV-2 in Washington State." *Science* 370, no. 6516 (2020): 571-575.

Betancourt, Walter Q., Bradley W. Schmitz, Gabriel K. Innes, Sarah M. Prasek, Kristen M. Pogreba Brown, Erika R. Stark, Aidan R. Foster et al. "COVID-19 Containment on a College Campus via Wastewater-Based Epidemiology, Targeted Clinical Testing and an Intervention." *Science of The Total Environment* 779 (2021): 146408.

Bibby, Kyle, Katherine Crank, Justin Greaves, Xiang Li, Zhenyu Wu, Ibrahim A. Hamza, and Elyse Stachler. "Metagenomics and the Development of Viral Water Quality Tools." *NPJ Clean Water* 2, no. 1 (2019): 1-13.

Bieler, Des. "'A Biological Bomb': Soccer Match in Italy Linked to Epicenter of Deadly Outbreak." *The Washington Post.* March 25, 2020.

Biobot Analytics. "How Many People Are Infected with COVID-19? Sewage Suggests That Number Is Much Higher than Officially Confirmed." *Medium.* April 8, 2020.

Brouwer, Andrew F., Joseph NS Eisenberg, Connor D. Pomeroy, Lester M. Shulman, Musa Hindiyeh, Yossi Manor, Itamar Grotto, James S. Koopman, and Marisa C. Eisenberg. "Epidemiology of the Silent Polio Outbreak in Rahat, Israel, Based on Modeling of Environmental Surveillance Data." *Proceedings of the National Academy of Sciences* 115, no. 45 (2018): E10625-E10633.

Brueck, Hilary. "COVID-19 Experts Say Omicron is Peaking in the US, Citing Data from Poop Samples." *Business Insider.* January 12, 2022.

Burgard, Daniel A., Jason Williams, Danielle Westerman, Rosie Rushing, Riley Carpenter, Addison LaRock, Jane Sadetsky et al. "Using Wastewater-Based Analysis to Monitor the Effects of Legalized Retail Sales on Cannabis Consumption in Washington State, USA." *Addiction* 114, no. 9 (2019): 1582-1590.

Choi, Phil M., Benjamin Tscharke, Saer Samanipour, Wayne D. Hall, Coral E. Gartner, Jochen F. Mueller, Kevin V. Thomas, and Jake W. O'Brien. "Social, Demographic, and Economic Correlates of Food and Chemical Consumption Measured by Wastewater-Based Epidemiology." *Proceedings of the National Academy of Sciences* 116, no. 43 (2019): 21864-21873.

Cima, Greg. "Pandemic Prevention Program Ending after 10 Years." *JAVMA News.* January 2, 2020.

Cohen, Elizabeth. "China Says Coronavirus Can Spread Before Symptoms Show—Calling into Question US Containment Strategy." *CNN.* January 26, 2020.

Crank, K., W. Chen, A. Bivins, S. Lowry, and K. Bibby. "Contribution of SARS-CoV-2 RNA Shedding Routes to RNA Loads in Wastewater." *Science of The Total Environment* 806 (2022): 150376.

Devoid, Alex. "Pima County Braces for Rise in COVID-19 Cases As Arizona Continues to See Increase." *Tucson.com.* October 27, 2020.

Endo, Norkio, Newsha Ghaeli, Claire Duvallet, Katelyn Foppe, Timothy B. Erickson, Mariana Matus, and Peter R. Chai. "Rapid Assessment of Opioid Exposure and Treatment in Cities Through Robotic Collection and Chemical Analysis of Wastewater." *Journal of Medical Toxicology* 16, no. 2 (2020): 195-203.

BIBLIOGRAPHY

Engelhart, Katie. "What Happened in Room 10?" *California Sunday*. August 23, 2020.

European Monitoring Centre for Drugs and Drug Addition. *European Drug Report 2016: Trends and Developments*. Luxembourg: Publications Office of the European Union, 2016.

Fink, Sheri, and Mike Baker. "'It's Just Everywhere Already': How Delays in Testing Set Back the U.S. Coronavirus Response." *New York Times*. March 10, 2020.

Giacobbo, Alexandre, Marco Antônio Siqueira Rodrigues, Jane Zoppas Ferreira, Andréa Moura Bernardes, and Maria Norberta de Pinho. "A Critical Review on SARS-CoV-2 Infectivity in Water and Wastewater. What Do We Know?" *Science of The Total Environment* 774 (2021): 145721.

Graham, Katherine E., Stephanie K. Loeb, Marlene K. Wolfe, David Catoe, Nasa Sinnott-Armstrong, Sooyeol Kim, Kevan M. Yamahara et al. "SARS-CoV-2 RNA in Wastewater Settled Solids is Associated with COVID-19 Cases in a Large Urban Sewershed." *Environmental Science & Technology* 55, no. 1 (2020): 488-498.

Grange, Zoë L., Tracey Goldstein, Christine K. Johnson, Simon Anthony, Kirsten Gilardi, Peter Daszak, Kevin J. Olival et al. "Ranking the Risk of Animal-to-Human Spillover for Newly Discovered Viruses." *Proceedings of the National Academy of Sciences* 118, no. 15 (2021).

Gundy, Patricia M., Charles P. Gerba, and Ian L. Pepper. "Survival of Coronaviruses in Water and Wastewater." *Food and Environmental Virology* 1, no. 1 (2009): 10-14.

Hess, Peter. "Scientists Can Tell How Wealthy You Are by Examining Your Sewage." *Inverse*. October 9, 2019.

Hjelmsø, Mathis Hjort, Sarah Mollerup, Randi Holm Jensen, Carlotta Pietroni, Oksana Lukjancenko, Anna Charlotte Schultz, Frank M. Aarestrup, and Anders Johannes Hansen. "Metagenomic Analysis of Viruses in Toilet Waste from Long Distance Flights—A New Procedure for Global Infectious Disease Surveillance." *PLoS One* 14, no. 1 (2019): e0210368.

Johnson, Gene. "Gee Whiz: Testing of Tacoma Sewage Confirms Rise in Marijuana Use." *The Seattle Times*. June 24, 2019.

Karimi, Faith, Mallika Kallingal, and Theresa Waldrop. "Second Coronavirus Death in Washington State as Number of Cases Rises to 13." *CNN*. March 1, 2020.

Kaufman, Rachel. "Sewage May Hold the Key to Tracking Opioid Abuse." *Smithsonian Magazine*. August 22, 2018.

Kim, Sooyeol, Lauren C. Kennedy, Marlene K. Wolfe, Craig S. Criddle, Dorothea H. Duong, Aaron Topol, Bradley J. White et al. "SARS-CoV-2 RNA Is Enriched by Orders of Magnitude in Primary Settled Solids Relative to Liquid Wastewater at Publicly Owned Treatment Works." *Environmental Science: Water Research & Technology* (2022).

Kirby, Amy E., Maroya Spalding Walters, Wiley C. Jennings, Rebecca Fugitt, Nathan LaCross, Mia Mattioli, Zachary A. Marsh et al. "Using Wastewater Surveillance Data to Support the COVID-19 Response—United States, 2020–2021." *Morbidity and Mortality Weekly Report* 70, no. 36 (2021): 1242.

Kitajima, Masaaki, Hannah P. Sassi, and Jason R. Torrey. "Pepper Mild Mottle Virus as a Water Quality Indicator." *NPJ Clean Water* 1, no. 1 (2018): 1-9.

BIBLIOGRAPHY

Kling, C., G. Olin, J. Fåhraeus, and G. Norlin. "Sewage as a Carrier and Disseminator of Poliomyelitis Virus. Part I. Searching for Poliomyelitis Virus in Stockholm Sewage." *Acta Medica Scandinavica* 112, no. 3-4 (1942): 217-49.

Kling, C., G. Olin, J. Fåhraeus, and G. Norlin. "Sewage as a Carrier and Disseminator of Poliomyelitis Virus. Part II. Studies on the Conditions of Life of Poliomyelitis Virus outside the Human Organism." *Acta Medica Scandinavica* 112, no. 3-4 (1942): 250-63.

Komar, Nicholas, Stanley Langevin, Steven Hinten, Nicole Nemeth, Eric Edwards, Danielle Hettler, Brent Davis, Richard Bowen, and Michel Bunning. "Experimental Infection of North American Birds with the New York 1999 Strain of West Nile Virus." *Emerging Infectious Diseases* 9, no. 3 (2003): 311.

La Rosa, Giuseppina, Marcello Iaconelli, Pamela Mancini, Giusy Bonanno Ferraro, Carolina Veneri, Lucia Bonadonna, Luca Lucentini, and Elisabetta Suffredini. "First Detection of SARS-CoV-2 in Untreated Wastewaters in Italy." *Science of The Total Environment* 736 (2020): 139652.

La Rosa, Giuseppina, Pamela Mancini, Giusy Bonanno Ferraro, Carolina Veneri, Marcello Iaconelli, Lucia Bonadonna, Luca Lucentini, and Elisabetta Suffredini. "SARS-CoV-2 Has Been Circulating in Northern Italy Since December 2019: Evidence from Environmental Monitoring." *Science of the Total Environment* 750 (2021): 141711.

Larsen, David A., and Krista R. Wigginton. "Tracking COVID-19 with Wastewater." *Nature Biotechnology* 38, no. 10 (2020): 1151-1153.

Lusk, Jayson L., and Ranveer Chandra. "Farmer and Farm Worker Illnesses and Deaths from COVID-19 and Impacts on Agricultural Output." *Plos One* 16, no. 4 (2021): e0250621.

Ma, Qiuyue, Jue Liu, Qiao Liu, Liangyu Kang, Runqing Liu, Wenzhan Jing, Yu Wu, and Min Liu. "Global Percentage of Asymptomatic SARS-CoV-2 Infections among the Tested Population and Individuals with Confirmed COVID-19 Diagnosis: A Systematic Review and Meta-Analysis." *JAMA Network Open* 4, no. 12 (2021): e2137257.

Macklin, Grace, Ousmane M. Diop, Asghar Humayun, Shohreh Shahmahmoodi, Zeinab A. El-Sayed, Henda Triki, Gloria Rey et al. "Update on Immunodeficiency-Associated Vaccine-Derived Polioviruses—Worldwide, July 2018–December 2019." *Morbidity and Mortality Weekly Report* 69, no. 28 (2020): 913.

Macklin, G. R., K. M. O'Reilly, N. C. Grassly, W. J. Edmunds, O. Mach, R. Santhana Gopala Krishnan, A. Voorman et al. "Evolving Epidemiology of Poliovirus Serotype 2 Following Withdrawal of the Serotype 2 Oral Poliovirus Vaccine." *Science* 368, no. 6489 (2020): 401-405.

Mancini, Pamela, Giusy Bonanno Ferraro, Elisabetta Suffredini, Carolina Veneri, Marcello Iaconelli, Teresa Vicenza, and Giuseppina La Rosa. "Molecular Detection of Human Salivirus in Italy Through Monitoring of Urban Sewages." *Food and Environmental Virology* 12, no. 1 (2020): 68-74.

McKinney, Kelly R., Yu Yang Gong, and Thomas G. Lewis. "Environmental Transmission of SARS at Amoy Gardens." *Journal of Environmental Health* 68, no. 9 (2006): 26.

BIBLIOGRAPHY

McMichael, Temet M., Dustin W. Currie, Shauna Clark, Sargis Pogosjans, Meagan Kay, Noah G. Schwartz, James Lewis et al. "Epidemiology of Covid-19 in a Long-Term Care Facility in King County, Washington." *New England Journal of Medicine* 382, no. 21 (2020): 2005-2011.

McNeil, Megan. "Wastewater Epidemiology Used to Stave off Lettuce Shortage." *KOLD News 13.* January 21, 2021.

Medema, Gertjan, Leo Heijnen, Goffe Elsinga, Ronald Italiaander, and Anke Brouwer. "Presence of SARS-Coronavirus-2 RNA in Sewage and Correlation with Reported COVID-19 Prevalence in the Early Stage of the Epidemic in The Netherlands." *Environmental Science & Technology Letters* 7, no. 7 (2020): 511-516.

Medrano, Kastalia. "Huge European Poop Study Shows Amsterdam's MDMA Is Strong and Spain Likes Cocaine." *Inverse.* May 31, 2016.

Melnick, Joseph L. "Poliomyelitis Virus in Urban Sewage in Epidemic and in Nonepidemic Times." *American Journal of Hygiene* 45, no. 2 (1947): 240-253.

Nelson, Bryn. "America Botched Coronavirus Testing. We're About to Find Out Just How Badly." *Daily Beast.* March 18, 2020.

Nelson, Bryn. "Coronavirus Patient Had Close Contact With 16 in Washington State." *Daily Beast.* January 22, 2020.

Nelson, Bryn. "Seattle's Covid-19 Lessons Are Yielding Hope." *BMJ* 369 (2020): m1389.

Nelson, Bryn. "The Next Coronavirus Nightmare Is Closer Than You Think." *Daily Beast.* January 29, 2020.

Nordahl Petersen, Thomas, Simon Rasmussen, Henrik Hasman, Christian Carøe, Jacob Bælum, Anna Charlotte Schultz, Lasse Bergmark et al. "Meta-Genomic Analysis of Toilet Waste from Long Distance Flights; A Step Towards Global Surveillance of Infectious Diseases and Antimicrobial Resistance." *Scientific Reports* 5, no. 1 (2015): 1-9.

O'Reilly, Kathleen M., David J. Allen, Paul Fine, and Humayun Asghar. "The Challenges of Informative Wastewater Sampling for SARS-CoV-2 Must Be Met: Lessons from Polio Eradication." *The Lancet Microbe* 1, no. 5 (2020): e189-e190.

Parasa, Sravanthi, Madhav Desai, Viveksandeep Thoguluva Chandrasekar, Harsh K. Patel, Kevin F. Kennedy, Thomas Roesch, Marco Spadaccini et al. "Prevalence of Gastrointestinal Symptoms and Fecal Viral Shedding in Patients with Coronavirus Disease 2019: A Systematic Review and Meta-Analysis." *JAMA Network Open* 3, no. 6 (2020): e2011335-e2011335.

Paul, John R., James D. Trask, and Sven Gard. "II. Poliomyelitic Virus in Urban Sewage." *The Journal of Experimental Medicine* 71, no. 6 (1940): 765-777.

Peccia, Jordan, Alessandro Zulli, Doug E. Brackney, Nathan D. Grubaugh, Edward H. Kaplan, Arnau Casanovas-Massana, Albert I. Ko et al. "Measurement of SARS-CoV-2 RNA in Wastewater Tracks Community Infection Dynamics." *Nature Biotechnology* 38, no. 10 (2020): 1164-1167.

Pineda, Paulina, and Rachel Leingang. "University of Arizona Wastewater Testing Finds Virus at Dorm, Prevents Outbreak." *Arizona Republic.* August 27, 2020.

"Record Rat Invasion in Stockholm." *Radio Sweden.* November 27, 2014.

BIBLIOGRAPHY

Sagan, Carl, and Ann Druyan. *The Demon-Haunted World: Science as a Candle in the Dark*. New York: Random House, 1996.

Sah, Pratha, Meagan C. Fitzpatrick, Charlotte F. Zimmer, Elaheh Abdollahi, Lyndon Juden-Kelly, Seyed M. Moghadas, Burton H. Singer, and Alison P. Galvani. "Asymptomatic SARS-CoV-2 infection: A Systematic Review and Meta-Analysis." *Proceedings of the National Academy of Sciences* 118, no. 34 (2021): e2109229118.

Seymour, Christopher. "Stockholm—The Rat Capital of Scandinavia?" *The Local*. October 8, 2008.

Shumaker, Lisa. "U.S. Shatters Coronavirus Record with over 77,000 Cases in a Day." *Reuters*. July 16, 2020.

"Sifting Through Garbage For Clues on American Life." *New York Times*. March 6, 1976.

Strubbia, Sofia, My VT Phan, Julien Schaeffer, Marion Koopmans, Matthew Cotten, and Françoise S. Le Guyader. "Characterization of Norovirus and other Human Enteric Viruses in Sewage and Stool Samples Through Next-Generation Sequencing." *Food and Environmental Virology* 11, no. 4 (2019): 400-409.

Suffredini, E., M. Iaconelli, M. Equestre, B. Valdazo-González, A. R. Ciccaglione, C. Marcantonio, S. Della Libera, F. Bignami, and G. La Rosa. "Genetic Diversity among Genogroup II Noroviruses and Progressive Emergence of GII. 17 in Wastewaters in Italy (2011–2016) Revealed by Next-Generation and Sanger Sequencing." *Food and Environmental Virology* 10, no. 2 (2018): 141-150.

Symonds, E. M., Karena H. Nguyen, V. J. Harwood, and Mya Breitbart. "Pepper Mild Mottle Virus: A Plant Pathogen with a Greater Purpose in (Waste) Water Treatment Development and Public Health Management." *Water Research* 144 (2018): 1-12.

Tai, Don Bambino Geno, Aditya Shah, Chyke A. Doubeni, Irene G. Sia, and Mark L. Wieland. "The Disproportionate Impact of COVID-19 on Racial and Ethnic Minorities in the United States." *Clinical Infectious Diseases* 72, no. 4 (2021): 703-706.

Vere Hodge, R. Anthony. "Meeting Report: 30th International Conference on Antiviral Research, in Atlanta, GA, USA." *Antiviral Chemistry & Chemotherapy 26* (2018): 2040206618783924.

Wu, Fuqing, Jianbo Zhang, Amy Xiao, Xiaoqiong Gu, Wei Lin Lee, Federica Armas, Kathryn Kauffman et al. "SARS-CoV-2 Titers in Wastewater are Higher Than Expected from Clinically Confirmed Cases." *Msystems* 5, no. 4 (2020): e00614-20.

Ye, Yinyin, Robert M. Ellenberg, Katherine E. Graham, and Krista R. Wigginton. "Survivability, Partitioning, and Recovery of Enveloped Viruses in Untreated Municipal Wastewater." *Environmental Science & Technology* 50, no. 10 (2016): 5077-5085.

Yong, Ed. "America Is Trapped in a Pandemic Spiral." *The Atlantic*. September 9, 2020.

Yu, Ignatius TS, Yuguo Li, Tze Wai Wong, Wilson Tam, Andy T. Chan, Joseph HW Lee, Dennis YC Leung, and Tommy Ho. "Evidence of Airborne Transmission of the Severe Acute Respiratory Syndrome Virus." *New England Journal of Medicine* 350, no. 17 (2004): 1731-1739.

Zhang, Tao, Mya Breitbart, Wah Heng Lee, Jin-Quan Run, Chia Lin Wei, Shirlena Wee Ling Soh, Martin L. Hibberd, Edison T. Liu, Forest Rohwer, and Yijun Ruan. "RNA Viral Community in Human Feces: Prevalence of Plant Pathogenic Viruses." *PLoS Biology* 4, no. 1 (2006): e3.

Zhang, Yawen, Mengsha Cen, Mengjia Hu, Lijun Du, Weiling Hu, John J. Kim, and Ning Dai. "Prevalence and Persistent Shedding of Fecal SARS-CoV-2 RNA in Patients with COVID-19 Infection: A Systematic Review and Meta-Analysis." *Clinical and Translational Gastroenterology* 12, no. 4 (2021): e00343.

Chapter Seven

Angelakis, E., D. Bachar, M. Yasir, D. Musso, Félix Djossou, B. Gaborit, S. Brah et al. "Treponema Species Enrich the Gut Microbiota of Traditional Rural Populations but Are Absent from Urban Individuals." *New Microbes and New Infections* 27 (2019): 14-21.

Aversa, Zaira, Elizabeth J. Atkinson, Marissa J. Schafer, Regan N. Theiler, Walter A. Rocca, Martin J. Blaser, and Nathan K. LeBrasseur. "Association of Infant Antibiotic Exposure with Childhood Health Outcomes." *Mayo Clinic Proceedings* 96, no. 1 (2021): 66-77.

Blaser, Martin J. "Antibiotic Use and its Consequences for the Normal Microbiome." *Science* 352, no. 6285 (2016): 544-545.

Blaser, Martin J. *Missing Microbes: How the Overuse of Antibiotics Is Fueling Our Modern Plagues*. New York: Henry Holt, 2014.

Cepon-Robins, Tara J., Theresa E. Gildner, Joshua Schrock, Geeta Eick, Ali Bedbury, Melissa A. Liebert, Samuel S. Urlacher et al. "Soil-Transmitted Helminth Infection and Intestinal Inflammation Among the Shuar of Amazonian Ecuador." *American Journal of Physical Anthropology* 170, no. 1 (2019): 65-74.

Chauhan, Ashish, Ramesh Kumar, Sanchit Sharma, Mousumi Mahanta, Sudheer K. Vayuuru, Baibaswata Nayak, and Sonu Kumar. "Fecal Microbiota Transplantation in Hepatitis B E Antigen-Positive Chronic Hepatitis B Patients: A Pilot Study." *Digestive Diseases and Sciences* 66, no. 3 (2021): 873-880.

Chou, Han-Hsuan, Wei-Hung Chien, Li-Ling Wu, Chi-Hung Cheng, Chen-Han Chung, Jau-Haw Horng, Yen-Hsuan Ni et al. "Age-Related Immune Clearance of Hepatitis B Virus Infection Requires the Establishment of Gut Microbiota." *Proceedings of the National Academy of Sciences* 112, no. 7 (2015): 2175-2180.

Clemente, Jose C., Erica C. Pehrsson, Martin J. Blaser, Kuldip Sandhu, Zhan Gao, Bin Wang, Magda Magris et al. "The Microbiome of Uncontacted Amerindians." *Science Advances* 1, no. 3 (2015): e1500183.

Cummings, J. H., W. Branch, D. J. A. Jenkins, D. A. T. Southgate, Helen Houston, and W. P. T. James. "Colonic Response to Dietary Fibre from Carrot, Cabbage, Apple, Bran, and Guar Gum." *The Lancet* 311, no. 8054 (1978): 5-9.

Curry, Andrew. "Piles of Ancient Poop Reveal 'Extinction Event' in Human Gut Bacteria." *Science*. May 12, 2021.

BIBLIOGRAPHY

Davido, B., R. Batista, H. Fessi, H. Michelon, L. Escaut, C. Lawrence, M. Denis, C. Perronne, J. Salomon, and A. Dinh. "Fecal Microbiota Transplantation to Eradicate Vancomycin-Resistant Enterococci Colonization in Case of an Outbreak." *Médecine Et Maladies Infectieuses* 49, no. 3 (2019): 214-218.

El-Salhy, Magdy, Jan Gunnar Hatlebakk, Odd Helge Gilja, Anja Bråthen Kristoffersen, and Trygve Hausken. "Efficacy of Faecal Microbiota Transplantation for Patients with Irritable Bowel Syndrome in a Randomised, Double-Blind, Placebo-Controlled Study." *Gut* 69, no. 5 (2020): 859-867.

Fauconnier, Alan. "Phage Therapy Regulation: From Night to Dawn." *Viruses* 11, no. 4 (2019): 352.

Furfaro, Lucy L., Matthew S. Payne, and Barbara J. Chang. "Bacteriophage Therapy: Clinical Trials and Regulatory Hurdles." *Frontiers in Cellular and Infection Microbiology* (2018): 376.

Gerson, Jacqueline, Austin Wadle, and Jasmine Parham. "Gold Rush, Mercury Legacy: Small-Scale Mining for Gold Has Produced Long-Lasting Toxic Pollution, from 1860s California to Modern Peru." *The Conversation.* May 28, 2020.

Ghorayshi, Azeen. "Her Husband Was Dying From A Superbug. She Turned To Sewer Viruses Collected By The Navy." *BuzzFeed News.* May 6, 2017.

Groussin, Mathieu, Mathilde Poyet, Ainara Sistiaga, Sean M. Kearney, Katya Moniz, Mary Noel, Jeff Hooker et al. "Elevated Rates of Horizontal Gene Transfer in the Industrialized Human Microbiome." *Cell* 184, no. 8 (2021): 2053-2067.

Iida, Toshiya, Moriya Ohkuma, Kuniyo Ohtoko, and Toshiaki Kudo. "Symbiotic Spirochetes in the Termite Hindgut: Phylogenetic Identification of Ectosymbiotic Spirochetes of Oxymonad Protists." *FEMS Microbiology Ecology* 34, no. 1 (2000): 17-26.

Lam, Nguyet-Cam, Patricia B. Gotsch, and Robert C. Langan. "Caring for Pregnant Women and Newborns with Hepatitis B or C." *American Family Physician* 82, no. 10 (2010): 1225-1229.

Laporte, Dominique. *History of Shit.* Translated by Nadia Benadbid and Rodolphe el-Khoury. Cambridge, Massachusetts: MIT Press, 2002.

Linden, S. K., P. Sutton, N. G. Karlsson, V. Korolik, and M. A. McGuckin. "Mucins in the Mucosal Barrier to Infection." *Mucosal Immunology* 1, no. 3 (2008): 183-197.

Louca, Stilianos, Patrick M. Shih, Matthew W. Pennell, Woodward W. Fischer, Laura Wegener Parfrey, and Michael Doebeli. "Bacterial Diversification Through Geological Time." *Nature Ecology & Evolution* 2, no. 9 (2018): 1458-1467.

Maizels, Rick M. "Parasitic Helminth Infections and the Control of Human Allergic and Autoimmune Disorders." *Clinical Microbiology and Infection* 22, no. 6 (2016): 481-486.

Matson, Richard G., and Brian Chisholm. "Basketmaker II Subsistence: Carbon Isotopes and Other Dietary Indicators from Cedar Mesa, Utah." *American Antiquity* 56, no. 3 (1991): 444-459.

Milhorance, Flávia. "Yanomami Beset by Violent Land-Grabs, Hunger and Disease in Brazil." *The Guardian.* May 17, 2021.

Mitchell, Piers D. "Human Parasites in the Roman World: Health Consequences of Conquering an Empire." *Parasitology* 144, no. 1 (2017): 48-58.

BIBLIOGRAPHY

Moayyedi, Paul, Michael G. Surette, Peter T. Kim, Josie Libertucci, Melanie Wolfe, Catherine Onischi, David Armstrong et al. "Fecal Microbiota Transplantation Induces Remission in Patients with Active Ulcerative Colitis in a Randomized Controlled Trial." *Gastroenterology* 149, no. 1 (2015): 102-109.

Nagpal, Ravinder, Tiffany M. Newman, Shaohua Wang, Shalini Jain, James F. Lovato, and Hariom Yadav. "Obesity-Linked Gut Microbiome Dysbiosis Associated with Derangements in Gut Permeability and Intestinal Cellular Homeostasis Independent of Diet." *Journal of Diabetes Research* 2018 (2018): 3462092.

Paraguassu, Lisandra, and Anthony Boadle. "Brazil to Deploy Special Force to Protect the Yanomami from Wildcat Gold Miners." *Reuters.* June 14, 2021.

Park, Young Jun, Jooyoung Chang, Gyeongsil Lee, Joung Sik Son, and Sang Min Park. "Association of Class Number, Cumulative Exposure, and Earlier Initiation of Antibiotics during the First Two-Years of Life with Subsequent Childhood Obesity." *Metabolism* 112 (2020): 154348.

Philips, Dom. "'Like a Bomb Going Off': Why Brazil's Largest Reserve is Facing Destruction." *The Guardian.* January 13, 2020.

Popescu, Medeea, Jonas D. Van Belleghem, Arya Khosravi, and Paul L. Bollyky. "Bacteriophages and the Immune System." *Annual Review of Virology* 8 (2021): 415-435.

Poyet, Mathilde, and Mathieu Groussin. "The 'Global Microbiome Conservancy'—Extending Species Conservation to Microbial Biodiversity." *Science for Society.* May 9, 2020.

Ren, Yan-Dan, Zhen-Shi Ye, Liu-Zhu Yang, Li-Xin Jin, Wen-Jun Wei, Yong-Yue Deng, Xiao-Xiao Chen et al. "Fecal Microbiota Transplantation Induces Hepatitis B Virus E-Antigen (HBeAg) Clearance in Patients with Positive HBeAg After Long-Term Antiviral Therapy." *Hepatology* 65, no. 5 (2017): 1765-1768.

Sabin, Susanna, Hui-Yuan Yeh, Aleks Pluskowski, Christa Clamer, Piers D. Mitchell, and Kirsten I. Bos. "Estimating Molecular Preservation of the Intestinal Microbiome via Metagenomic Analyses of Latrine Sediments from Two Medieval Cities." *Philosophical Transactions of the Royal Society B* 375, no. 1812 (2020): 20190576.

Santiago-Rodriguez, Tasha M., Gino Fornaciari, Stefania Luciani, Scot E. Dowd, Gary A. Toranzos, Isolina Marota, and Raul J. Cano. "Gut Microbiome of an 11th century AD Pre-Columbian Andean Mummy." *PloS One* 10, no. 9 (2015): e0138135.

Schooley, Robert T., Biswajit Biswas, Jason J. Gill, Adriana Hernandez-Morales, Jacob Lancaster, Lauren Lessor, Jeremy J. Barr et al. "Development and Use of Personalized Bacteriophage-Based Therapeutic Cocktails to Treat a Patient with a Disseminated Resistant *Acinetobacter baumannii* Infection." *Antimicrobial Agents and Chemotherapy* 61, no. 10 (2017): e00954-17.

Shaffer, Leah. "Old Friends: The Promise of Parasitic Worms." *Undark.* December 20, 2016.

Sharma, Sapna, and Prabhanshu Tripathi. "Gut Microbiome and Type 2 Diabetes: Where We Are and Where To Go?" *The Journal of Nutritional Biochemistry* 63 (2019): 101-108.

Shillito, Lisa-Marie, John C. Blong, Eleanor J. Green, and Eline N. van Asperen. "The What, How and Why of Archaeological Coprolite Analysis." *Earth-Science Reviews* 207 (2020): 103196.

BIBLIOGRAPHY

Singh, Madhu V., Mark W. Chapleau, Sailesh C. Harwani, and Francois M. Abboud. "The Immune System and Hypertension." *Immunologic Research* 59, no. 1 (2014): 243-253.

Skoulding, Lucy. "We Visited a London Curiosity Shop and Found Vintage McDonald's Toys, a Mermaid, and Kylie Minogue's Poo." *MyLondon*. November 30, 2019.

Sonnenburg, Erica D., Samuel A. Smits, Mikhail Tikhonov, Steven K. Higginbottom, Ned S. Wingreen, and Justin L. Sonnenburg. "Diet-Induced Extinctions in the Gut Microbiota Compound over Generations." *Nature* 529, no. 7585 (2016): 212-215.

South Park. 2019. Season 23, Episode 8 "Turd Burglars." Directed by Trey Parker. Aired November 27, 2019 on Comedy Central.

Statovci, Donjete, Mònica Aguilera, John MacSharry, and Silvia Melgar. "The impact of Western Diet and Nutrients on the Microbiota and Immune Response at Mucosal Interfaces." *Frontiers in Immunology* 8 (2017): 838.

Stephen, Alison M., and J. H. Cummings. "The Microbial Contribution to Human Faecal Mass." *Journal of Medical Microbiology* 13, no. 1 (1980): 45-56.

Strathdee, Steffanie, Thomas Patterson, and Teresa Barker. *The Perfect Predator: A Scientist's Race to Save Her Husband from a Deadly Superbug.* New York: Hachette, 2019.

Tall, Alan R., and Laurent Yvan-Charvet. "Cholesterol, Inflammation and Innate Immunity." *Nature Reviews Immunology* 15, no. 2 (2015): 104-116.

Urlacher, Samuel S., Peter T. Ellison, Lawrence S. Sugiyama, Herman Pontzer, Geeta Eick, Melissa A. Liebert, Tara J. Cepon-Robins, Theresa E. Gildner, and J. Josh Snodgrass. "Tradeoffs Between Immune Function and Childhood Growth Among Amazonian Forager-Horticulturalists." *Proceedings of the National Academy of Sciences* 115, no. 17 (2018): E3914-E3921.

Verhagen, Lilly M., Renzo N. Incani, Carolina R. Franco, Alejandra Ugarte, Yeneska Cadenas, Carmen I. Sierra Ruiz, Peter WM Hermans et al. "High Malnutrition Rate in Venezuelan Yanomami Compared to Warao Amerindians and Creoles: Significant Associations with Intestinal Parasites and Anemia." *PLoS One* 8, no. 10 (2013): e77581.

Wibowo, Marsha C., Zhen Yang, Maxime Borry, Alexander Hübner, Kun D. Huang, Braden T. Tierney, Samuel Zimmerman et al. "Reconstruction of Ancient Microbial Genomes from the Human Gut." *Nature* 594, no. 7862 (2021): 234-239.

Wu, Katherine J. "In Collecting Indigenous Feces, A Slew of Sticky Ethics." *Undark*. April 6, 2020.

"Xawara: Tracing the Deadly Path of Covid-19 and Government Negligence in the Yanomami Territory." Yanomami and Ye'kwana Leadership Forum and the Pro-Yanomami and Ye'kwana Network, November 2020.

Xie, Yurou, Zhangran Chen, Fei Zhou, Ligang Chen, Jianquan He, Chuanxing Xiao, Hongzhi Xu, Jianlin Ren, and Xiang Zhang. "IDDF2018-ABS-0201 Faecal Microbiota Transplantation Induced HBSAG Decline in HBEAG Negative Chronic Hepatitis B Patients After Long-Term Antiviral Therapy." (2018): A110-A111.

Yatsunenko, Tanya, Federico E. Rey, Mark J. Manary, Indi Trehan, Maria Gloria Dominguez-Bello, Monica Contreras, Magda Magris et al. "Human Gut

Microbiome Viewed Across Age and Geography." *Nature* 486, no. 7402 (2012): 222-227.

Ye, Jianyu, and Jieliang Chen. "Interferon and Hepatitis B: Current and Future Perspectives." *Frontiers in Immunology* 12 (2021): 733364.

Yeh, Hui-Yuan, Aleks Pluskowski, Uldis Kalējs, and Piers D. Mitchell. "Intestinal Parasites in a Mid-14th Century Latrine from Riga, Latvia: Fish Tapeworm and the Consumption of Uncooked Fish in the Medieval Eastern Baltic Region." *Journal of Archaeological Science* 49 (2014): 83-89.

Yeh, Hui-Yuan, Kay Prag, Christa Clamer, Jean-Baptiste Humbert, and Piers D. Mitchell. "Human Intestinal Parasites from a Mamluk Period Cesspool in the Christian Quarter of Jerusalem: Potential Indicators of Long Distance Travel in the 15th Century AD." *International Journal of Paleopathology* 9 (2015): 69-75.

Yong, Ed. *I Contain Multitudes: The Microbes Within Us and a Grander View of Life.* New York: Ecco, 2016.

Chapter Eight

Arvin, Jariel. "Norway Wants to Lead on Climate Change. But First It Must Face Its Legacy of Oil and Gas." *Vox.* January 15, 2021.

Asbjørnsen, Peter Christen & Jørgen Engebretsen Moe. *The Complete Norwegian Folktales and Legends of Asbjørnsen & Moe.* Translated by Simon Roy Hughes. 2020.

Baalsrud, Kjell. "Pollution of the Outer Oslofjord." *Water Science and Technology* 24, no. 10 (1991): 321-322.

Beckwith, Martha. *Hawaiian Mythology.* Honolulu: University of Hawai'i Press, 1970.

Bergmo, Per E. S., Erik Lindeberg, Fridtjof Riis, and Wenche T. Johansen. "Exploring Geological Storage Sites for CO2 from Norwegian Gas Power Plants: Johansen Formation." *Physics Procedia* 1, no. 1 (2009): 2945-2952.

Bernton, Hal. "Giant Landfill in Tiny Washington Hamlet Turns Trash to Natural Gas, as Utilities Fight for a Future." *The Seattle Times.* March 4, 2021.

Bevanger, Lars. "First-World Problem? Norway and Sweden Battle over Who Gets to Burn Waste." *DW.* November 23, 2015.

Björn, Annika, Sepehr Shakeri Yekta, Ryan M. Ziels, Karl Gustafsson, Bo H. Svensson, and Anna Karlsson. "Feasibility of OFMSW Co-Digestion with Sewage Sludge for Increasing Biogas Production at Wastewater Treatment Plants." *Euro-Mediterranean Journal for Environmental Integration* 2, no. 1 (2017): 1-10.

Bond, Tom, and Michael R. Templeton. "History and Future of Domestic Biogas Plants in the Developing World." *Energy for Sustainable Development* 15, no. 4 (2011): 347-354.

Cambi. "How Does Thermal Hydrolysis Work?" Accessed April 20, 2022, https://www .cambi.com/what-we-do/thermal-hydrolysis/how-does-thermal-hydrolysis -work/.

Campbell, Kristina. "The Science on Gut Microbiota and Intestinal Gas: Everything You Wanted to Know but Didn't Want to Ask." *ISAPP Science Blog.* March 2, 2020.

BIBLIOGRAPHY

Chaudhary, Prem Prashant, Patricia Lynne Conway, and Jørgen Schlundt. "Methanogens in Humans: Potentially Beneficial or Harmful for Health." *Applied Microbiology and Biotechnology* 102, no. 7 (2018): 3095-3104.

Day, Adrienne. "Waste Not: Addressing the Sanitation and Fuel Need" *DEMAND*. April 3, 2016.

Defenders of Wildlife. "Take Refuge: Tualatin River National Wildlife Refuge." August 2, 2011.

de Souza, Sandro Maquiné, Thierry Denoeux, and Yves Grandvalet. "Recycling experiments for sludge monitoring in waste water treatment." In *2004 IEEE International Conference on Systems, Man and Cybernetics* (IEEE Cat. No. 04CH37583) 2 (2004): 1342-1347.

Doyle, Amanda. "CCS Pilot Phase Successfully Completed on Norwegian Waste-to-Energy Plant." *The Chemical Engineer.* May 20, 2020.

Elliott, Douglas C., Patrick Biller, Andrew B. Ross, Andrew J. Schmidt, and Susanne B. Jones. "Hydrothermal Liquefaction of Biomass: Developments from Batch to Continuous Process." *Bioresource Technology* 178 (2015): 147-156.

European Biogas Initiative. "The Contribution of the Biogas and Biomethane Industries to Medium-Term Greenhouse Gas Reduction Targets and Climate Neutrality by 2050." April 2020.

FirstGroup. "New Bio-Methane Gas Bus Filling Station Opens in Bristol." News release. February 10, 2020, https://www.firstgroupplc.com/news-and-media/latest-news/2020/10-02-20b.aspx.

FlixBus. "Biogas in Detail: What's Behind Bio-CNG and Bio-LNG?" News release, June 29, 2021. https://corporate.flixbus.com/biogas-in-detail-whats-behind-bio-cng-and-bio-lng/.

Franklin Institute, The. "Benjamin Franklin's Inventions." Accessed April 20, 2022, https://www.fi.edu/benjamin-franklin/inventions.

Geneco. "Case Study: Bio-Bus." Accessed April 20, 2022, https://www.geneco.uk.com/case-studies/bio-bus.

Gonzalez, Ahtziri. "Beyond Bans: Toward Sustainable Charcoal Production in Kenya." *Forests News.* October 30, 2020.

He, Pin Jing. "Anaerobic Digestion: An Intriguing Long History in China." *Waste Management* 30, no. 4 (2010): 549-550.

Hede, Karyn. "The Path to Renewable Fuel Just Got Easier." Pacific Northwest National Laboratory. News release. February 2, 2022, https://www.pnnl.gov/news-media/path-renewable-fuel-just-got-easier.

Hutkins, Robert. "Got Gas? Blame It on Your Bacteria." *ISAPP Science Blog.* September 8, 2016.

"ISAPP Board Members Look Back in Time to Respond to Benjamin Franklin's Suggestion on How to Improve "Natural Discharges of Wind from Our Bodies." *ISAPP News.* January 29, 2021.

Ishaq, Suzanne L., Peter L. Moses, and André-Denis G. Wright. "The Pathology of Methanogenic Archaea in Human Gastrointestinal Tract Disease" In *The Gut Microbiome—Implications for Human Disease.* London: IntechOpen: 2016.

BIBLIOGRAPHY

International Biochar Initiative. "Biochar Production and By-Products." Accessed April 20, 2022, https://biochar.international/the-biochar-opportunity/biochar -production-and-by-products/.

Jain, Sarika. "Global Potential of Biogas." World Biogas Association. June 2019.

Jensen, Michael. "Cheap Jet Fuel from Biogas." *LinkedIn*. May 3, 2017.

Kirk, Esben. "The Quantity and Composition of Human Colonic Flatus." *Gastroenterology* 12, no. 5 (1949):782–794.

Klackenberg, Linus. "Biomethane in Sweden—Market Overview and Policies." Swedish Gas Association. March 16, 2021.

Kuenen, J. Gijs. "Anammox Bacteria: From Discovery to Application." *Nature Reviews Microbiology* 6, no. 4 (2008): 320-326.

Levaggi, Laura, Rosella Levaggi, Carmen Marchiori, and Carmine Trecroci. "Waste-to-Energy in the EU: The Effects of Plant Ownership, Waste Mobility, and Decentralization on Environmental Outcomes and Welfare." *Sustainability* 12, no. 14 (2020): 5743.

Metro Vancouver. "Hydrothermal Processing Biocrude Oil for Low Carbon Fuel." May 2020, https://mvupdate.metrovancouver.org/issue-62/hydrothermal-processing -biocrude-oil-for-low-carbon-fuel/.

National Archives. "From Benjamin Franklin to the Royal Academy of Brussels [After 19 May 1780]." Founders Online. Accessed April 20, 2022, https://founders .archives.gov/documents/Franklin/01-32-02-0281.

National Oceanic and Atmospheric Administration. "Despite Pandemic Shutdowns, Carbon Dioxide and Methane Surged in 2020." April 7, 2021.

Nikel, David. "Norway's Climate Plan to Halve Emissions by 2030." *Life In Norway*. January 8, 2021.

Norsk Folkemuseum. "Hygiene." Accessed April 20, 2022, https://norskfolkemu seum.no/en/hygiene.

Oregon Health Authority, Public Health Division. "Climate and Health in Oregon 2020." Accessed April 20, 2022. https://www.oregon.gov/oha/PH/HEALTHY ENVIRONMENTS/CLIMATECHANGE/Pages/profile-report.aspx.

Price, Toby. "Scandinavia Boasts 'World's First' Biogas-Powered Train." *Renewable Energy Magazine*. November 29, 2011.

Rehkopf Smith, Jill. "An Unexpected River Runs Through Western Washington County." *The Oregonian*. October 1, 2009.

Rudek, Joe, and Stefan Schwietzke. "Not All Biogas Is Created Equal." *EDF Blogs*. April 15, 2019.

Sahakian, Ara B., Sam-Ryong Jee, and Mark Pimentel. "Methane and the Gastrointestinal Tract." *Digestive Diseases and Sciences* 55, no. 8 (2010): 2135-2143.

Scania. "Premiere for the First International Biogas Bus." Press release. June 30, 2021, https://www.prnewswire.com/news-releases/premiere-for-the-first-international -biogas-bus-301322876.html.

Scanlan, Pauline D., Fergus Shanahan, and Julian R. Marchesi. "Human Methanogen Diversity and Incidence in Healthy and Diseased Colonic Groups Using mcrA Gene Analysis." *BMC Microbiology* 8, no. 1 (2008): 1-8.

Schindler, David W. "Eutrophication and Recovery in Experimental Lakes: Implications for Lake Management." *Science* 184, no. 4139 (1974): 897-899.

Schuster-Wallace, C.J., C. Wild, and C. Metcalfe. "Valuing Human Waste as an Energy Resource: A Research Brief Assessing the Global Wealth in Waste." United Nations University Institute for Water, Environment and Health. 2015.

Shaw Street, Erin. "A 'Poop Train' from New York Befouled a Small Alabama Town, Until the Town Fought Back." *The Washington Post*. April 20, 2018.

Sola, Phosiso, and Paolo Omar Cerutti. "Kenya Has Been Trying to Regulate the Charcoal Sector: Why It's Not Working." *The Conversation*. February 23, 2021.

Solli, Hilde. "Oslo's New Climate Strategy." KlimaOslo. News release, June 10, 2020, https://www.klimaoslo.no/2020/06/10/oslos-new-climate-strategy/.

Staalstrøm, André, and Lars Petter Røed. "Vertical Mixing and Internal Wave Energy Fluxes in a Sill Fjord." *Journal of Marine Systems* 159 (2016): 15-32.

Suarez, F. L., J. Springfield, and M. D. Levitt. "Identification of Gases Responsible for the Odour of Human Flatus and Evaluation of a Device Purported to Reduce This Odour." *Gut* 43, no. 1 (1998): 100-104.

"Thousands Lose Power in Northern California Amid Roll Out of PG&E Blackouts." *KXTV*. September 7, 2020.

Venkatesh, Govindarajan. "Wastewater Treatment in Norway: An Overview." *Journal AWWA* 105, no. 5 (2013): 92-97.

Villadsen, Sebastian NB, Philip L. Fosbøl, Irini Angelidaki, John M. Woodley, Lars P. Nielsen, and Per Møller. "The Potential of Biogas; The Solution to Energy Storage." *ChemSusChem* 12, no. 10 (2019): 2147-2153.

Wang, Hailong, Sally L. Brown, Guna N. Magesan, Alison H. Slade, Michael Quintern, Peter W. Clinton, and Tim W. Payn. "Technological Options for the Management of Biosolids." *Environmental Science and Pollution Research-International* 15, no. 4 (2008): 308-317.

Wiig Sørensen, Benedikte. "Sustainable Waste Management for a Carbon Neutral Europe." KlimaOslo. News release, February 26, 2021, https://www.klimaoslo.no/2021/02/26/the-klemetsrud-carbon-capture-project/.

Williams, Chris. "Mythbusters: Top 25 Moments." *Discovery Channel*. June 16, 2010.

World Bank. "Zero Routine Flaring by 2030 (ZRF) Initiative." Accessed April 20, 2022, https://www.worldbank.org/en/programs/zero-routine-flaring-by-2030/about.

"Yamauba." Yokai.com. Accessed April 20, 2022, https://yokai.com/yamauba/.

Chapter Nine

Alewell, Christine, Bruno Ringeval, Cristiano Ballabio, David A. Robinson, Panos Panagos, and Pasquale Borrelli. "Global Phosphorus Shortage Will Be Aggravated by Soil Erosion." *Nature Communications* 11, no. 1 (2020): 1-12.

Anawar, Hossain M., Zed Rengel, Paul Damon, and Mark Tibbett. "Arsenic-Phosphorus Interactions in the Soil-Plant-Microbe System: Dynamics of Uptake, Suppression and Toxicity to Plants." *Environmental Pollution* 233 (2018): 1003-1012.

BIBLIOGRAPHY

Barragan-Fonseca, Karol B., Marcel Dicke, and Joop JA van Loon. "Nutritional Value of the Black Soldier Fly (Hermetia illucens L.) and its Suitability as Animal Feed– A Review." *Journal of Insects as Food and Feed* 3, no. 2 (2017): 105-120.

Bhattacharya, Preeti Tomar, Satya Ranjan Misra, and Mohsina Hussain. "Nutritional Aspects of Essential Trace Elements in Oral Health and Disease: An Extensive Review." *Scientifica* 2016 (2016): 5464373.

Brown, Sally. "Connections: Compost + Cannabis." *BioCycle.* January 7, 2020.

Brown, Sally, Laura Kennedy, Mark Cullington, Ashley Mihle, and Maile Lono-Batura. "Relating Pharmaceuticals and Personal Care Products in Biosolids to Home Exposure." *Urban Agriculture & Regional Food Systems* 4, no. 1 (2019): 1-14.

Brown, Sally, Rufus Chaney, and David M. Hill. "Biosolids Compost Reduces Lead Bioavailability in Urban Soils." *BioCycle* 44, no. 6 (2003): 20-24.

Brown, Sally L., Rufus L. Chaney, and Ganga M. Hettiarachchi. "Lead in Urban Soils: A Real or Perceived Concern for Urban Agriculture?" *Journal of Environmental Quality* 45, no. 1 (2016): 26-36.

Brown, Sally L., Rufus L. Chaney, J. Scott Angle, and James A. Ryan. "The Phytoavailability of Cadmium to Lettuce in Long-Term Biosolids-Amended Soils." *Journal of Environmental Quality* 27, no. 5 (1998): 1071-1078.

Burke Museum. "Traditional Coast Salish Foods List." September 14, 2013, https://www.burkemuseum.org/news/traditional-coast-salish-foods-list.

Clague, John J. "Cordilleran Ice Sheet." In *Encyclopedia of Paleoclimatology and Ancient Environments.* Dordrecht: Springer, 2009.

Cohen, Lindsay. " 'Crappy' Solution to Soil Shortage: U.S. Open Human Waste." *KOMO News.* June 29, 2015.

Collivignarelli, Maria Cristina, Alessandro Abbà, Andrea Frattarola, Marco Carnevale Miino, Sergio Padovani, Ioannis Katsoyiannis, and Vincenzo Torretta. "Legislation for the Reuse of Biosolids on Agricultural Land in Europe: Overview." *Sustainability* 11, no. 21 (2019): 6015.

Crunden, E. A. "For Waste Industry, PFAS Disposal Leads to Controversy, Regulation, Mounting Costs." *SEJournal Online* 5, no. 42. November 18, 2020.

Defoe, Phillip Peterson, Ganga M. Hettiarachchi, Christopher Benedict, and Sabine Martin. "Safety of Gardening on Lead- and Arsenic-Contaminated Urban Brownfields." *Journal of Environmental Quality* 43, no. 6 (2014): 2064-2078.

Defoe, Phillip Peterson. "Urban Brownfields to Gardens: Minimizing Human Exposure to Lead and Arsenic." PhD diss. Kansas State University, 2014.

De Groote, Hugo, Simon C. Kimenju, Bernard Munyua, Sebastian Palmas, Menale Kassie, and Anani Bruce. "Spread and Impact of Fall Armyworm (*Spodoptera frugiperda* JE Smith) in Maize Production Areas of Kenya." *Agriculture, Ecosystems & Environment* 292 (2020): 106804.

Doughty, Christopher E., Andrew J. Abraham, and Joe Roman. "The Sixth R: Revitalizing the Natural Phosphorus Pump." *EcoEvoRxiv.* March 18, 2020.

Driscoll, Matt. "A Happy Ending for Tacoma Community College's Beloved Garden." *The News Tribune.* March 15, 2016.

BIBLIOGRAPHY

Driscoll, Matt. "How Can We Save the Community Garden at Tacoma Community College?" *The News Tribune.* September 28, 2015.

Duwamish Tribe. "Our History of Self Determination." Accessed April 20, 2022, https://www.duwamishtribe.org/history.

Flavell-White, Claudia. "Fritz Haber and Carl Bosch—Feed the World." *The Chemical Engineer.* March 1, 2010.

Gerling, Daniel Max. "American Wasteland: A Social and Cultural History of Excrement 1860-1920." Doctoral dissertation, University of Texas, 2012.

Gross, Daniel A. "Caliche: The Conflict Mineral that Fuelled the First World War." *The Guardian.* June 2, 2014.

Hilbert, Klaus, and Jens Soentgen. "From the '*Terra Preta de Indio*' to the 'Terra Preta do Gringo': A History of Knowledge of the Amazonian Dark Earths." In *Ecosystem and Biodiversity of Amazonia.* London: InTechOpen, 2021.

Historic England. "The Great Stink—How the Victorians Transformed London to Solve the Problem of Waste." Accessed April 20, 2022, https://historicengland .org.uk/images-books/archive/collections/photographs/the-great-stink/.

Johnson, Steven. *The Ghost Map: The Story of London's Most Terrifying Epidemic—and How it Changed Science, Cities, and the Modern World.* New York: Penguin, 2006.

Kawa, Nicholas C., Yang Ding, Jo Kingsbury, Kori Goldberg, Forbes Lipschitz, Mitchell Scherer, and Fatuma Bonkiye. "Night Soil: Origins, Discontinuities, and Opportunities for Bridging the Metabolic Rift." *Ethnobiology Letters* 10, no. 1 (2019): 40-49.

Kimmerer, Robin Wall. *Braiding Sweetgrass: Indigenous Wisdom, Scientific Knowledge and the Teachings of Plants.* Minneapolis: Milkweed Editions, 2013.

King County. "DNRP Carbon Neutral." December 17, 2019, https://kingcounty.gov /depts/dnrp/about/beyond-carbon-neutral.aspx.

King County Wastewater Treatment Division. "King County Biosolids Program Strategic Plan 2018-2037." June 2018, https://kingcounty.gov/~/media/services/environment /wastewater/resource-recovery/plans/1711_KC-WTD-Biosolids-2018-2037 -Strategic-Plan-rev2.ashx?la=en.

Kolbert, Elizabeth. "Head Count." *The New Yorker.* October 14, 2013.

Krietsch Boerner, Leigh. "Industrial Ammonia Production Emits More CO2 Than Any Other Chemical-Makin Reaction. Chemists Want to Change That." *Chemical & Engineering News.* June 15, 2019.

Laporte, Dominique. *History of Shit.* Translated by Nadia Benadbid and Rodolphe el-Khoury. Cambridge, Massachusetts: The MIT Press, 2002.

Lehman, J. "*Terra Preta* Nova—Where to from Here?" In *Amazonian Dark Earths: Wim Sombroek's Vision.* Dordrecht: Springer, 2009.

Matthews, Todd. "Outside Pacific Plaza: A Garden Grows High Above Downtown Tacoma." *Tacoma Daily Index,* Accessed April 20, 2022.

National Institute for Occupational Safety and Health. "Guidance For Controlling Potential Risks To Workers Exposed to Class B Biosolids." June 6, 2014.

Orozco-Ortiz, Juan Manuel, Clara Patricia Peña-Venegas, Sara Louise Bauke, Christian Borgemeister, Ramona Mörchen, Eva Lehndorff, and Wulf Amelung. "Terra

BIBLIOGRAPHY

Preta Properties in Northwestern Amazonia (Colombia)." *Sustainability* 13, no. 13 (2021): 7088.

Perkins, Tom. "Biosolids: Mix Human Waste with Toxic Chemicals, Then Spread on Crops." *The Guardian.* October 5, 2019.

Philpott, Tom. "Our Other Addiction: The Tricky Geopolitics of Nitrogen Fertilizer." *Grist.* February 12, 2010.

Piccolo, Alessandro. "Humus and Soil Conservation." In *Humic Substances in Terrestrial Ecosystems.* 225-264. Amsterdam: Elsevier, 1996.

Rockefeller Foundation. "Black Soldier Flies: Inexpensive and Sustainable Source for Animal Feed." November 10, 2020.

Rolph, Amy. "What Was Washington State Like During the Last Ice Age?" *KUOW.* August 10, 2017.

Ross, Rachel. "The Science Behind Composting." *LiveScience.* September 12, 2018.

Rout, Hemant Kumar. "India's First Coal Gasification Based Fertiliser Plant on Track." *The New Indian Express.* January 10, 2021.

Santana-Sagredo, Francisca, Rick J. Schulting, Pablo Méndez-Quiros, Ale Vidal-Elgueta, Mauricio Uribe, Rodrigo Loyola, Anahí Maturana-Fernández et al. "'White Gold' Guano Fertilizer Drove Agricultural Intensification in the Atacama Desert from AD 1000." *Nature Plants* 7, no. 2 (2021): 152-158.

Schmidt, Hans-Peter. "Terra Preta—Model of a Cultural Technique." *The Biochar Journal.* 2014.

Science History Institute. "Fritz Haber." December 7, 2017, https://www.science history.org/historical-profile/fritz-haber.

Sellars, Sarah, and Vander Nunes. "Synthetic Nitrogen Fertilizer in the U.S." *farmdoc daily* 11 (2021): 24.

Shumo, Marwa, Isaac M. Osuga, Fathiya M. Khamis, Chrysantus M. Tanga, Komi KM Fiaboe, Sevgan Subramanian, Sunday Ekesi, Arnold van Huis, and Christian Borgemeister. "The Nutritive Value of Black Soldier Fly Larvae Reared on Common Organic Waste Streams in Kenya." *Scientific Reports* 9, no. 1 (2019): 1-13.

Specter, Michael. "Ocean Dumping Is Ending, but Not Problems; New York Can't Ship, Bury or Burn Its Sludge, but No One Wants a Processing Plant." *New York Times.* June 29, 1992.

Tajima, Kayo. "The Marketing of Urban Human Waste in the Early Modern Edo/ Tokyo Metropolitan Area." *Environnement Urbain/Urban Environment* 1 (2007).

Thrush, Coll-Peter. "The Lushootseed Peoples of Puget Sound Country." University of Washington Libraries Digital Collections, Accessed April 20, 2022, https:// content.lib.washington.edu/aipnw/thrush.html.

Tong, Ziya. *The Reality Bubble: Blind Spots, Hidden Truths, and the Dangerous Illusions that Shape Our World.* London: Allen Lane, 2019.

Tulalip Tribes of Washington. "Lushootseed Encyclopedia." Accessed April 20, 2022, https://tulaliplushootseed.com/encyclopedia/.

United Nations Environment Programme. "Food Waste Index Report 2021." Accessed April 20, 2022, https://www.unep.org/resources/report/unep-food-waste-index -report-2021.

Washington State Department of Ecology. "Tacoma Smelter Plume Project." Accessed April 20, 2022, https://ecology.wa.gov/Spills-Cleanup/Contamination -cleanup/Cleanup-sites/Tacoma-smelter.

"Wastewater Treatment Plant to Recycle Nutrients into 'Green' Fertilizer." *Water-World*. September 29, 2008.

Wilfert, Philipp, Prashanth Suresh Kumar, Leon Korving, Geert-Jan Witkamp, and Mark CM Van Loosdrecht. "The Relevance of Phosphorus and Iron Chemistry to the Recovery of Phosphorus from Wastewater: A Review." *Environmental Science & Technology* 49, no. 16 (2015): 9400-9414.

Winick, Stephen. "Ostara and the Hare: Not Ancient, but Not As Modern As Some Skeptics Think." *Folklife Today*. April 28, 2016.

Xue, Yong. " 'Treasure Nightsoil As If It Were Gold:' Economic and Ecological Links between Urban and Rural Areas in Late Imperial Jiangnan." *Late Imperial China* 26, no. 1 (2005): 41-71.

Chapter Ten

Agyei, Dominic, James Owusu-Kwarteng, Fortune Akabanda, and Samuel Akomea-Frempong. "Indigenous African Fermented Dairy products: Processing Technology, Microbiology and Health Benefits." *Critical Reviews in Food Science and Nutrition* 60, no. 6 (2020): 991-1006.

Alley, William M., and Rosemarie Alley. *The Water Recycling Revolution: Tapping into the Future*. London: Rowman & Littlefield, 2022.

Apicella, Coren L., Paul Rozin, Justin TA Busch, Rachel E. Watson-Jones, and Cristine H. Legare. "Evidence from Hunter-Gatherer and Subsistence Agricultural Populations for the Universality of Contagion Sensitivity." *Evolution and Human Behavior* 39, no. 3 (2018): 355-363.

Awerbuch, Leon, and Corinne Tromsdorff. "From Seawater to Tap or from Toilet to Tap? Joint Desalination and Water Reuse Is the Future of Sustainable Water Management." *IWA*. September 14, 2016.

Bastiaanssen, Thomaz FS, Caitlin SM Cowan, Marcus J. Claesson, Timothy G. Dinan, and John F. Cryan. "Making Sense of...the Microbiome in Psychiatry." *International Journal of Neuropsychopharmacology* 22, no. 1 (2019): 37-52.

Beal, Colin M., F. Todd Davidson, Michael E. Webber, and Jason C. Quinn. "Flare Gas Recovery for Algal Protein Production." *Algal Research* 20 (2016): 142-152.

Bested, Alison C., Alan C. Logan, and Eva M. Selhub. "Intestinal Microbiota, Probiotics and Mental Health: From Metchnikoff to Modern Advances: Part I–Autointoxication Revisited." *Gut Pathogens* 5, no. 1 (2013): 1-16.

Borenstein, Seth. "Cheers! Crew Drinks up Recycled Urine in Space." *Associated Press*. May 20, 2009.

Boxall, Bettina. "L.A.'s Ambitious Goal: Recycle All of the City's Sewage into Drinkable Water." *Los Angeles Times*. February 22, 2019.

Buck, Chris, and Jennifer Lee, directors. *Frozen II*. Disney, 2019.

BIBLIOGRAPHY

Buendia, Justin R., Yanping Li, Frank B. Hu, Howard J. Cabral, M. Loring Bradlee, Paula A. Quatromoni, Martha R. Singer, Gary C. Curhan, and Lynn L. Moore. "Regular Yogurt Intake and Risk of Cardiovascular Disease Among Hypertensive Adults." *American Journal of Hypertension* 31, no. 5 (2018): 557-565.

Calysta. "Calysta Announces $39 Million Investment to Fund Global Expansion Plans." Press release. September 9, 2021, https://calysta.com/calysta-announces-39-million-investment-to-fund-global-expansion-plans/.

Calysta. "FeedKind Protein Can Enable Blue Economy and Increase Global Food Security." Press release. October 4, 2017, https://calysta.com/feedkind-protein-can-enable-blue-economy-and-increase-global-food-security/.

Campana, Raffaella, Saskia van Hemert, and Wally Baffone. "Strain-Specific Probiotic Properties of Lactic Acid Bacteria and Their Interference with Human Intestinal Pathogens Invasion." *Gut Pathogens* 9, no. 1 (2017): 1-12.

Casadevall, Arturo, Dimitrios P. Kontoyiannis, and Vincent Robert. "On the Emergence of Candida auris: Climate Change, Azoles, Swamps, and Birds." *MBio* 10, no. 4 (2019): e01397-19.

Chang, Chin-Feng, Yu-Ching Lin, Shan-Fu Chen, Enrique Javier Carvaja Barriga, Patricia Portero Barahona, Stephen A. James, Christopher J. Bond, Ian N. Roberts, and Ching-Fu Lee. "Candida theae sp. nov., a New Anamorphic Beverage-Associated Member of the Lodderomyces Clade." *International Journal of Food Microbiology* 153, no. 1-2 (2012): 10-14.

Chen, Mu, Qi Sun, Edward Giovannucci, Dariush Mozaffarian, JoAnn E. Manson, Walter C. Willett, and Frank B. Hu. "Dairy Consumption and Risk of Type 2 Diabetes: 3 Cohorts of US Adults and an Updated Meta-Analysis." *BMC Medicine* 12, no. 1 (2014): 1-14.

Cuthbert, M. O., Tom Gleeson, Nils Moosdorf, Kelvin M. Befus, A. Schneider, Jens Hartmann, and B. Lehner. "Global Patterns and Dynamics of Climate–Groundwater Interactions." *Nature Climate Change* 9, no. 2 (2019): 137-141.

Daly, Luke, Martin R. Lee, Lydia J. Hallis, Hope A. Ishii, John P. Bradley, Phillip Bland, David W. Saxey et al. "Solar Wind Contributions to Earth's Oceans." *Nature Astronomy* 5, no. 12 (2021): 1275-1285.

De Roos, Nicole M., and Martijn B. Katan. "Effects of Probiotic Bacteria on Diarrhea, Lipid Metabolism, and Carcinogenesis: A Review of Papers Published between 1988 and 1998." *The American Journal of Clinical Nutrition* 71, no. 2 (2000): 405-411.

Farré-Maduell, Eulàlia, and Climent Casals-Pascual. "The Origins of Gut Microbiome Research in Europe: From Escherich to Nissle." *Human Microbiome Journal* 14 (2019): 100065.

Fields, R. Douglas. "Raising the Dead: New Species of Life Resurrected from Ancient Andean Tomb." *Scientific American.* February 19, 2012.

Fishman, Charles. *The Big Thirst: The Secret Life and Turbulent Future of Water.* New York: Free Press, 2011.

Fox, Michael J., Kiran DK Ahuja, Iain K. Robertson, Madeleine J. Ball, and Rajaraman D. Eri. "Can Probiotic Yogurt Prevent Diarrhoea in Children on Antibiotics? A Double-Blind, Randomised, Placebo-Controlled Study." *BMJ Open* 5, no. 1 (2015): e006474.

BIBLIOGRAPHY

Gao, Xing Wang, Mohamed Mubasher, Chong Yu Fang, Cheryl Reifer, and Larry E. Miller. "Dose–Response Efficacy of a Proprietary Probiotic Formula of Lactobacillus acidophilus CL1285 and Lactobacillus casei LBC80R for Antibiotic-Associated Diarrhea and Clostridium difficile-Associated Diarrhea Prophylaxis in Adult Patients." *American Journal of Gastroenterology* 105, no. 7 (2010): 1636-1641.

Gates, Bill. "Janicki Omniprocessor." *YouTube.* January 5, 2015, https://www.youtube.com/watch?v=bVzppWSIFU0.

Gates, Bill. "This Ingenious Machine Turns Feces into Drinking Water." *GatesNotes.* January 5, 2015.

Goldenberg, Joshua Z., Christina Yap, Lyubov Lytvyn, Calvin Ka-Fung Lo, Jennifer Beardsley, Dominik Mertz, and Bradley C. Johnston. "Probiotics for the Prevention of Clostridium difficile-Associated Diarrhea in Adults and Children." *Cochrane Database of Systematic Reviews* 12 (2017).

Gorman, Steve. "U.S. High-Tech Water Future Hinges on Cost, Politics." *Reuters.* March 11, 2009.

Gross, Terry. "The Worldwide 'Thirst' For Clean Drinking Water." *NPR.* April 11, 2011.

Gunaratnam, Sathursha, Carine Diarra, Patrick D. Paquette, Noam Ship, Mathieu Millette, and Monique Lacroix. "The Acid-Dependent and Independent Effects of Lactobacillus acidophilus CL1285, Lacticaseibacillus casei LBC80R, and Lacticaseibacillus rhamnosus CLR2 on Clostridioides difficile R20291." *Probiotics and Antimicrobial Proteins* 13, no. 4 (2021): 949-956.

Haefele, Marc B., and Anna Sklar. "Revisiting 'Toilet to Tap.'" *Los Angeles Times.* August 26, 2007.

Hansman, Heather. "A New Efficient Filter Helps Astronauts Drink Their Own Urine." *Smithsonian Magazine.* September 11, 2015.

Harris-Lovett, Sasha, and David Sedlak. "Protecting the Sewershed." *Science* 369, no. 6510 (2020): 1429-1430.

Hurlimann, Anna, and Sara Dolnicar. "When Public Opposition Defeats Alternative Water Projects–The Case of Toowoomba Australia." *Water Research* 44, no. 1 (2010): 287-297.

Inyang, Mandu, and Eric RV Dickenson. "The Use of Carbon Adsorbents for the Removal of Perfluoroalkyl Acids from Potable Reuse Systems." *Chemosphere* 184 (2017): 168-175.

Jones, Anthony. "Bill Gates Drinks Glass of Water That Was Human Feces Minutes Earlier." *Business 2 Community.* January 7, 2015.

Jumrah, Wahab. "The 1962 Johor-Singapore Water Agreement: Lessons Learned." *The Diplomat.* September 30, 2021.

Kambale, Richard Mbusa, Fransisca Isia Nancy, Gaylord Amani Ngaboyeka, Joe Bwija Kasengi, Laure B. Bindels, and Dimitri Van der Linden. "Effects of Probiotics and Synbiotics on Diarrhea in Undernourished Children: Systematic Review with Meta-Analysis." *Clinical Nutrition* 40, no. 5 (2021): 3158-3169.

Kim, Jungbin, Kiho Park, Dae Ryook Yang, and Seungkwan Hong. "A Comprehensive Review of Energy Consumption of Seawater Reverse Osmosis Desalination Plants." *Applied Energy* 254 (2019): 113652.

BIBLIOGRAPHY

Kisan, Bhagwat Sameer, Rajender Kumar, Shelke Prashant Ashok, and Ganguly Sangita. "Probiotic Foods for Human Health: A Review." *Journal of Pharmacognosy and Phytochemistry* 8, no. 3 (2019): 967-971.

Kort, Remco. "A Yogurt to Help Prevent Diarrhea?" *On Biology*. December 8, 2015.

Kort, Remco, and Wilbert Sybesma. "Probiotics for Every Body." *Trends in Biotechnology* 30, no. 12 (2012): 613-615.

Kort, Remco, Nieke Westerik, L. Mariela Serrano, François P. Douillard, Willi Gottstein, Ivan M. Mukisa, Coosje J. Tuijn et al. "A Novel Consortium of Lactobacillus rhamnosus and Streptococcus thermophilus for Increased Access to Functional Fermented Foods." *Microbial Cell Factories* 14, no. 1 (2015): 1-14.

Lee, Hannah, and Thai Pin Tan. "Singapore's Experience with Reclaimed Water: NEWater." *International Journal of Water Resources Development* 32, no. 4 (2016): 611-621.

Mackie, Alec. "California's Water History: The Origin of 'Toilet-to-Tap.'" *CWEA*. Accessed April 20, 2022, https://www.cwea.org/news/whats-the-origin-of-toilet-to-tap/.

Marco, Maria L., Mary Ellen Sanders, Michael Gänzle, Marie Claire Arrieta, Paul D. Cotter, Luc De Vuyst, Colin Hill et al. "The International Scientific Association for Probiotics and Prebiotics (ISAPP) Consensus Statement on Fermented Foods." *Nature Reviews Gastroenterology & Hepatology* 18, no. 3 (2021): 196-208.

Marron, Emily L., William A. Mitch, Urs von Gunten, and David L. Sedlak. "A Tale of Two Treatments: The Multiple Barrier Approach to Removing Chemical Contaminants During Potable Water Reuse." *Accounts of Chemical Research* 52, no. 3 (2019): 615-622.

Martín, Rebeca, Sylvie Miquel, Leandro Benevides, Chantal Bridonneau, Véronique Robert, Sylvie Hudault, Florian Chain et al. "Functional Characterization of Novel *Faecalibacterium prausnitzii* Strains Isolated from Healthy Volunteers: A Step Forward in the Use of *F. prausnitzii* as a Next-Generation Probiotic." *Frontiers in Microbiology* (2017): 1226.

Mellen, Greg. "From Waste to Taste: Orange County Sets Guinness Record for Recycled Water." *The Orange County Register*. February 18, 2018.

Michels, Karin B., Walter C. Willett, Rita Vaidya, Xuehong Zhang, and Edward Giovannucci. "Yogurt Consumption and Colorectal Cancer Incidence and Mortality in the Nurses' Health Study and the Health Professionals Follow-up Study." *The American Journal of Clinical Nutrition* 112, no. 6 (2020): 1566-1575.

Nagpal, Ravinder, Shaohua Wang, Shokouh Ahmadi, Joshua Hayes, Jason Gagliano, Sargurunathan Subashchandrabose, Dalane W. Kitzman, Thomas Becton, Russel Read, and Hariom Yadav. "Human-Origin Probiotic Cocktail Increases Short-Chain Fatty Acid Production via Modulation of Mice and Human Gut Microbiome." *Scientific Reports* 8, no. 1 (2018): 1-15.

NASA. "NASA Awards Grants for Technologies That Could Transform Space Exploration." Press release, August 14, 2015, https://www.nasa.gov/press-release/nasa-awards-grants-for-technologies-that-could-transform-space-exploration.

National Research Council. *Water Reuse: Potential for Expanding the Nation's Water Supply Through Reuse of Municipal Wastewater*. Washington, DC: The National Academies Press, 2012.

BIBLIOGRAPHY

Nemeroff, Carol, and Paul Rozin. "The Contagion Concept in Adult Thinking in the United States: Transmission of Germs and of Interpersonal Influence." *Ethos* 22, no. 2 (1994): 158-186.

O'Connell, Todd. "1,4-Dioxane: Another Forever Chemical Plagues Drinking-Water Utilities." *Chemical & Engineering News.* November 9, 2020.

Piani, Laurette, Yves Marrocchi, Thomas Rigaudier, Lionel G. Vacher, Dorian Thomassin, and Bernard Marty. "Earth's Water May Have Been Inherited from Material Similar to Enstatite Chondrite Meteorites." *Science* 369, no. 6507 (2020): 1110-1113.

Plunkett, Luke. "Bill Gates Drinks Water Made From Human Poop." *Kotaku.* January 7, 2015.

"Reclaimed Wastewater Meets 40% of Singapore's Water Demand." *WaterWorld.* January 24, 2017.

Reynolds, Ross, and Kate O'Connell. "Poop Water: Why You Should Drink It." *KUOW.* March 19, 2015.

Rivard, Ry. "A Brief History of Pure Water's Pure Drama." *Voice of San Diego.* September 17, 2019.

Rotch, Thomas Morgan, and John Lovett Morse. "Report on Pediatrics." *Boston Medical and Surgical Journal* 153 (1905): 724-727.

Rozin, Paul, Brent Haddad, Carol Nemeroff, and Paul Slovic. "Psychological Aspects of the Rejection of Recycled Water: Contamination, Purification and Disgust." *Judgment and Decision Making* 10, no. 1(2015): 50-63.

Rubio, Raquel, Anna Jofré, Belén Martín, Teresa Aymerich, and Margarita Garriga. "Characterization of Lactic Acid Bacteria Isolated from Infant Faeces as Potential Probiotic Starter Cultures for Fermented Sausages." *Food Microbiology* 38 (2014): 303-311.

Rubio, Raquel, Anna Jofré, Teresa Aymerich, Maria Dolors Guàrdia, and Margarita Garriga. "Nutritionally Enhanced Fermented Sausages as a Vehicle for Potential Probiotic Lactobacilli Delivery." *Meat Science* 96, no. 2 (2014): 937-942.

Ruiz-Moyano, Santiago, Alberto Martín, María José Benito, Francisco Pérez Nevado, and María de Guía Córdoba. "Screening of Lactic Acid Bacteria and Bifidobacteria for Potential Probiotic Use in Iberian Dry Fermented Sausages." *Meat Science* 80, no. 3 (2008): 715-721.

Sanitation Technology Platform. "Preparing for Commercial Field Testing of the Janicki Omni Processor." November 2019, https://gatesopenresearch.org/documents/4-181.

Singapore Ministry of Foreign Affairs. "Water Agreements." Accessed April 20, 2022, https://www.mfa.gov.sg/SINGAPORES-FOREIGN-POLICY/Key-Issues/Water-Agreements.

Slovic, Paul. "Talking About Recycled Water—And Stigmatizing It." Decision Research. February 28, 2009.

Stefan, Mihaela I., and James R. Bolton. "Mechanism of the Degradation of 1, 4-Dioxane in Dilute Aqueous Solution Using the UV/Hydrogen Peroxide Process." *Environmental Science & Technology* 32, no. 11 (1998): 1588-1595.

Steinberg, Lisa M., Rachel E. Kronyak, and Christopher H. House. "Coupling of Anaerobic Waste Treatment to Produce Protein-and Lipid-Rich Bacterial Biomass." *Life Sciences in Space Research* 15 (2017): 32-42.

St. Fleur, Nicholas. "The Water in Your Glass Might Be Older Than the Sun." *New York Times*. April 15, 2016.

Sun, Fengting, Qingsong Zhang, Jianxin Zhao, Hao Zhang, Qixiao Zhai, and Wei Chen. "A Potential Species of Next-Generation Probiotics? The Dark and Light Sides of *Bacteroides fragilis* in Health." *Food Research International* 126 (2019): 108590.

Tan, Audrey, and Ng Keng Gene. "Linggiu Reservoir, Singapore's Main Water Source in Malaysia, Back at Healthy Levels for First Time Since 2016." *The Straits Times*. February 4, 2021.

Tan, Thai Pin, and Stuti Rawat. "NEWater in Singapore." Global Water Forum. January 15, 2018.

Vikhanski, Luba. *Immunity: How Elie Metchnikoff Changed the Course of Modern Medicine*. Chicago: Chicago Review, 2016.

Wastyk, Hannah C., Gabriela K. Fragiadakis, Dalia Perelman, Dylan Dahan, Bryan D. Merrill, B. Yu Feiqiao, Madeline Topf et al. "Gut-Microbiota-Targeted Diets Modulate Human Immune Status." *Cell* 184, no. 16 (2021): 4137-4153.

Westerik, Nieke, Arinda Nelson, Alex Paul Wacoo, Wilbert Sybesma, and Remco Kort. "A Comparative Interrupted Times Series on the Health Impact of Probiotic Yogurt Consumption Among School Children From Three to Six Years Old in Southwest Uganda." *Frontiers in Nutrition* (2020): 303.

Zhang, Ting, Qianqian Li, Lei Cheng, Heena Buch, and Faming Zhang. "*Akkermansia muciniphila* Is a Promising Probiotic." *Microbial Biotechnology* 12, no. 6 (2019): 1109-1125.

Chapter Eleven

Al-Azzawi, Mohammed, Les Bowtell, Kerry Hancock, and Sarah Preston. "Addition of Activated Carbon into a Cattle Diet to Mitigate GHG Emissions and Improve Production." *Sustainability* 13, no. 15 (2021): 8254.

Altieri, Miguel A., and Fernando R. Funes-Monzote. "The Paradox of Cuban Agriculture." *Monthly Review*. January 1, 2012.

American Chemical Society. "Sewage—Yes, Poop—Could Be a Source of Valuable Metals and Critical Elements." Press release. March 23, 2015.

Arden, Amanda. "'Kidneys of the Earth.' Wetlands Filter and Cool Wash. Co. Wastewater." *KOIN*. September 14, 2021.

Augustin, Ed, and Fraces Robles. "Cuba's Economy Was Hurting. The Pandemic Brought a Food Crisis." *New York Times*. September 20, 2020.

Baisre, Julio A. "Assessment of Nitrogen Flows into the Cuban Landscape." *Biogeochemistry* 79, no. 1 (2006): 91-108.

BIBLIOGRAPHY

BC Salmon Farmers Association. "Small Business Week Profile—Salish Soils." Press release. October 24, 2014, https://www.3blmedia.com/news/small-business-week -profile-salish-soils.

Cato, M. Porcius, and M. Terentius Varro. *On Agriculture (Loeb Classical Library No. 283).* Translated by W. D. Hooper and Harrison Boyd Ash. Loeb Classical Library 283. Cambridge, Massachusetts: Harvard University Press, 1934.

Cederholm, C. J., D. H. Johnson, R. E. Bilby, L.G. Dominguez, A. M. Garrett, W. H. Graeber, E. L. Greda, et al. "Pacific Salmon and Wildlife—Ecological Contexts, Relationships, and Implications for Management." Washington Department of Fish and Wildlife, 2000.

Costello, Christopher, Ling Cao, Stefan Gelcich, Miguel Á. Cisneros-Mata, Christopher M. Free, Halley E. Froehlich, Christopher D. Golden et al. "The Future of Food from the Sea." *Nature* 588, no. 7836 (2020): 95-100.

Cui, Liqiang, Matt R. Noerpel, Kirk G. Scheckel, and James A. Ippolito. "Wheat Straw Biochar Reduces Environmental Cadmium Bioavailability." *Environment International* 126 (2019): 69-75.

Doubilet, David, and Jennifer Hayes. "Cuba's Underwater Jewels Are in Tourism's Path." *National Geographic.* November 2016.

Feinstein, Dianne. "Feinstein, Toomey, Menendez, Collins Introduce Bipartisan Bill to Repeal Ethanol Mandate." Press release, July 20, 2021. https://www.feinstein .senate.gov/public/index.cfm/2021/7/feinstein-toomey-menendez-collins -introduce-bipartisan-bill-to-repeal-ethanol-mandate.

Food and Agriculture Organization of the United Nations. "Fertilizer Use by Crop in Cuba." 2003. https://www.fao.org/3/y4801e/y4801e00.htm.

Food and Agriculture Organization of the United Nations. "Tackling Climate Change Through Livestock: A Global Assessment of Emissions and Mitigation Opportunities." 2013.

Gale, Mark, Tu Nguyen, Marissa Moreno, and Kandis Leslie Gilliard-AbdulAziz. "Physiochemical Properties of Biochar and Activated Carbon from Biomass Residue: Influence of Process Conditions to Adsorbent Properties." *ACS Omega* 6, no. 15 (2021): 10224-10233.

Galford, Gillian L., Margarita Fernandez, Joe Roman, Irene Monasterolo, Sonya Ahamed, Greg Fiske, Patricia González-Díaz, and Les Kaufman. "Cuban Land Use and Conservation, from Rainforests to Coral Reefs." *Bulletin of Marine Science* 94, no. 2 (2018): 171-191.

Genchi, Giuseppe, Maria Stefania Sinicropi, Graziantonio Lauria, Alessia Carocci, and Alessia Catalano. "The Effects of Cadmium Toxicity." *International Journal of Environmental Research and Public Health* 17, no. 11 (2020): 3782.

Gewin, Virginia. "How Corn Ethanol for Biofuel Fed Climate Change." *Civil Eats.* February 14, 2022.

Hansen, H. H., IML Drejer Storm, and A. M. Sell. "Effect of Biochar on in Vitro Rumen Methane Production." *Acta Agriculturae Scandinavica, Section A–Animal Science* 62, no. 4 (2012): 305-309.

BIBLIOGRAPHY

Hoegh-Guldberg, Ove, Catherine Lovelock, Ken Caldeira, Jennifer Howard, Thierry Chopin, and Steve Gaines. "The Ocean as a Solution to Climate Change: Five Opportunities for Action." Washington, DC: World Resources Institute, 2019.

Hoffman, Jeremy S., Vivek Shandas, and Nicholas Pendleton. "The Effects of Historical Housing Policies on Resident Exposure to Intra-Urban Heat: A Study of 108 US Urban Areas." *Climate* 8, no. 1 (2020): 12.

Hu, Winnie. "Please Don't Flush the Toilet. It's Raining." *New York Times*. March 2, 2018.

International Biochar Initiative. "Profile: Using Biochar for Water Filtration in Rural Southeast Asia." October 2012.

Kearns, Joshua, Eric Dickenson, Myat Thandar Aung, Sarangi Madhavi Joseph, Scott R. Summers, and Detlef Knappe. "Biochar Water Treatment for Control of Organic Micropollutants with UVA Surrogate Monitoring." *Environmental Engineering Science* 38, no. 5 (2021): 298-309.

Kilcoyne, Clodagh, and Conor Humphries. "Ireland Looks to Seaweed in Quest to Curb Methane from Cows." *Reuters*. November 17, 2021.

Lark, Tyler J., Nathan P. Hendricks, Aaron Smith, Nicholas Pates, Seth A. Spawn-Lee, Matthew Bougie, Eric G. Booth, Christopher J. Kucharik, and Holly K. Gibbs. "Environmental Outcomes of the US Renewable Fuel Standard." *Proceedings of the National Academy of Sciences* 119, no. 9 (2022): e2101084119.

Lee, Uisung, Hoyoung Kwon, May Wu, and Michael Wang. "Retrospective Analysis of the US Corn Ethanol Industry for 2005–2019: Implications for Greenhouse Gas Emission Reductions." *Biofuels, Bioproducts and Biorefining* 15, no. 5 (2021): 1318-1331.

Leng, R. A., Sangkhom Inthapanya, and T. R. Preston. "Biochar Lowers Net Methane Production from Rumen Fluid in Vitro." *Livestock Research for Rural Development* 24, no. 6 (2012): 103.

Lim, XiaoZhi. "Can Microbes Save Us from PFAS?" *Chemical & Engineering News*. March 22, 2021.

McNerthney, Casey. "Heat Wave Broils Western Washington, Shattering Seattle and Regional Temperature Records on June 28, 2021." *HistoryLink.org*. July 1, 2021.

MesoAmerican Research Center. "Milpa Cycle." Accessed April 20, 2022, https://www.marc.ucsb.edu/research/maya-forest-is-a-garden/maya-forest-gardens/milpa-cycle.

Murphy, Andi. "Meet the Three Sisters Who Sustain Native America." *PBS*. November 16, 2018.

Nelson, Amy. "An Oasis in the Most Unlikely Place." *Biohabitats*. August 3, 2015.

Newman, Andy. "2 Months of Rain in a Day and a Half: New York City Sets Records." *New York Times*. August 23, 2021.

Newtown Creek Alliance. "Combined Sewer Overflow." Accessed April 20, 2022, http://www.newtowncreekalliance.org/combined-sewer-overflow/.

Newtown Creek Alliance. "The History and Geography of Newtown Creek." Accessed April 20, 2022, https://storymaps.arcgis.com/stories/4d38389f05a94d5e8bb67ef7e5b03b32.

BIBLIOGRAPHY

New York City Department of Environmental Protection. "Combined Sewer Overflow Long Term Control Plan for Newtown Creek." June 2017.

Norris, Charlotte E., G. Mac Bean, Shannon B. Cappellazzi, Michael Cope, Kelsey LH Greub, Daniel Liptzin, Elizabeth L. Rieke, Paul W. Tracy, Cristine LS Morgan, and C. Wayne Honeycutt. "Introducing the North American Project to Evaluate Soil Health Measurements." *Agronomy Journal* 112, no. 4 (2020): 3195-3215.

Odell, Jenny. *How to Do Nothing: Resisting the Attention Economy.* Brooklyn: Melville House, 2020.

Oro Loma Sanitary District. "Horizontal Levee Project." Accessed April 20, 2022, https://oroloma.org/horizontal-levee-project/.

Pandey, Avaneesh. "Poop Gold: Study Finds Human Feces Contain Precious Metals Worth Millions." *International Business Times.* March 24, 2015.

Plumer, Brad, and Nadja Popovich. "How Decades of Racist Housing Policy Left Neighborhoods Sweltering." *New York Times.* August 24, 2020.

Ramirez, Rachel. "Report: Utilities Are Less Likely to Replace Lead Pipes in Low-Income Communities of Color." *Grist.* March 12, 2020.

"Record Rainfall Floods The City: Sewer Overflow Swamps Eight Hotels and Times Square Subway Station." *New York Times.* July 29, 1913.

Richardson, Dana Erin, and Sarah Zentz, directors. "Back to Eden." ProVisions Productions, 2011.

Richter, Brent. "Lehigh's Sechelt' Mine Wins Provincial Recognition." *Coast Reporter.* October 8, 2010.

Ryan, John. "Extreme Heat Cooks Shellfish Alive on Puget Sound Beaches." KUOW. July 7, 2021.

"Salish Soils Takes Leading Role in Community." *The Local Weekly.* June 19, 2013.

"Sewage Yields More Gold than Top Mines." *Reuters.* January 30, 2009.

Shandas, Vivek, Jackson Voelkel, Joseph Williams, and Jeremy Hoffman. "Integrating Satellite and Ground Measurements for Predicting Locations of Extreme Urban Heat." *Climate* 7, no. 1 (2019): 5.

Shindell, Drew, Yuqiang Zhang, Melissa Scott, Muye Ru, Krista Stark, and Kristie L. Ebi. "The Effects of Heat Exposure on Human Mortality Throughout the United States." *GeoHealth* 4, no. 4 (2020): e2019GH000234.

Smith, Kathleen S., Geoffrey Plumlee, and Philip L. Hageman. "Mining for Metals in Society's Waste." *The Conversation.* October 1, 2015.

Sughis, Muhammad, Joris Penders, Vincent Haufroid, Benoit Nemery, and Tim S. Nawrot. "Bone Resorption and Environmental Exposure to Cadmium in Children: A Cross-Sectional Study." *Environmental Health* 10, no. 1 (2011): 1-6.

"The Ocean As Solution, Not Victim." *Living on Earth.* April 2, 2021.

U.S. Environmental Protection Agency. "Case Summary: Settlement Reached at Newtown Creek Superfund Site." Accessed April 20, 2022, https://www.epa.gov/enforcement/case-summary-settlement-reached-newtown-creek-superfund-site.

Westerhoff, Paul, Sungyun Lee, Yu Yang, Gwyneth W. Gordon, Kiril Hristovski, Rolf U. Halden, and Pierre Herckes. "Characterization, Recovery Opportunities, and

Valuation of Metals in Municipal Sludges from US Wastewater Treatment Plants Nationwide." *Environmental Science & Technology* 49, no. 16 (2015): 9479-9488.

Winders, Thomas M., Melissa L. Jolly-Breithaupt, Hannah C. Wilson, James C. MacDonald, Galen E. Erickson, and Andrea K. Watson. "Evaluation of the Effects of Biochar on Diet Digestibility and Methane Production from Growing and Finishing Steers." *Translational Animal Science* 3, no. 2 (2019): 775-783.

Yearwood, Burl, Cho Cho Aung, Ridima Pradhan, and Jennifer Vance. "Toxins in Newtown Creek." *World Environment* 5, no. 2 (2015): 77-79.

Chapter Twelve

American Society of Civil Engineers. "2021 Infrastructure Report Card." Accessed April 20, 2022, www.infrastructurereportcard.org.

Barth, Brian. "Humanure: The Next Frontier in Composting." *Modern Farmer.* March 7, 2017.

BBC website; *A History of the World*, "Moule's Mechanical Dry Earth Closet," 2014.

Bertschi School. "Where Science Lives." Accessed April 20, 2022, https://www .bertschi.org/science-wing.

Bill & Melinda Gates Foundation. "Reinvent the Toilet: A Brief History." Accessed April 20, 2022, https://www.gatesfoundation.org/our-work/programs/global-growth -and-opportunity/water-sanitation-and-hygiene/reinvent-the-toilet-challenge -and-expo.

Clivus Multrum. "About Us." Accessed April 20, 2022, https://www.clivusmultrum .eu/about-us/.

Cosgrove, Anne. "Restrooms That Recapture Water and Waste." *Facility Executive.* December 2019.

De Ceuvel. "Sustainability." Accessed April 20, 2022, https://deceuvel.nl/en/about /sustainable-technology/.

DELVA. "De Ceuvel—Amsterdam." Accessed April 20, 2022, https://delva.la/projecten /de-ceuvel/.

"Earth Closets." *OldandInteresting.com* August 15, 2007. http://www.oldandinteresting .com/earth-closet.aspx.

Engineering for Change. "Tiger Toilet." Accessed April 20, 2022, https://www .engineeringforchange.org/solutions/product/tiger-toilet/.

ENR California. "Green Project Best Project: Arch | Nexus SAC." October 5, 2017.

Haukka, J. K. "Growth and Survival of *Eisenia fetida* (Sav.)(Oligochaeta: Lumbricidae) in Relation to Temperature, Moisture and Presence of *Enchytraeus albidus* (Henle)(Enchytraeidae)." *Biology and Fertility of Soils* 3, no. 1 (1987): 99-102.

Hennigs, Jan, Kristin T. Ravndal, Thubelihle Blose, Anju Toolaram, Rebecca C. Sindall, Dani Barrington, Matt Collins et al. "Field Testing of a Prototype Mechanical Dry Toilet Flush." *Science of the Total Environment* 668 (2019): 419-431.

Historic England. "The Story of London's Sewer System." *The Historic England Blog.* March 28, 2019, https://heritagecalling.com/2019/03/28/the-story-of-londons -sewer-system/.

BIBLIOGRAPHY

Holland, Oscar. "Has the Wooden Skyscraper Revolution Finally Arrived?" *CNN*. February 19, 2020.

Hugo, Victor. *Les Misérables*. Translated by Lee Fahnestock and Norman MacAfee. London: Penguin, 2013.

International Living Future Institute. "Bertschi Living Building Science Wing." Accessed April 20, 2022, https://living-future.org/lbc/case-studies/bertschi-living-building-science-wing/.

International Living Future Institute. "Perkins SEED Classroom." Accessed April 20, 2022, https://living-future.org/lbc/case-studies/perkins-seed-classroom/.

Jenkins, Joseph. *The Humanure Handbook*. Grove City, Pennsylvania: Joseph Jenkins, 1994.

Kenter, Peter. "A Community That Will Last." *Municipal Sewer & Water*. September 2016.

Lalander, Cecilia, Stefan Diener, Maria Elisa Magri, Christian Zurbrügg, Anders Lindström, and Björn Vinnerås. "Faecal Sludge Management with the Larvae of the Black Soldier Fly (*Hermetia illucens*)—From a Hygiene Aspect." *Science of the Total Environment* 458 (2013): 312-318.

Leich, Harold H. "The Sewerless Society." *The Bulletin of the Atomic Scientists*. November 1975.

Lewis-Hammond, Sarah. "Composting Toilets: A Growing Movement in Green Disposal." *The Guardian*. July 23, 2014.

LIXIL. "LIXIL to Pilot Household Reinvented Toilets in Partnership with the Gates Foundation." Press release. November 6, 2018, https://www.lixil.com/en/news/pdf/181106_BMGF_E.pdf.

Mackinnon, Eve. "Eco-Friendly Composting Toilets Already Bring Relief to Big Cities—Just Ask London's Canal Boaters." *Independent*. May 16, 2018.

Montesano, Jin Song. "Refreshing our Sanitation Targets, Standing Firm on Our Commitments." LIXIL press release. November 15, 2019, https://www.lixil.com/en/stories/stories_16/.

Nelson, Bryn. "A Building Not Just Green, but Practically Self-Sustaining." *New York Times*. April 2, 2013.

Nelson, Bryn. "In Rural Minnesota, A 70-Acre Lab for Sustainable Living." *New York Times*. January 11, 2013.

New York Academy of Medicine Center for History (blog); "A Different Kind of Flush," by Johanna Goldberg, November 19, 2013.

Nierenberg, Jacob. "SEED Classroom Is the Learning Space of a Greener, Better Future." *The Section Magazine*. April 7, 2020.

O'Neill, Meaghan. "The World's Tallest Timber-Framed Building Finally Opens Its Doors." *Architectural Digest*. March 22, 2019.

Perrone, Jane. "To Pee or Not to Pee." *The Guardian*. November 13, 2009.

Progress on Household Drinking Water, Sanitation and Hygiene 2000-2020: Five Years into the SDGs. Geneva: World Health Organization (WHO) and the United Nations Children's Fund (UNICEF), 2021.

Rolston, Kortny. "CSU Research Is in the Toilet, Literally." *Source*. March 9, 2015.

Schuster-Wallace, C.J., C. Wild, and C. Metcalfe. "Valuing Human Waste as an Energy Resource: A Research Brief Assessing the Global Wealth in Waste." United Nations University Institute for Water, Environment and Health. 2015.

BIBLIOGRAPHY

Sedlak, David. "The Solution to Cities' Water Problems Has Been Hiding in Rural Areas This Whole Time." *Quartz*. November 14, 2018.

Sky City Cultural Center and Haak'u Museum. "Virtual Tour." Accessed April 20, 2022, https://beta.acomaskycity.org/page/virtual_tour.

TBF Environmental Solutions. "The Tiger Toilet." Accessed April 20, 2022, https://www.tbfenvironmental.in/the-tiger-toilet.html.

UNICEF. "Billions of People Will Lack Access to Safe Water, Sanitation and Hygiene in 2030 Unless Progress Quadruples." Press release. July 1, 2021, https://www.unicef.org/press-releases/billions-people-will-lack-access-safe-water-sanitation-and-hygiene-2030-unless.

Ward, Barbara J., Tesfayohanes W. Yacob, and Lupita D. Montoya. "Evaluation of Solid Fuel Char Briquettes from Human Waste." *Environmental Science & Technology* 48, no. 16 (2014): 9852-9858.

WHO/UNICEF Joint Monitoring Programme for Water Supply, Sanitation and Hygiene (JMP). "Open Defecation." Accessed April 20, 2022, https://washdata.org/monitoring/inequalities/open-defecation.

INDEX

INDEX

INDEX

B RYN NELSON, PHD, IS AN AWARD-WINNING SCIENCE WRITER and former microbiologist who decided he'd much rather write about microbes than do research on them. Since receiving his doctoral degree from the University of Washington and completing a graduate program in science writing at the University of California at Santa Cruz, he has accumulated more than two decades of journalism experience. As a staff writer on the *Newsday* science desk, he covered genetics, stem cell research, evolution, ecology, and conservation, among other fields.

Nelson has written for dozens of other news outlets as well, including the *New York Times*, NBCNews.com, the Daily Beast, *Nature*, *Mosaic*, the *BMJ*, and *Science News for Students*. He was a contributing editor for two chapters on microbiology and food safety in the best-selling *Modernist Cuisine: The Art and Science of Cooking* and has won awards from the Association of Health Care Journalists and the Deadline Club of New York, among other honors. In his spare time, he enjoys photography, singing, travel, and gardening in Seattle, where he lives with his husband, Geoff, and their energetic boxador, Piper.